T0327571

Seafood and Aquaculture Marketing Handbook

Seafood and Aquaculture Marketing Handbook

Carole R. Engle

Kwamena K. Quagrainie

Madan M. Dey

SECOND EDITION

WILEY Blackwell

This [second] edition first published 2017 © 2017 by John Wiley & Sons, Ltd
Edition history: Blackwell Publishing Ltd (1e, 2006)

Registered Office
John Wiley & Sons, Ltd, The Atrium, Southern Gate, Chichester, West Sussex, PO19 8SQ, UK

Editorial Offices
9600 Garsington Road, Oxford, OX4 2DQ, UK
The Atrium, Southern Gate, Chichester, West Sussex, PO19 8SQ, UK
111 River Street, Hoboken, NJ 07030-5774, USA

For details of our global editorial offices, for customer services and for information about how to
apply for permission to reuse the copyright material in this book please see our website at
www.wiley.com/wiley-blackwell

Library of Congress Cataloging-in-Publication data applied for

ISBN: 9781118845509

A catalogue record for this book is available from the British Library.

Set in 9.5/13pt Meridien by SPi Global, Pondicherry, India

1   2017

Contents

About the authors

Carole R. Engle, Ph.D., is Owner/Manager of Engle-Stone Aquatic$ LLC, a consulting company that specializes in the marketing and economics of aquaculture. She has more than 35 years of experience in the analysis of economics and marketing issues of aquaculture in the U.S. and around the world. She has worked in 20 different countries on all major continents, using her combined understanding of aquaculture and economics to benefit aquaculture businesses and to assist aquaculture research and extension scientists. She previously worked for 27 years as Full Professor, Director of the Aquaculture/Fisheries Center, and Chairperson of the Department of Aquaculture and Fisheries at the University of Arkansas at Pine Bluff. Through Engle-Stone Aquatic$ LLC, she continues to provide expertise to aquaculture researchers, extension aquaculture scientists, and aquaculture businesses on strategic planning, marketing, and financial analysis. She has served as President of the International Association of Aquaculture Economics and Management, the U.S. Aquaculture Society is a Chapter of the World Aquaculture Society, and continues to serve on both the IAAEM and WAS boards. She is a three-time recipient of the Joseph P. McCraren Award of the National Aquaculture Association, Researcher-of-the Year Award from the Catfish Farmers of America and two-time recipient of Service Awards from the Catfish Farmers of Arkansas. She also serves as Executive Editor of the Journal of the World Aquaculture Society and as Editor-in-Chief of *Aquaculture Economics & Management*.

Kwamena K. Quagrainie, Ph.D., is Associate Professor in Aquaculture Economics and Marketing in the Department of Agricultural Economics at Purdue University, Indiana. He leads Purdue University's engagement efforts to the aquaculture industry in Indiana as well as the efforts of Illinois-Indiana Sea Grant Program on aquaculture development in Illinois. He has over 20 years of industry and research experience in economics and marketing issues related to both agriculture and aquaculture. In addition to his work in the U.S., he works in Asia and Africa on economics and marketing issues in agriculture, aquaculture, and fisheries. He leads a team of interdisciplinary researchers from universities across the U.S. that has developed an internationally recognized aquaculture program in Africa with funding from USAID. He serves as a director on the U.S. Aquaculture Society board, and chairs the board's promotion and membership committee. He is an Associate Editor of the *Journal of Applied Aquaculture* and coordinates the review of submitted manuscripts with focus on economics and social sciences. He also serves on the editorial board of *Applied Aquaculture Economics & Management*.

Madan M. Dey is the Chair of the Department of Agriculture and Full Professor of Agricultural Economics/Agribusiness at Texas State University. He served as a Full Professor of Aquaculture/Seafood Economics and Marketing at the Aquaculture/Fisheries Center of Excellence, University of Arkansas at Pine Bluff (UAPB), USA, from August 2007 to June 2016. A Ph.D in Agricultural Economics, he has more than 25 years of post Ph.D. experience in research, teaching, and governance in a variety of agricultural and natural resources contexts. Before joining UAPB in 2007, he held various senior-level, progressively responsible, research and leadership positions at three different centers of the CGIAR Consortium (formerly known as Consultative Group of International Agricultural Research), namely International Rice Research Institute, International Food Policy Research Institute, and WorldFish Center (WorldFish) for more than 15 years. He has published widely in several fields of agricultural/natural resource economics with more than 120 journal articles and book chapters. For his research contribution, he received the "Best Research Achievement Award" from the WorldFish Board of Trustees in 1997 and again in 2005. He received the Chancellor's "Distinguished Research, Scholarship, and Creative Activity Award" from UAPB in 2014. He has held posts in the United States, Malaysia, the Philippines and Bangladesh, and led research and development projects in more than 20 other countries of Asia, Africa, and the Pacific.

CHAPTER 1

Seafood and aquaculture markets

This introductory chapter will provide an overview of seafood and aquaculture markets worldwide, the global supply of major seafood and aquaculture species, the location of major markets, and international trade volumes and partners. The chapter continues with a discussion of characteristics of aquaculture products and the market competition between wild-caught and farmed fish. The chapter concludes by summarizing trends in consumption of seafood and aquaculture products. Practical examples from aquaculture are included throughout.

Global trends in seafood and aquaculture markets

Successful industries must be successful in marketing their products yet marketing is not well understood by many aquaculturists. This book both defines and explains many key marketing concepts and components of theory fundamental to a thorough understanding of marketing that is necessary for aquaculture businesses to successfully develop effective marketing plans and strategies. A market can be defined in a number of ways. It can be a location, such as the Fulton Fish Market in New York City or the Tsukiji Market in Tokyo, Japan, a product such as the jumbo shrimp market, a time such as the Lenten season market in the United States or the European Christmas market, or a level such as the retail or wholesale market.

This chapter will focus mostly on geographic markets but will touch on several other levels of markets. Chapter 3 presents more specific information on fundamental marketing terms and concepts.

A frieze in an Egyptian tomb dated to 2500 B.C. shows the harvest of cultured tilapia (Bardach et al. 1972). While this date places aquaculture as an ancient technology, it is still quite young when compared to terrestrial agriculture. Diamond (1999) shows that domesticated species of both crops and animals were

Seafood and Aquaculture Marketing Handbook, Second Edition. Carole R. Engle, Kwamena K. Quagrainie and Madan M. Dey.

Table 1.1 Dates of domestication of various plant and animal crops important in the cultural development of humans.

Area	Domesticated		Earliest attested date of domestication
	Plants	Animals	
Independent origins of domestication			
Southwest Asia	Wheat, pea, olive	Sheep, goat	8500 B.C.
China	Rice, millet	Pig, silkworm	By 7500 B.C.
Mesoamerica	Corn, beans, squash	Turkey	By 3500 B.C.
Andes and Amazonia	Potato, manioc	Llama, guinea pig	By 3500 B.C.
Eastern U.S.	Sunflower, goosefoot	None	2500 B.C.
Sahel	Sorghum, African rice	Guinea fowl	By 5000 B.C.
Tropical West Africa	African yams, oil palm	None	By 3000 B.C.
Ethiopia	Coffee, tea	None	Unknown
New Guinea	Sugar cane, banana	None	7000 B.C.
Local demonstration following arrival of founder crops from elsewhere			
Western Europe	Poppy, oat	None	6000–3500 B.C.
Indus Valley	Sesame, eggplant	Humped cattle	7000 B.C.
Egypt	Sycamore fig, chufa	Donkey, cat	6000 B.C.

Source: Diamond (1999).

being cultivated by 8500 B.C. (Table 1.1). Southwest Asia and China served as the birthplace for many types of terrestrial agriculture and aquatic crops. Diamond theorized that areas with sparse game would provide greater returns to the effort in developing farming technologies. For most species of fish, scarcities due to overfishing have become evident only in the latter part of the 1900s. Thus, strong incentives to explore and invest in widespread domesticated production of aquatic plants and animals have been of comparatively recent origin. The ensuing level of scientific and technological development of aquaculture in the 1900s has resulted in a dramatic blossoming of aquaculture industries.

Continued growth in the global economy and in the world's population has resulted in increasing demand for seafood. However, the volume of seafood supplied from capture fisheries across the world has leveled off since about 1994, while the quantity of aquaculture production supplied worldwide has continued to increase (Fig. 1.1). The global supply from capture fisheries increased most rapidly during the late 1950s through the end of the 1960s. From that point, capture fisheries continued to increase, but at a slower rate, reaching slightly more than 95 million metric tons in 1996. Since then, world capture fisheries have fluctuated from 86.8 million to 94.8 million metric tons, averaging about 92 million metric tons. It is clear that most of the increase in the world supply of fish and seafood has been due to the expansion of aquaculture production.

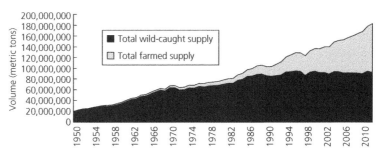

Fig. 1.1 Volume of wild-caught and farmed supply of seafood, 1950–2012. Source: FAO (2014).

Global aquaculture production has increased more than 40-fold, from 2 million metric tons in 1960 to 90.4 million metric tons in 2012 (FAO 2014), while chicken meat production increased by a factor of 10 and beef production doubled (Thornton 2010). From 2008 to 2012, the annual growth rate of cultured finfish and shellfish production averaged 4%. Capture fisheries production has declined by 3% from 1996 to 2012.

All aquatic farming combined represented a 3% share of the world harvest of fish, shellfish, and seaweeds in 1950 (FAO 2014). By 2012, this share had increased to 49.4% and consisted of a record 90.4 million metric tons of total farmed aquatic production. Of this, the greatest increase was for freshwater diadromous fishes (41.97 million metric tons), aquatic plants (23.78 million metric tons), and mollusks (15.17 million metric tons). The total value of aquaculture production worldwide increased to $144.3 billion in 2012.

The relative costs of capture fisheries have increased over time while those of aquaculture production have decreased. In the United States, the Magnuson Fishery Conservation and Management Act established a 200 nautical mile (370 km) Exclusive Economic Zone (EEZ) for commercial fisheries. The U.S. Magnuson Act, combined with declining abundance of many types of fish stocks, requires trawlers to travel greater distances to find supplies of fish. In other parts of the world, countries such as Chile, Ecuador, and Peru have also claimed rights to 200 nautical mile zones for fishing. However, a few countries, such as Papua New Guinea and Anguilla, still use a 5-km limit, while others have moved to a 12 nautical mile limit. Costs of capture fisheries are likely to continue to increase over time. At the same time, aquaculture costs have declined as new technologies have been developed and refined. According to a 2013 World Bank study (World Bank 2013; Kobayashi et al. 2015), global fish supply is projected to rise to 187 million metric tons by 2030. Capture production is expected to remain fairly stable over the 2000–2030 period, with a projected supply of about 93.2 million metric tons in 2030. In contrast, global aquaculture projection is likely to maintain its steady rise, reaching 93.6 million metric tons by 2030. In terms of food fish production, the World Bank study predicts that aquaculture will contribute 62% of the global supply by 2030.

Where are most aquaculture crops produced?

Asia is the birthplace of early aquaculture production technology and continues to be the world's leading aquaculture region. Production in Asia reached 46.7 million metric tons in 2012, accounting for 91% of the world's output (Fig. 1.2). Next to Asia, the Americas was the second leading aquaculture producing region, but with only 4% of total world production. Europe followed closely at 3% of total world production, and Africa at 2%.

The nation that leads the world in aquaculture production is China (Fig. 1.3). Of the top 10 countries in aquaculture production, eight are located in Asia (China, Indonesia, India, Vietnam, The Philippines, Bangladesh, Republic of Korea, and Thailand). Norway and Chile are the only non-Asian countries in the top 10 (ranking eighth and tenth, respectively, in terms of quantity produced). While aquaculture's contribution to world aquatic production averaged 35% in 2002, it reached 66% to 77% in some of the top aquaculture producing countries (China, India).

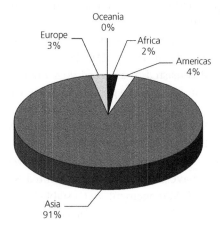

Fig. 1.2 World aquaculture production by region, 2012. Source: FAO (2014).

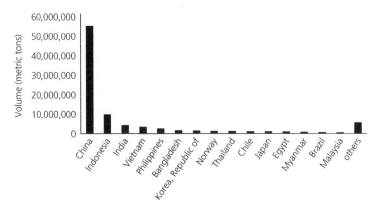

Fig. 1.3 Volume of global aquaculture production by country, 2012. Source: FAO (2014).

Much of the aquaculture production in the world occurs in lesser-developed nations (FAO 2014). Of the top 20 aquaculture producing nations, only three, Japan, Norway, and the U.S., are considered developed nations by the FAO. Moreover, much of the increase in aquaculture production has been from low-income food deficit countries, such as China.

Global aquaculture production has grown at an annual rate of approximately 10% (FAO 2014). Aquaculture production in China grew at an annual rate of about 5%, down from 14% in previous decades. However, the rate of growth of aquaculture in Indonesia was 21% annually from 2000 to 2012, and 17% in Vietnam. By comparison, Africa had the greatest annual percentage increases in production at 12% per year for 2001 to 2012. The Americas and Asia averaged 7%, Europe 3%, and Oceania 4% per year over this same time period.

Global fish production will further concentrate in Asia toward 2030 (World Bank 2013). China is expected to account for an overwhelming 37% of the world's fish production by 2030. Fish supply from other Asian countries/regions (including India and Southeast Asia) will also likely expand. Latin America and Caribbean countries are projected to experience large aquaculture growth over the next 20 years or so (World Bank 2013; Kobayashi et al. 2015).

What are the major species cultured worldwide?

Worldwide, the greatest volume produced of an aquaculture product in 2001 was that of Eucheuma seaweeds (*Eucheuma* spp.), followed by Japanese kelp (*Undaria* spp.), grass carp (*Ctenopharyngodon idellus*), silver carp (*Hypophthalmichthys molitrix*), various cupped oysters (*Crassostrea* spp.), common carp (*Cyprinus carpio*), Japanese carpet shell (*Ruditapes philippinarum*), Nile tilapia (*Oreochromis tilapia*), whitelegged shrimp (*Litopenaeus vannamei*), bighead carp (*Hypophthalmichthys nobilis*), various aquatic plants, catla (*Catla catla*), Crucian carp (*Carassius carassius*), wakame (*Undaria pinnatifida*), and Elkhorn sea moss (*Kappaphycus alvarezii*) (Fig. 1.4). The various carp species combined represent the major volume of finfish harvested, by several orders of magnitude. The top three finfish species harvested, by volume, are all different species of carp, and carp are the only finfish other than tilapia included in the list of the top 10 aquaculture products (by volume).

The aquaculture species that generated the greatest value in 2012 was the whitelegged shrimp, followed by Atlantic salmon, grass carp, silver carp, and catla (Fig. 1.5). These top five species in terms of value were followed in descending order by Nile tilapia, common carp, Chinese mitten crab, giant tiger prawn, bighead carp, rainbow trout, Japanese carpet shell, roho labeo, red swamp crawfish, and Crucian carp. Of the top 15, six were carp. However, the overall rankings of the top five valued species have changed dramatically over time. Whitelegged shrimp was not in the top 15 in 2002 but accounted for the highest value in 2012. Atlantic salmon increased from fourth place to second and Nile

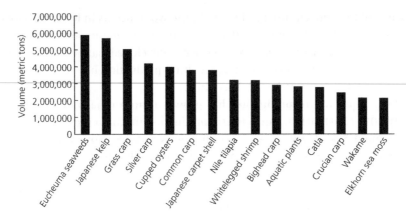

Fig. 1.4 Global aquaculture production of the top 15 species (2012). Source: FAO (2014). nei, not elsewhere indicated.

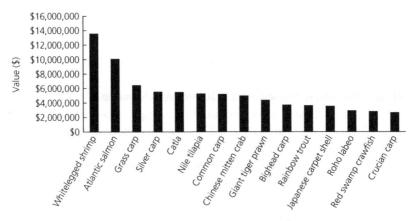

Fig. 1.5 Value of the top 15 farmed species, 2012. Source: FAO (2014).

tilapia increased to the sixth highest value from fifteenth in 2002. Shrimp, salmon, and tilapia combined composed 45% of the total value of aquaculture supplied.

Over the next 20 years or so, further growth in supply is expected for tilapia, carp, and *Pangasius* (World Bank 2013; Kobayashi et al. 2015). Production of some high-value species (such as shrimp and salmon) is also likely to grow over the period. However, only marginal growth in supply is expected for species with limited aquaculture potential.

Real prices of all fish aquaculture species are projected to increase modestly by about 10% during the 2010–30 period (World Bank 2013). However, the real prices of fishmeal, fish oil, and capture fisheries products that are used for these ingredients are expected to rise substantially more than those of fish for direct consumption.

What are the major finfish species caught and supplied to world markets?

The Peruvian anchovy constitutes the greatest volume of worldwide capture fisheries (Fig. 1.6). The primary use of anchovies is for fishmeal production, not as a food product. The second greatest catch is that of pollock. Pollock is used commonly in fish sandwiches, fish sticks, and other popular frozen and breaded preparations. It is also used for production of surimi in many countries. Following pollock are several other types of tuna, herring, and mackerel. Croakers and drums occupied fifteenth place in 2012.

If the volumes of worldwide aquaculture production (Fig. 1.4) are compared with those of worldwide capture fisheries, it is clear that more grass or silver carp are produced worldwide than any single marine species used for direct food consumption by humans[1]. There was also more common carp produced from aquaculture (3.8 million metric tons) than of the next largest volume of wild-caught foodfish, pollock (3.27 million metric tons).

While aquaculture production is approximately equal to that of capture fisheries, culture techniques have been developed for only a limited number of finfish species. In contrast, a large number of different freshwater and marine species are caught and sold, many for production of fishmeal and not for direct human consumption. Thus, there is a great deal of potential for future growth of aquaculture as new culture techniques are developed for other species.

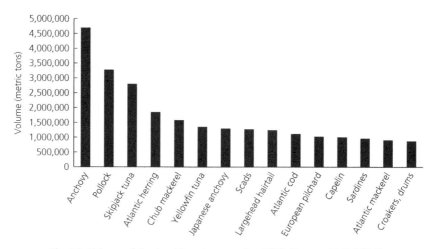

Fig. 1.6 Volume of the top 15 capture species, 2012. Source: FAO (2014).

---

[1] Grass carp volume was 3.6 million metric tons in 2001 and the volume of Alaskan pollock was 3.1 million metric tons.

What countries are the major markets for seafood and aquaculture?

Per capita consumption of seafood by world region[2] averaged 12–48 kg/capita (Table 1.2) (FAO 2014). However, per capita consumption varied tremendously, even from 0.3 to more than 140 kg/capita within the same region of the world. For example, in the North American region, Greenland averaged per capita seafood consumption of 84.1 kg, while seafood consumption in the U.S. was 22.7 kg/capita. Oceania ranked second, followed by the Far East, and then the Caribbean. Table 1.3 presents the top five countries in terms of highest per capita consumption of seafood for 2001. The country with the highest per capita consumption of fish and seafood in the world, the Maldives, is located in the Far East world region. However, this same region includes countries such as Mongolia (0.1 kg/capita) and Nepal (1.0 kg/capita). In terms of the percentage of countries within a region that consumed more than 25 kg/capita, there were 46% of the countries in the Far East region, 65% in Oceania, and 22% in Europe.

Table 1.3 presents the top five countries in terms of total volume of consumption of fish and seafood in 2007–2009 (NOAA-NMFS 2011). The total amount is clearly related to the combination of per capita consumption and total population. Topping the list was China that has both a high per capita consumption rate and the highest population in the world, resulting in consumption of over 40 million metric tons. Japan followed, with total consumption of 7.2 million metric tons with the U.S. third with 7.1 million metric tons. While per capita consumption in India is among the lowest in the world, it still ranks fourth in total consumption

Table 1.2 Average per capita consumption of fish and shellfish by world region, 2007–09.

Region	Mean ± SD	Maximum	Minimum
		kg/capita	
Africa	13 ± 14	68	0.2
Caribbean	27 ± 14	55	0.57
Europe	20 ± 20	90	0.3
Far East	35 ± 29	141	0.3
Latin America	12 ± 9	35	1.4
Near East	12 ± 9	29	0.0
North America	48 ± 30	86	22.7
Oceania	37 ± 17	74	2.5

Source: NOAA-NMFS (2011).

---

[2] FAO defines world regions as Africa, the Caribbean, Europe, the Far East, Latin America, the Near East, North America, and Oceania.

Table 1.3 Top five countries worldwide with highest per capita consumption and highest total consumption of fish and seafood, 2007–09.

Country	Per capita consumption (kg/capita)	Total population (million people)	Total consumption of fishery products (metric tons)
Highest per capita consumption			
Maldives	140.8	317,280	44,673
Iceland	89.8	326,340	29,305
Faroe Islands	87.7	48,359	4,241
Greenland	86.1	56,483	4,863
Kiribati	73.8	106,461	7,857
Countries with highest consumption of fish and seafood			
China	30.5	1,365,500,000	41,647,750
Japan	55.9	127,090,000	7,104,331
U.S.	22.7	318,360,000	7,226,772
India	5.5	1,246,460,000	6,855,530
Indonesia	24.7	252,164,800	6,228,471

Source: NOAA-NMFS (2011).

due to its large population. Indonesia completed the top five countries in total consumption of fish and seafood in 2012.

Trade in seafood and aquaculture

Approximately 38% (live weight equivalent) of world fish production was traded internationally in 2010 (FAO 2014). The continued increase in aquaculture production results in continued increases in the total supply of fishery products worldwide.

Are aquaculture products different from agriculture products?

Characteristics of aquaculture products

Aquaculture is a unique form of food production. Most cultured species of fish are not substantially different from wild-caught species. While common carp, with 2000 years of culture, has been bred selectively into strains of fish recognizably different from wild-caught fish, this is not the case for most other cultured aquatic species. Genetic advances may change this situation rapidly, but unlike animal and row crop agriculture, aquaculture growers find themselves competing in the marketplace with wild-caught seafood products. In many cases, wild-caught product still dominates the market and has a major effect on price. Some segments of the aquaculture industry have been more successful than others in differentiating their product from wild-caught supplies.

Aquaculture products offer distinct advantages in terms of control over the product. Many aquaculture products can be supplied year-round. In contrast, most wild-caught seafood is characterized by seasonal fluctuations related to weather and fishing regulations that can result in dramatic price swings. The domination of seafood markets by wild-caught species has resulted in a tendency towards high volatility. While aquaculture products offer the advantage of controlled year-round supply, these products must compete within the volatile seafood market.

Controlled production techniques also allow the aquaculture grower to produce a consistent product. Consistency in supply refers to size, quality, and other product characteristics in addition to consistency in the quantity supplied. Consistently supplied aquaculture products would be expected to lend some stability to the seafood market as the market share of aquaculture products continues to grow over time. Enhanced reliability and regularity in supply of farmed product should enable producers to negotiate better prices (Asche 2001). Theoretically, buyers would be willing to pay higher prices to compensate for reduction in the financial risk that results from supply problems. Market sectors, such as the retail sector, that prefer fresh product, might be expected to prefer farmed supplies (Young et al. 1993). Fresh product requires a short re-order period. Supply chains of captured fisheries products are more fixed due to seasonality of supply and cannot respond readily to changes in retail demand.

Consumers in many countries and for many years have exhibited strong preferences for the freshness of seafood. By contrast, one rarely hears an emphasis on the freshness of beef, pork, or chicken. This strong consumer preference for fresh seafood likely derives from the perishability of seafood as compared to other products. Technological advances enable processors to produce quality frozen and preserved seafood products. However, the preferences for fresh seafood have driven some retail grocers to purchase frozen product, thaw it, and sell it as fresh.

It is easier to trace farmed product back to its original source than wild-caught product. The complexity of market channels for wild-caught product may obscure steps in the supply chain and make tracing products to their source difficult (Asche 2001). Some wild-caught seafood is marked, logged, and stored separately, but this is the exception. The greater traceability of aquaculture products should become increasingly advantageous especially in the U.S. with its country-of-origin labeling laws that require certification of product origin. Individual states in the U.S. also have enacted state laws related to notification of the origin of the seafood sold. Aquaculture suppliers should find compliance less onerous than suppliers of wild-caught seafood.

The potential to control attributes and their levels in a product can offer an opportunity for farmers to target specific consumer segments (Asche 2001).

For example, producing the exact fat content to produce a particular smoked flavor or production of fish of a given size may provide aquaculture growers a significant marketing advantage over capture fisheries. In most cases, additional research will be required to develop cost-effective means of producing these attributes.

Fish and other aquaculture production allows for reliable delivery schedules to comply with contractual agreements to supply fish of a given size and quality grade. The uncertainty of what species, size, and, to some extent, quality of fish will be caught is an important characteristic that can be used to differentiate farm-raised from wild-caught seafood.

The management required for successful aquaculture businesses can be used to reassure consumers of the safety of the product. Consumers increasingly desire assurances that products are free of chemicals, pesticides, and other undesirable additives. This concern can include assurance that the product has not been modified genetically.

A survey of consumers in 2007 showed increasing concerns in the U.S. over food safety (Brewer and Rojas 2008). Greatest concerns were expressed about pesticide residues and hormones in poultry and meat. These concerns have been extended to seafood. The particular concerns for seafood are related to concentrations of dioxin and mercury in seafood products and the status of menhaden and other pelagics used for fishmeal in fish diets (Millar 2001), and levels of metal ions such as mercury in seafood (Petroczi and Naughton 2009).

There has been growing resistance to aquaculture products by some activist groups. There are groups who consider aquaculture as unnatural and detrimental to the environment. In some areas of the U.S., for example, farmed salmon is considered less desirable than wild-caught salmon. On the other hand, some consumers may be convinced to pay a premium price for environmentally sustainable products. Farm-raised catfish is preferred to wild-caught catfish in southern states for a variety of reasons, but primarily for the consistency of flavor, quality, and the certainty that it is free of contaminants and adulterations. U.S. farm-raised tilapia, catfish, trout, and hybrid striped bass are listed as environmentally acceptable seafood choices by the Monterey Bay Aquarium (Seafood Watch 2014).

A major disadvantage of aquaculture products as compared to wild-caught seafood is the price. Costs of production have frequently been higher for aquaculture products than for wild-caught seafood. However, as wild fish stocks have declined and boats have had to travel farther on fewer fishing days, costs of capture fisheries have increased. At the same time, research and development have reduced costs of producing a number of aquaculture species. Thus, there is a greater number of farmed species for which production costs are competitive with those of wild-caught species than before. However, the consistent production and supply of aquaculture products results in more consistent costs and prices. Buyers who are accustomed to waiting for periods of abundant supply and low prices of wild-caught seafood may be reluctant to pay a consistently higher price for aquaculture products.

Market opportunities have developed for aquaculture species when declining stocks of similar wild-caught species resulted in higher prices. This has been the case for hybrid striped bass in the U.S., cultured turbot, halibut, and other species even though framed turbot and halibut are considered inferior to wild-caught product (Asche 2001).

Market competition between wild-caught and farmed finfish

Prices for several aquacultured species such as Atlantic salmon, rainbow trout, sea bass, and sea bream have fallen as production has increased. These finfish species have grown in importance in seafood markets in the European Union and in the U.S. (Asche 2001). Atlantic salmon, rainbow trout, sea bass, and sea bream were high-value species before aquaculture production became significant. The increased supplies from aquaculture have been accompanied by lower prices.

A farmed product that competes in a large market will face limited price effects from increased aquaculture production. As long as supplies of the farmed species are low in comparison with wild-caught species, the impact of the farmed quantity supplied on price will be small.

When the supply of the farmed species is high, farm-level production is likely to determine market price because of the greater control that farmers have over the production process (Asche 2001). Salmon (Asche et al. 1999), catfish (Quagrainie and Engle 2002), tilapia, carp, shrimp, oysters, and mussels are examples of seafood markets that are dominated by farmed production. With few or no substitutes, it may be more difficult for the industry to grow because farmers will then have to create and promote the market for their product.

U.S. catfish was a low-value species prior to development of the catfish farming industry. While price in recent years has been low, there is no clear long-term trend. From 1993 to 2000, the U.S. catfish industry successfully moved its product into new markets, sustaining price ($0.748±0.03/lb) even with consistent growth (4% increase per year from 1993 to 2000) in volumes produced and sold. New market development was predicated upon changing consumer attitudes towards what had been regarded as an inferior, scavenging fish.

Most seafood demand studies show that the seafood market is highly segmented. Farmed species seem to compete mainly with similar, wild species, but not with other species (Asche 2001). However, Dey et al. (2014) showed that, at the retail level in the U.S., there is substitution among species, but the substitutability varied by region, product form, and ethnicity of buyers. Aquaculture growers are capturing market share even though demand studies have not determined clearly what market is being captured. Aquaculture products may create new market segments and may win parts of market shares from a variety of goods such that the effects on individual goods are not measurable (Asche 2001).

Consumption trends in seafood and aquaculture markets, expenditures, effects of income, and at-home versus away-from-home purchases

Until the development of advanced transportation and refrigeration and freezing technologies, the only seafood available was what could be caught locally. There remains a strong tendency for consumers to prefer species that live in nearby water. Many people are conservative and traditional about the fish and seafood that they eat. Consumer preferences typically are based on what they, their family, and their friends have been able to catch or gather from their hometown areas. For example, Engle et al. (1990) asked consumers nationwide what their most preferred type of finfish was. The preferred finfish on the Pacific Coast of the U.S. was salmon. Consumers in the Mountain region preferred trout that is caught in the mountain streams in the region. Catfish was most preferred in the West South Central and East South Central regions where catfish are abundant in the Mississippi River and its tributaries in the south. Catfish was also most preferred by consumers in the West North Central region through which the Mississippi River flows but also has a large number of inhabitants who have moved there from the south. The East North Central region has a tradition of Friday night fish fries that are based on the catch of locally available yellow perch. The Middle and South Atlantic regions have provided consumers with an abundant flounder fishery, and the 1989 survey showed preferences by Middle and South Atlantic consumers for flounder. Haddock was most preferred by consumers in the New England region.

European research showed that fish were associated with the natural environment in which they were found (i.e., the sea, rivers, lagoons, and ponds), leading to regional preferences for fish in Europe as in the U.S. (Gabriel 1990). Kinnucan et al. (1993) supported this by showing that preferences for fish products were influenced to a large degree by source availability.

Preparation methods also vary by region and the associated culinary traditions. Northern Europeans, for example, prefer fish fried, in breadcrumbs, soused, smoked, or cooked in foil (Gabriel 1990). In central Europe, French cuisine dominates and fish are steamed, poached, fried, smoked, simmered, or wrapped in foil. In southern Europe, fish is most often fried, grilled, simmered, or eaten dried.

Consumer tastes and preferences change over time. In the U.S., for example, beef consumption has declined while consumption of poultry has increased. Increasing health concerns and choices of lower-fat protein sources have been credited with the increased consumption of poultry products. However, declines in the cost of producing chicken in the U.S. and the resulting lower prices of chicken as compared to beef, no doubt have contributed to increased consumption of chicken. Pork and seafood consumption patterns, on the other hand, have changed little. Quality and flavor perceptions often have the greatest impact on preferences (Kinnucan et al. 1993). Other variables such as price, household size, coupon value, household income, geographic region, urbaniza-tion, race, and seasonality have been shown to explain the variation in house-hold expenditures on fresh and frozen seafood commodities (Cheng and Capps 1988).

Dey et al. (2014), used retail-level scanner data in the U.S. to examine mar-ket trends in seafood sales across 52 cities. Frozen seafood sales in supermarkets were found to increase by 6% per year from 2005 to 2010. Retail prices and volume of sales varied considerably by product form, ethnic characteristics of market area, and geographic region. Patterns of substitute and complementary seafood products also varied by region. Thus, it has become more important in recent years to design differentiated marketing strategies that target specific segments of targeted market regions.

Older consumers tend to eat more seafood, particularly if the consumer is health conscious and views seafood as a convenient choice (Olsen 2003). In Belgium, fish was consumed more frequently by women and consumption fre-quency increased with age (Verbeke and Vackier 2005). However, regional dif-ferences were also identified.

The most promising customers for at-home sales were shown to be older, well-educated (four or more years of college), higher-income (more than $30,000), non-white urban-suburban residents in families without young chil-dren (age 10 or under) present (Rauniyar et al. 1997). New England households were significantly more likely to be frequent purchasers for at-home use as com-pared to households in the West North Central and West South Central regions.

Frequent purchasers at restaurants were more likely to have annual incomes above $20,000, and especially above $40,000 (Hanson et al. 1994). The role of income, race, seasonality, few small children and adherence to the catholic faith were found important to restaurant consumption. The recognition in all con-sumer profiles of fish as a nutritious and healthful product represented an advantage for future marketing strategies in aquaculture.

Aquaculture market synopsis: tilapia

Tilapia (*Oreochromis* spp.; *Tilapia* spp.) is the eighth most important aquaculture crop worldwide in terms of volume (Fig. 1.4) and sixth in terms of value (Fig. 1.5). It is the fourth most important in terms of volume of all finfish and fifth most important in terms of value. World tilapia production has climbed steadily over the last half a century, with a marked increase in the rate of growth beginning in the 1990s (Fig. 1.7). Total worldwide production of tilapia and cich-lids exceeded 4.5 million metric tons in 2012. Average annual growth in tilapia production averaged 12.3% from the 1990s to 2012.

There has been a major shift in the countries leading the supply of tilapia over the years. In 1971, for example, the five leading tilapia producing countries

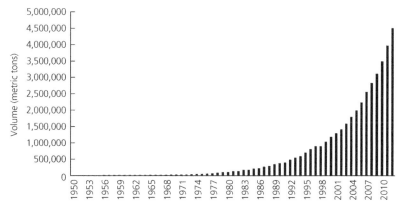

Fig. 1.7 Global tilapia and cichlid production, 1950–2012. Source: FAO (2014).

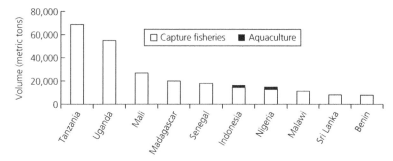

Fig. 1.8 Top ten tilapia producing (capture fisheries and aquaculture) countries, 1971. Source: FAO (2014).

(Tanzania, Uganda, Mali, Madagascar, and Senegal) were all African countries with endemic tilapia populations (Fig. 1.8). All of this supply was from capture fisheries. Only Indonesia and Nigeria registered measurable amounts of tilapia production from aquaculture and these were negligible. By 2012, only one of the five leading tilapia producing countries (China, Egypt, Indonesia, Brazil, and The Philippines) was an African country (Fig. 1.9). Of these countries, only Egypt and Indonesia have endemic populations of tilapia whereas tilapia were introduced into the other countries. Moreover, the supply of tilapia had shifted heavily to production from aquaculture.

China emerged as the dominant world producer of tilapia in the late 1990s. Over the 19-year period from 1994 to 2012, tilapia production increased by 558% with an average annual increase of 29%/yr (Fig. 1.10). Some of this production is exported while other portions of the production are consumed in the domestic market.

The major species of tilapia farmed worldwide is the Nile tilapia (*Oreochromis niloticus*), with 71% of total world production in 2012 (Fig. 1.11). Other, unspecified tilapia composed 20% of global production. The blue tilapia (*Oreochromis*

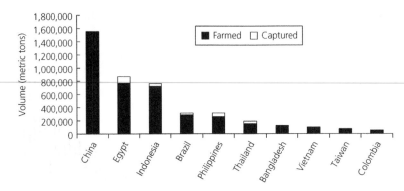

Fig. 1.9 Top ten tilapia producing (farmed and capture) countries, 2012. Source: FAO (2014).

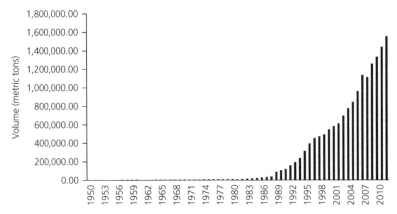

Fig. 1.10 Tilapia production in China, 1950–2012. Source: FAO (2014).

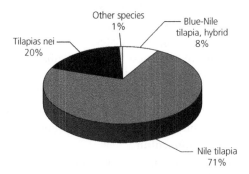

Fig. 1.11 Major species of farmed tilapia worldwide, 2012. Source: FAO (2014).

aureus) – Nile tilapia hybrid accounted for 8% of global production, and a variety of other species composed 1% of world production.

Much of the growth in tilapia aquaculture is a result of the development of improved production practices and both domestic and export market development (Engle 2006). Key technological developments in reproductive control led to rapid growth of commercial-scale aquaculture production (Kumar 2015). Sex reversal technology (Phelps and Popma 2000) was eventually replaced by development of genetically male tilapias through selective breeding (Mair et al. 1997). The GIFT (Genetically Improved Farm Tilapia) program of the WorldFish Center, Penang, Malaysia, has been the most widely adopted (Ponzoni et al. 2008). Dey et al. (2000a, b, c) showed that production costs were lower with GIFT strains than non-GIFT strains and benefited both producers and consumers. Development of intensive raceway/tank production in Central and South America led to further growth of large-scale tilapia production (Engle 1997).

The availability of supply of high-quality fillets and marketing expertise has resulted in the successful introduction of fresh and frozen tilapia fillets into the U.S. and European markets. The development of export markets has resulted in a change in the major tilapia production centers and a shift from a dominance of tilapia from capture fisheries to tilapia produced on farms.

The U.S. is the major export market for tilapia. Imports of tilapia into the U.S. have grown rapidly, particularly since 2000. The majority of this growth has been in the form of imported fresh and frozen fillets. Tilapia are also imported as frozen whole fish, but these volumes have not increased as rapidly as the imported volumes of fresh and frozen tilapia fillets.

The major suppliers of fresh tilapia fillets to the U.S. in 2003 were Costa Rica, Ecuador, and Honduras. Tilapia from Costa Rica and Honduras originate primarily from farms designed to specialize in tilapia production while, in Ecuador, shrimp farmers have begun to diversify into tilapia production. The pond and processing infrastructure in Ecuador allowed shrimp farmers to move quickly into tilapia production as shrimp disease problems escalated.

Indonesia has been the major supplier of frozen tilapia fillets into the U.S. for many years. In more recent years, though, Taiwan has begun to increase exports of frozen fillets in addition to export of lower-priced, frozen whole tilapia. Taiwan continues to be the major supplier to the U.S. of frozen whole tilapia. The U.S. tilapia production industry has targeted sales of live tilapia to Asian and Hispanic grocery stores. Large cities such as New York, Toronto, Chicago, and San Francisco have historically been the major targets for the U.S. industry, but other markets have been developed successfully in smaller cities throughout the U.S.

Tilapia continue to be raised for subsistence purposes. In subsistence farming areas, tilapia are consumed whole, gutted, scaled, and either fried or roasted. Tilapia is now accepted in many national dishes around the world and is popular in many forms, including smoked, as sashimi, and even as fried tilapia skins. Whole dressed tilapia are common in many open-air markets around the world. Export markets, however, require primarily filleted products although there is also international trade in frozen whole tilapia. Frozen whole tilapia imported into the U.S. are targeted towards Asian grocery stores throughout the U.S. Taiwan has dominated the supply of frozen whole tilapia to the U.S. for many years but China increased the export volume of frozen whole tilapia to the U.S. in the early 2000s.

Large commercial tilapia ventures began to emerge in the 1990s. These businesses developed techniques that led to the production of export-quality fresh and frozen tilapia fillets.

Tilapia have been introduced from their native ranges in Africa and spread widely across the world (FAO 1997). The early introductions of tilapia (1950s to 1970s) were part of development projects targeted towards increasing the availability of animal protein in subsistence farming areas. Surplus tilapia were sold as a means of generating cash income.

While the growth of the global market for tilapia has been an undisputed success story in aquaculture, challenges are emerging that may begin to threaten the high rate of growth of tilapia sales. First, controversy emerged in the late 1990s over the use of carbon monoxide by some tilapia processing plants (*SeaFood Business* 2001–2003). Carbon monoxide treatment results in a deep red color to the fillets that is considered desirable. Second, tilapia fillets have a lower dress out ratio (fillet weight: live weight of fish) than do fillets of other fish species. This results in a higher relative meat cost at the processing plant for the same farm-gate price of fish that dress out at higher ratios. Third, tilapia growers have recently come under criticism by buyers of organic supermarkets in the U.S. for use of the hormone methyltestosterone to sex reverse young tilapia. Sex reversal has allowed tilapia growers to achieve higher yields and growth rates by stocking the faster-growing all-male populations of tilapia.

A more significant challenge to tilapia production worldwide may come from environmentalist groups. Some commercial-scale tilapia ventures depend upon high flows of surface water for the discharge of waste products. Increased awareness of environmental effects of effluent discharges may result in additional regulations. Also, concern globally over the introduction of exotic species is growing rapidly. Tilapia have become established in natural waters in many countries with tropical climates and are increasingly being labeled as an invasive species.

The tilapia industry can likely adapt to these challenges as it has to others over time. Challenges such as these arise as an industry matures and attracts increasing attention. The success in market development that has led to the growth of the tilapia industry will provide incentives to continue to adapt to new challenges that arise.

Summary

Much of the increased total fishery production worldwide is from aquaculture. Aquaculture costs of production have declined as the cost of capture fisheries has increased. The result has been an increase in the proportion of fish and seafood supplies from aquaculture as compared to capture fisheries. The majority of aquaculture products in the world are produced in Asia. Kelp, oysters, and carps are the major aquaculture species produced and sold. Japan and the U.S. are the major seafood markets worldwide, while the leading seafood exporter is Thailand. Aquaculture products, as compared to wild-caught fisheries products, offer advantages such as: (1) greater control over the product and its consistency; (2) freshness; (3) traceability; and (4) enhanced food safety. Nevertheless, some activist groups consider farmed product undesirable and unsustainable, while others prefer farm-raised product for its positive attributes.

Study and discussion questions

1 What percentage of the total world supply of fish and seafood was from aquaculture in 2012?
2 From a marketing perspective, how do aquaculture products differ from wild-caught products?
3 What are some of the reasons that aquaculture has grown so rapidly in recent years?
4 What are the most important farmed and wild-caught species worldwide? List and describe the five most important farmed and the five most important wild-caught species worldwide.
5 Describe the major aquaculture producing countries in terms of volumes, types of products produced, and target markets.
6 Describe the major world markets for seafood and aquaculture.
7 Discuss the controversies related to aquaculture and the various points of view.
8 How does consumption of seafood compare with that of other protein products in the U.S.?
9 Describe some important consumption trends related to seafood and aquaculture products.
10 How has the market for tilapia changed from the 1970s to recent times? (Remember that the term "market" includes both demand and supply considerations.)

References

Asche, F. 2001. Testing the effect of an anti-dumping duty: the US salmon market. *Empirical Economics* 26:343–355.
Asche, F., H. Bremnes, and C.R. Wessells. 1999. Product aggregation, market integration and relationships between prices: an application to world salmon markets. *American Journal of Agricultural Economics* 81:568–581.
Bardach, J.E., J.H. Ryther, and W.O. McLarney. 1972. *Aquaculture: The Farming and Husbandry of Freshwater and Marine Organisms*. Wiley-Interscience, New York.

Brewer, M.S. and M. Rojas. 2008. Consumer attitudes towards issues in food safety. *Journal of Food Safety* 28:1–22.

Cheng, H. and O. Capps, Jr. 1988. Demand analysis of fresh and frozen finfish and shellfish in the United States. *American Journal of Agricultural Economics* 70:533–542.

Dey, M.M., A.E. Eknath, S. Li, M.G. Hussain, M.T. Tran, V.H. Nguyen, S. Ayyappa, and N. Pongthana. 2000a. Performance and nature of genetically improved farmed tilapia. A bio-economic analysis. *Aquaculture Economics & Management* 4:83–106.

Dey, M.M., G.B. Bimbao, L. Yong, P.B. Regaspi, A.K.M. Kohinoor, N. Pongthana, and F.J. Paraguas. 2000b. Current status of production and consumption of tilapia in selected Asian countries. *Aquaculture Economics & Management* 4:107–115.

Dey, M.M. and M.V. Gupta. Socioeconomics of disseminating genetically improved Nile tilapia in Asia: An introduction. 2000c. *Aquaculture Economics & Management* 4:116–128.

Dey, M.M., A. Rabbani, K. Singh, and C.R. Engle. 2014. Determinants of retail price and sales volume of catfish products in the United States: an application of retail scanner data. *Aquaculture Economics & Management* 18:120–148.

Diamond, J. 1999. *Guns, Germs, and Steel: The Fates of Human Societies*. W.W. Norton & Company, New York.

Engle, C.R. 1997. Economics of tilapia aquaculture. In: Costa-Pierce, B.A. and J.E. Rakocy, eds. *Tilapia Aquaculture in the Americas*, vol. 1, pp. 18–33. World Aquaculture Society, Baton Rouge, Louisiana.

Engle, C.R. 2006. Marketing and economics. In: Lim, C. and C.D. Webster, eds. *Tilapia: Biology, Culture, and Nutrition*, pp. 619–644. Food Products Press, New York.

Engle, C.R., O. Capps, Jr., L. Dellenbarger, J. Dillard, U. Hatch, H. Kinnucan, and R. Pomeroy. 1990. The U.S. Market for Farm-Raised Catfish: An Overview of Consumer, Supermarket and Restaurant Surveys. Arkansas Agricultural Experiment Station, Division of Agriculture, Bulletin 925.

Food and Agricultural Organization of the United Nations (FAO). 1997. FAO database on introduced aquatic species. FAO, Rome. Available at www.fao.org.

Food and Agricultural Organization of the United Nations (FAO). 2014. FISHSTATPlus. FAO, Rome. Available at www.fao.org.

Gabriel, R. 1990. A market study of the portion-sized trout in Europe. Final Report, Federation of European Salmon and Trout Growers, Belgium.

Hanson, G.D., G.P. Rauniyar, and R.O. Herrmann. 1994. Using consumer profiles to increase the U.S. market for seafood: implications for aquaculture. *Aquaculture* 127:303–316.

Kinnucan, H.W., R.G. Nelson, and J. Hiariay. 1993. U.S preferences for fish and seafood: an evoked set analysis. *Marine Resource Economics* 8:273–291.

Kumar, G. 2015. Economics and risk of alternative production systems in channel catfish aquaculture. Ph.D. dissertation, University of Arkansas at Pine Bluff, Pine Bluff, Arkansas.

Kobayashi, M., S. Msangi, M. Batka, S. Vannuccini, M.M. Dey, and J.L. Anderson. 2015. Fish to 2030: The role and opportunity for aquaculture. *Aquaculture Economics & Management* 19:282–300.

Mair, G.C., J.S. Abucay, D.F. Skibinski, T.A. Abella, and J.A. Beardmore. 1997. Genetic manipulation of sex ratio for the large-scale production of all-male tilapia Oreochromis niloticus. *Canadian Journal of Fisheries and Aquatic Science* 54:396–404.

Millar, S. 2001. Salmon farmers braced for clampdown on toxins. *The Observer*, January 7.

National Marine Fisheries Service (NMFS). 2004. Imports and exports of fishery products annual summary, 2003. Can be found at www.st.nmfs.noaa.gov. Accessed July 2014.

National Oceanic and Atmospheric Administration and National Marine Fisheries Service (NOAA-NMFS). 2011. Fisheries of the United States. U.S. Department of Commerce, Washington, D.C.

Olsen, S.O. 2003. Understanding the relationship between age and seafood consumption: the mediating role of attitude, health involvement and convenience. *Food Quality and Preference* 14(3):199–209.

Petroczi, A. and D.P. Naughton. 2009. Mercury, cadmium and lead contamination in seafood: a comparative study to evaluate the usefulness of Target Hazard Quotients. *Food and Chemical Toxicology* 47:298–302.

Phelps, R.P. and T.J. Popma. 2000. Sex reversal of tilapia. In: Costa-Pierce, B.A. and J.E. Rakocy, eds. *Tilapia Aquaculture in the Americas*, vol. 2. The World Aquaculture Society, Baton Rouge, Louisiana.

Ponzoni, R.W., N.H. Nguyen, H.L. Khaw, N. Kamaruzzaman, A. Hamzeh, K.R.A. Bakar, and Y. Yee. 2008. Genetic improvement of Nile tilapia (*Oreochromis niloticus*) – present and future. In: Proceedings of the 8th International Symposium on Tilapia in Aquaculture, Cairo, Egypt, 2005, pp. 33–52.

Quagrainie, K. and C.R. Engle. 2002. Analysis of catfish pricing and market dynamics: the role of imported catfish. *Journal of the World Aquaculture Society* 33:389–397.

Rauniyar, G.P., R.O. Herrmann, and G.D. Hanson. 1997. Identifying frequent purchasers of seafood for at-home and restaurant consumption. *The Southern Business & Economic Journal* 20:114–129.

Seafood Business. 2001–2003. News archives. www.seafoodbusiness.com/archives. Accessed October 2014.

Seafood Watch. 2014. Monterey Bay Aquarium. www.seafoodwatch.org. Accessed July 23, 2014.

Thornton, P.K. 2010. Livestock production: recent trends, future prospects. *Philosophical Transactions of The Royal Society* 365:2853–2867.

Verbeke, W. and I. Vackier. 2005. Individual determinants of fish consumption: application of the theory of planned behaviour. *Appetite* 44:67–82.

World Bank. 2013. Fish to 2030: prospects for fisheries and aquaculture. Agriculture and environmental services discussion paper; no. 3. World Bank Group, Washington, D.C.

Young, J.A., S.L. Burt, and J.F. Muir. 1993. Study of the marketing of fisheries and aquaculture. Products in the European Community. European Commission Directorate-General for Maritime Affairs and Fisheries, Brussels.

CHAPTER 2

Demand and supply: basic economic premises

The concepts of demand and supply are fundamental to all aquaculture market-ing efforts. This chapter will provide definitions of these concepts along with a series of related aquaculture examples. Supply and demand together determine the price that is paid in the market for the quantity of product that is sold. Thus, understanding how supply and demand affect prices and quantities is critical to understanding markets for seafood and aquaculture. The types of factors that determine demand and supply relationships are explained along with the effects of changes in the levels of these factors. The chapter concludes with a discussion of special supply and demand conditions related to the interaction between wild-caught and aquacultured seafood. The salmon market synopsis describes one of the highest-volume aquaculture markets worldwide and one in which farm-raised production volumes have surpassed wild-caught volumes.

What is economics?

Many people view economics as a field that focuses entirely on money. While economists do spend a great deal of time estimating monetary values, economics is much more than a study of money.

Most people understand that an economy includes both production of goods and services by producers and consumption of goods and services by consumers. Consumers "demand" goods and services and producers "supply" the goods and services that consumers "demand." However, if it were that simple, there would be no need for an entire field of study such as economics. The problem is that there is no end to the wants and desires of human beings and no one has every-thing that they would like to have. Yet the resources needed to supply goods and services are often scarce. The fundamental problem addressed by economics is how to allocate scarce resources to meet unlimited wants and needs of human beings.

Seafood and Aquaculture Marketing Handbook, Second Edition. Carole R. Engle, Kwamena K. Quagrainie and Madan M. Dey.
© 2017 John Wiley & Sons, Ltd. Published 2017 by John Wiley & Sons, Ltd.

Fig. 2.1 A market includes the supply obtained from aquaculture growers and all the functions, transactions, transformations, and exchanges required to meet the demand of consumers.

Demand represents the needs and wants of human beings and supply represents the scarce resources that have been converted into goods that are needed and wanted by human beings (Fig. 2.1). The allocation process takes place within what is referred to as a "market." While many people are familiar with markets as places for consumers to purchase goods, a market is a much broader concept than a location where sales take place. The market encompasses the entire relationships of demand and supply but also includes the transformation of scarce resources into goods and the transactions that occur throughout the value chain from producers to processors, to wholesalers, and to the end consumer.

Where does the price of a product come in? The quantities of a product that are bought and sold at different prices send messages about how consumers value different products and what products producers can put on the market at various prices. Thus, the price of the good is the signal that sends information between producers and consumers about the extent and relative scarcity of resources used to produce that particular good and the extent to which consumers need and want that particular good. Thus, demand represents what people want and are willing to pay at different prices, and supply represents what producers can make available to the market at different prices. These forces of demand and supply interact in the marketplace until an "equilibrium" is achieved at which buyers and sellers agree to exchange a particular quantity at a particular price. This is the level that demonstrates "agreement" between producers and consumers about the value of the product and the terms of exchange. When does money enter this discussion? Money is simply the medium of exchange that is used in many transactions.

Demand

Demand represents a relationship between price and quantity of any particular product. More formally, consumer demand represents the various quantities of a commodity that consumers are willing and able to purchase as the price varies, when all other factors that affect demand are held constant (*ceteris paribus*[1]). Since economics involves the interaction of a number of parameters that continue to change, it is necessary to first hold many of these parameters constant. By doing so, it is possible to determine the fundamental relationships among different key parameters. This is what is referred to by the term "*ceteris paribus*." Once the relationship between key parameters is understood, then other parameter values can be varied and their effects analyzed. This type of analysis is referred to as a partial budget equilibrium analysis because changes in price in the market under consideration do not have dramatic effects on prices in other markets. Partial equilibrium analysis assumes that each market is independent and that it is self-contained. This concept is similar to the experimental designs used by aquaculture researchers to control for all but one or two variables in aquaculture experiments. Once the individual relationships between specific variables are understood, then more complex models can be built that allow several variables to vary at a time to begin to understand how these variables interact with each other. An analysis that encompasses the entire economy, including households, firms, markets, and income, would be classified as a general equilibrium analysis.

An analysis that seeks to identify a profit-maximizing management strategy, such as the optimal stocking or feeding rate for a fish farm, may assume a market price. Such an analysis would be a partial equilibrium analysis because an assumption is made that changes in the fish price will not result in changes in the prices of other goods (such as tractors or land) associated with fish production. However, an analysis that would evaluate the impact of new environmental regulations on fish farming would likely take into consideration that increased costs on fish farms may affect the quantity of product supplied and the market price of the product. As the price of that particular type of fish product changes, the quantity demanded of other, similar types of fish products is likely to change. This type of analysis would require a general equilibrium analysis.

The word "demand" is often misused, and it is important to understand the different contexts within which the term is used. For example, existing demand represents the quantities that would be purchased of a particular product for a range of specific prices. In other words, existing demand measures the quantity of fish taken from the market under all the imperfections that exist in any specific situation. Demand cannot be measured if there are no historic sales data from which to measure prices and quantities involved in the transactions. However, many people refer to potential demand when they use the term "demand." Potential demand represents the quantities that would be purchased of a new aquaculture product if it were available. Different analyses must be

---

[1] "*Ceteris paribus*" is a term that is used to mean "all else being equal," or "all else held constant."

used depending upon whether the intent is to analyze the relationship of quantities purchased at various prices or products already in the market or if the intent is to assess what consumers might do in terms of quantities they would purchase at various prices of new products.

Formally, demand is represented as a relationship between the quantities of product that consumers are willing and able to take off the market at all alternative prices of that product. Economists typically represent this relationship as a graph (Fig. 2.2). As price (P) goes up, the quantity demanded (Q) by consumers of that product typically goes down. Likewise, as the price (P) goes down, the quantity demanded (Q) by consumers of that product goes up. Demand can also be viewed as the maximum quantity of a product that is desired by consumers who are able to purchase the good at a given price. Demand can also be viewed

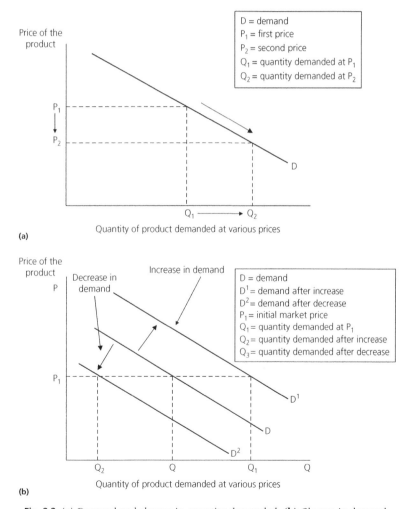

Fig. 2.2 (a) Demand and change in quantity demanded. (b) Change in demand.

as the maximum price that people are willing and able to pay for a given quantity of the good.

The resulting rate of change in the quantity demanded is represented by the slope of the curve in the graph (Fig. 2.2a). In aquaculture markets, this "law of demand" can be seen in salmon markets. As the market price of salmon dropped in the 1990s, the quantity demanded by U.S. consumers increased. Because quantity demanded will always move in the opposite direction from its price, the demand relationship between the quantity demanded and the price of the good is said to be negative, or inverse. A negative relationship is depicted graphically as a line that slopes downwards to the right. Figure 2.2a depicts a classic demand curve. It is important to identify whether changes in the marketplace are due to a change in the price of a good or if there has been a fundamental shift in the entire set of price and quantity changes. The above discussion refers to a change in the quantity demanded in response to a change in price with such a change tracked along the same demand curve, moving either up or down depending upon the direction of the price change.

A change in demand refers to a shift of the entire demand curve that represents the total relationship between price and quantity demanded (Fig. 2.2b). For example, salmon market prices (not adjusted for inflation) have trended upwards since 2003 in spite of increased volumes of production. Such a trend implies that other changes may be occurring in the market for salmon that may have resulted in a change in demand as compared to a change in the quantity demanded.

Changes in demand can occur when one of the following determinants of demand changes:
- population size and distribution;
- consumer income and distribution;
- consumer tastes and preferences;
- prices of other, related goods;
- availability of substitutes.

When these factors are held constant, *ceteris paribus*, then we can discuss and analyze the relationships and interactions of price and quantity demanded. However, when any of these factors change, then the demand curve itself will change as illustrated in Fig. 2.2b. As demand increases, the demand curve shifts to the right. When this happens, the quantity demanded is increased for all prices. If demand decreases, the demand curve shifts to the left. Consequently, for every price, then, the quantity demanded is less.

We can represent demand algebraically as follows:

$$Q_d = f\left(P_d, Pop_d, I, T, P_R\right)$$

where
Q_d = quantity demanded;
P_d = price of the product under investigation;

Pop_d=size of the population;
I=income;
T=tastes and preferences;
P_R=prices of related goods.
Traditional demand models use the quantity demanded as the dependent variable in a multiple regression analysis using econometric (statistical methods developed for economics analyses) methods. The independent variables frequently used are those listed above as factors or determinants of demand. Thus, the product's own price, population levels, consumers' incomes, consumer tastes and preferences, and prices of related products are frequently selected as independent variables.

A good example of the contrast between increases in quantity demanded and increasing demand is provided by shrimp consumption in the U.S. Shrimp was once a product consumed only occasionally and in certain consumer segments. Increasing quantities supplied of shrimp resulted in lower shrimp prices that resulted in a shift along the demand curve and increases in quantity demanded. In recent years, environmentalist groups opposed to shrimp production have developed advertising campaigns to convince consumers not to purchase shrimp due to the alleged environmental and social injustices related to shrimp production. If these advertising campaigns successfully change consumer tastes and preferences for shrimp, a decrease in demand could result. This decrease in demand would cause the demand curve to shift to the left, and quantities demanded would decrease at all prices. The factors affecting demand are further described below.

Population

The world's population is projected to grow to 8.3 billion by 2030 from 5 billion in 2003. Even with stable per capita seafood consumption, world demand for seafood would increase from 143 to 186 million metric tons (MMT) by 2030 from population increases alone. Aquaculture production will need to increase from 55 MMT in 2009 to 93 MMT by 2030 (World Bank 2013).

Income

Income levels of consumers also affect the demand for a good, but different types of goods are affected in different ways as income levels change. A good is classified as a necessity or a luxury good depending on the nature of the changes in expenditures on that good as a result of changes in income. For a good that is a necessity, the change in expenditure is less than proportionate to the change in price (income elasticity between 0 and 1), but for a luxury good, the change in expenditure is proportionately greater than the change in income.

For example, in many developing countries, fish such as tilapia are considered "poor people's food" (Neira et al. 2003). In such cases, as people's incomes rise, their consumption habits may change by substituting a higher-priced source of protein for what they perceive to be less desirable types of fish. Thus, consumers may begin to purchase large tilapia fillets instead of small, whole tilapia, or purchase shrimp or other higher-valued species instead of small, whole tilapia. Consumers might switch to filet mignon or some other type of protein altogether. A product that is considered to be "poor people's food" is classified by economists as an inferior good. An inferior good is defined as one for which demand would decrease with an increase in income levels.

Normal goods are those for which demand increases as income levels increase. An example might be imported farmed salmon in China. In this case, as income levels increase, consumers in China will desire to eat imported salmon as compared to lower-valued carps raised domestically. Some aquaculture products may be considered as superior goods in certain markets. Superior goods typically are quite expensive, luxury types of products.

This relationship between demand and income levels of consumers can change over the life of the product. The demand for tilapia over time provides a good example. Historically, tilapia were viewed as a low-valued product and were consumed mostly by limited-resource farmers around the world. However, demand for large, export-quality tilapia fillets has increased over time and the relationship has changed to where increasing income levels now typically result in increased demand for large, export-quality tilapia fillets.

Consumer tastes and preferences

Consumer tastes and preferences affect demand over time, and demand for different products will change as tastes and preferences change. A good example of this in seafood markets would be the demand for fresh salmon and tuna. With an increasingly health-conscious consumer population and information on the benefits of eating fish with high content of omega-3 fatty acids, the demand for tuna steaks and salmon products has increased. Demand for whole-dressed finfish has decreased over time as consumers' preferences have turned towards fresh and frozen fillets rather than whole-dressed fish.

Consumer behavior

Consumer expectations of the future will also affect demand for different goods. If consumers expect the economy to grow and individuals expect to enjoy increasing salaries and wages over the coming years, they are likely to spend more money on luxury goods. Since many seafood products in the U.S. are considered by consumers to be luxury products, expectations of a healthy, growing economy often will result in increasing demand for seafood products. However, if consumers expect poor economic growth, or are concerned over the safety of seafood products, demand may decrease.

A relationship between certain products in the minds of consumers may affect the demand for one of those products. Consumers will substitute some products for others; thus, substitutes can be considered to be competing or rival products. A seafood example of a substitute product is that consumers are likely to purchase either sea bass or trout to prepare for dinner, but they are unlikely to choose both sea bass and trout for the same meal. If the price of trout has recently gone up, someone considering trout might choose to purchase sea bass instead and the overall quantity demanded of sea bass will go up as the price of the substitute product (trout in this case) goes up.

Products can also be complements. Complementary products are those products that consumers tend to consume together at the same meal. For example, many people serve lemon wedges with fish; thus, lemons could be considered as complements to fish. If the price of fish goes down such that consumers increase the quantity demanded of fish, the quantity demanded of lemons would be expected to increase as well.

Supply

Supply is the amount of goods and services that producers are willing and able to offer in the marketplace at specific prices. Formally, supply is represented as a relationship between the quantity of product that producers are willing and able to place on the market at all alternative prices of the product. Economists typically represent this relationship as a graph (Fig. 2.3a). The data used are referred to as a "schedule." The supply schedule includes the alternative quantities (Q) offered for sale at different prices (P). As price goes up, the quantity supplied by growers typically goes up (Fig. 2.3a). Likewise, as the price goes down, the quantity supplied by growers goes down (Fig. 2.3a). This classic relationship can be seen in the hybrid striped bass market in the U.S. in the 1980s to 1990s. As the price of hybrid striped bass increased in this time period, the quantity supplied by U.S. growers increased. Because quantity supplied will typically move in the same direction as price, the relationship between the quantity supplied and the price of the good is said to be positive, or direct. A positive relationship is depicted graphically as a line that slopes upward to the right. Figure 2.3a depicts a classic style of supply curve.

As the price of a good changes, then the quantity supplied changes. To identify the change in quantity supplied, this change would be tracked along the same supply curve, moving either up or down depending upon the direction of the price change. Thus, a change in price results in a movement along the supply curve and generates a change in the quantity supplied. A change in the quantity supplied is distinct from a change in supply, just as a change in quantity demanded is distinct from a change in demand. Since supply represents the total relationship between price and quantity supplied, a change in supply reflects a shift in the entire supply curve.

Figure 2.3b shows that a shift to the right of the supply curve will result in increased quantities supplied at all prices. A decrease in supply will shift the supply curve to the left and will result in lower quantities supplied at all prices.

Price of the product

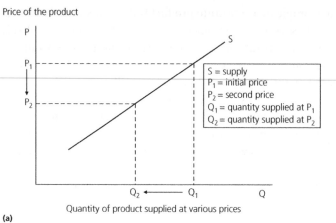

(a)

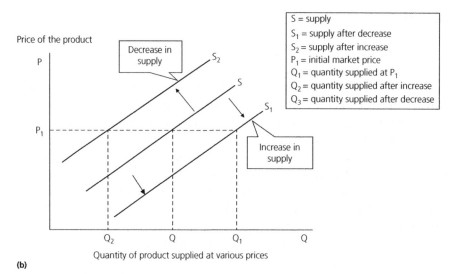

(b)

Fig. 2.3 **(a)** Supply and change in quantity supplied. **(b)** Change in supply.

Any given supply relationship between price and quantity supplied is based on holding constant the factors that affect supply. These include:
- changes in price of inputs;
- changes in price of related products;
- changes in production technology;
- changes in price of joint products;
- institutional and environmental changes – government regulations and programs.

When these factors are held constant, *ceteris paribus*, then the relationships and interactions of price and quantity supplied can be analyzed. However, when any of these factors change, then the supply curve itself will change.

We can represent supply algebraically as follows:

$$Q_s = f\left(\begin{array}{l} P_1, Price\ of\ inputs, Price\ of\ related\ commodities, Production\ technology, \\ Price\ of\ joint\ products, \text{and } Institutional\ and\ environmental\ changes \end{array} \right)$$

where
Q_s = quantity supplied;
P = price of the product under investigation;
Price of inputs = price of feed, labor, electricity, etc.;
Price of related commodities = price of other types of finfish or shellfish that consumers would consider switching to;
Production technology = change in the way the product is produced;
Price of joint products = price of a product that is produced in the same production system;
Institutional and environmental changes.

The farm-raised shrimp industry provides an example of the differences between changes in the quantity supplied versus changes in supply. When market prices fall, shrimp farmers produce less. Those farmers whose prices fall below their costs of production will go out of business. This situation represents a decrease in the quantity supplied and represents a movement downward along the supply curve. In contrast, improved feed formulations and the development of hatchery techniques to consistently and reliably supply shrimp seed (post-larvae) were major technological breakthroughs that resulted in increases in supply. Thus, the supply curve shifted to the right, and greater quantities were produced at all prices. An example of an institutional change would be the loss of a market due to regulations that prohibit the sale of a particular species due to its over-exploitation, with clear effects related to decreasing its supply.

Costs of production

Increases in production costs will cause supply to decrease. For example, increased costs of feed, labor, or utilities will decrease the quantity that producers can supply for any given price. Kouka and Engle (1998) showed that supply of food-size catfish would decrease by 2% with a 20% increase in the cost of feed, *ceteris paribus*.

Technology

Improved technologies can also cause supply to increase. Technologies that improve productivity by increasing the output per unit of input will result in increased supply. For example, the development of improved shrimp diet formulations and pellets that remained intact longer in the water contributed greatly to expansion of the shrimp industry (Csavas 1994). The improved feeds allowed farmers to increase feed efficiency and increase yields. Production costs declined as yields increased, and the result contributed to an increase in the supply of farm-raised shrimp. The development of efficient aerators in the 1980s resulted in a similar increase in catfish supply. With a consistent, reliable, and low-cost source of oxygen, farmers could stock and feed at higher rates and increase yields with lower yield risk. Supply increased as a result.

Price determination

The amount and price of a product are determined in the marketplace by the interactions between supply and demand (Fig. 2.4a, b, c). If producers seek too high a price, there will be fewer buyers who are willing and able to purchase the product at that price. Thus, the quantity demanded in the market will be less at higher prices. Not all the product offered will be removed from the market and either some producers will have to offer lower quantities or some producers will go out of business. In order to move their product, sellers will have to lower the price. At lower prices, fewer sellers will be able to sell at a profitable level and some will go out of business. However, at lower prices, the quantity demanded by consumers increases. At some point, equilibrium is reached in which the quantity demanded equals the quantity supplied for a given price. This is called market equilibrium and it is described by the equilibrium price and the equilibrium quantity.

When demand or supply changes, the equilibrium price and quantity change. For example, in Fig. 2.4a, the demand curve has shifted to the right, demonstrating an increase in demand. This increase in demand causes the equilibrium price and quantity to increase. If demand would decrease (shifting downwards to the left), then the equilibrium price and quantity would both decrease.

Similar effects on price and quantity occur with changes in supply. If supply decreases (shifts to the left), then the quantity decreases, but the price increases (Fig. 2.4b). If supply increases (shifts to the right), then the quantity will increase and the price will decrease.

Elasticity

The concepts of demand and supply are so fundamental to any discussion of market forces and effects that it is important to further explore the characteristics of demand and supply relationships. An important concept to any discussion of demand and supply is that of elasticity. Elasticity concepts measure the percentage change in the quantity demanded or supplied with a given percentage change in market price or income. The following will define and discuss various types of elasticity.

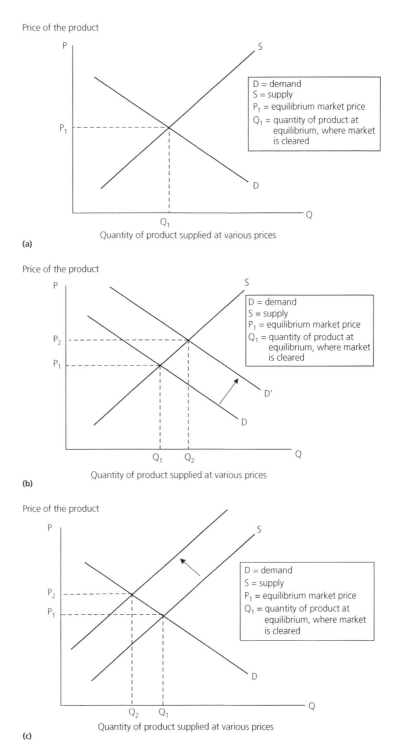

Fig. 2.4 (a) Price determination. **(b)** Price determination with an increase in demand. **(c)** Price determination with a decrease in supply.

Demand elasticity

Elasticity of demand is the degree of responsiveness of quantity demanded to a given change in price. Thus, elasticity is a measure of changes relative to a single demand curve, not changes in the determinants of demand that result in a shift in the demand curve itself. Elasticity of demand measures the change in quantity demanded as a result of a given change in price or the percentage change in the quantity demanded that results from a 1% change in one of the independent variables in the estimated demand equation. If the product's own price is used to calculate elasticity, this measure is also referred to as price elasticity of demand, or its own-price elasticity. It is measured as the percentage change in quantity demanded due to a percentage change in price, *ceteris paribus*.

Price elasticity of demand (E_d) is measured by the following equation:

$$E_d = \frac{Percentage\ change\ in\ quantity\ demanded}{Percentage\ change\ in\ price} = \frac{\%\Delta Q}{\%\Delta P}$$

It is important to note that, at this point, the discussion refers to the percentage change in price of the good itself that results from a 1% change in its "own" price. Later on, elasticities related to changes in price of a related good will be discussed.

It is important to note that the price elasticity of demand is not the same along the entire length of the demand curve and can vary with the nature of the curve itself. Price elasticity of demand is more elastic at demand relationships with higher prices and lower quantities whereas it becomes more inelastic as price decreases and quantity demanded increases. The midpoint along a demand curve corresponds to unitary elasticity. An example of how demand changes from being inelastic to elastic can be found in the cod market. The cod fishery in the 1800s exhibited characteristics of a product with highly inelastic demand. The abundance of cod resulted in lower prices that enabled cod to be transported around the world and to become a staple commodity for over a century. However, as cod stocks diminished and price increased, buyers began to substitute other types of fish for cod and quantities demanded decreased.

Elasticity does not measure the slope of the curve. If it did, it would not change along the length of the demand curve (straight lines have constant slopes). It does not determine the shape of the demand curve, but there are some shape relationships. For example, if the coefficient for the price elasticity of demand is greater than the absolute value of 1, then demand is considered to be elastic. This means that the percentage change in quantity demanded is greater than the percentage change in price. Thus, the quantity demanded is very responsive to price changes, or the relationship is very "elastic." In simplistic terms, the quantity demanded "stretches" a great deal with a small change in price when demand is elastic. Goods that are price elastic tend to be goods with many substitutes. If price goes up and there are other types of fish that can readily be substituted by consumers, buyers are likely to switch quickly to the other types of fish. Thus, the quantity demanded will decrease quickly as price increases if there are many substitutes.

However, if the absolute value of the coefficient of the price elasticity of demand is less than 1 ($E_d < 1$), then demand is considered to be inelastic. With inelastic demand, the percentage change in quantity demanded is less than the percentage change in price and the quantity demanded is not responsive to changes in price.

Figures 2.5a and 2.5b present the two extremes of perfectly elastic demand and perfectly inelastic demand, respectively. Products with perfectly elastic demand have very large numbers of other products that are very similar and among which consumers substitute readily. In markets where consumers readily substitute among a number of different types of fish fillets, a price change of an individual species will trigger substitution to other species. This ready substitutability may derive from the nature of seasonal catches of various species, such

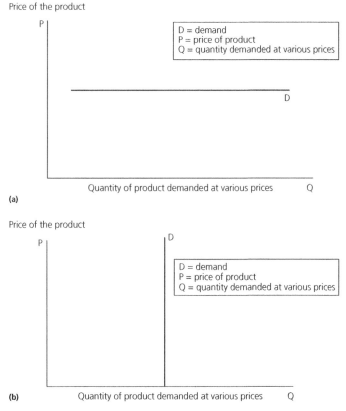

Fig. 2.5 (a) Elastic demand: elasticity is from 0 to infinity. **(b)** Perfectly inelastic demand.

that consumers become accustomed to changing to whatever species is available at that point in time. At the other extreme, if the quantity demanded of a product does not change regardless of changes in price, then demand is perfectly inelastic. Carp served for traditional Christmas Eve dinner in European countries would be expected to have highly inelastic own prices. If people believe that they need to pay whatever price to have the appropriate meal for an important holiday or ceremony, they are likely to purchase the same quantity regardless of its price.

The price elasticity of demand determines the extent to which the good is considered a necessity. Goods that are basic necessities will have inelastic demand. If a particular item is truly necessary for survival, consumers will purchase it regardless of changes in price. Thus, the quantity demanded will change very little even if price changes by a great deal. More formally, a 1% change in price results in less than a 1% change in quantity demanded. The opposite will also be true. If a good is a luxury good, then the price elasticity of demand will be highly elastic. The quantity demanded of a luxury good will vary greatly with relatively small changes in price. Fish in some developing countries is considered a necessity, therefore demand for fish in those markets is inelastic. Shrimp at one time was a high-priced seafood product with highly elastic demand. As price has decreased, shrimp demand has become less elastic, as the product has become less of a luxury item.

Price elasticity of demand for seafood products varies greatly. For example, Dey et al. (2008) estimated price elasticities for a wide variety of freshwater fish species in eight different countries in Asia. Elasticities varied by fish price levels, by consumer income levels, by species, and by country.

Cross-price elasticity

Cross-price elasticity measures the responsiveness of quantity demanded in one good to changes in price of a related good. It is measured by the following equation:

$$E_{x,y} = \frac{Percentage\ change\ in\ quantity\ demanded\ of\ good\ x}{Percentage\ change\ in\ price\ of\ good\ y} = \frac{\%\Delta Q_{dx}}{\%\Delta P_y}$$

where

$E_{x,y}$ = cross-price elasticity of good x with respect to changes in the price of y
Q_d = the quantity demanded of good x
P_y = price of good y
Δ = small change.

Cross price elasticity is used to measure the degree of substitutability of goods, or the degree to which goods compete in the same market. It measures the effect on quantity demanded of one product as a result of changes in price of another product. Completely unrelated goods have a zero cross elasticity. A negative sign shows an inverse relationship that indicates that the two goods are complements.

Substitute goods would have a positive relationship between the change in price of one and the quantity demanded of the other.

The availability of substitute goods will also determine the price elasticity of demand. The more substitutes that are available, the easier it will be for consumers to switch to another good when price increases. Thus, demand tends to be more price elastic, as more substitutes are available. However, if there are no close substitutes for a good, those consumers who really wish to purchase it will find it necessary to pay whatever the market price is. Thus, if the price increases, all else being equal and if substitutes are available, consumers will switch to the cheaper good. However, if seafood is not considered to be a good substitute for red meat and poultry consumption, then the demand for seafood will be more price inelastic.

The literature on substitutes among types of seafood is not clear. However, it would be reasonable to suppose that different species of marine fish fillets would substitute for each other. Thus, an increase in the price of orange roughy fillets might result in an increased quantity demanded of red snapper as an example. Shrimp and cocktail sauce could be considered complements. Decreasing prices of shrimp would be expected to increase quantity demanded of shrimp and also of cocktail sauce to accompany the shrimp.

Economists refer to income elasticity as a measure of the response of the quantity demanded to changes in income, *ceteris paribus*. Specifically, income elasticity is measured by the following equation:

$$E_I = \frac{Percentage\ change\ in\ quantity\ demanded}{Percentage\ change\ in\ income} = \frac{\%\Delta Q}{\%\Delta I}$$

Income elasticity will vary with the proportion of income spent on the product. Goods are classified as a necessity or luxury depending on the income elasticity. For a necessity, a 1% change in income results in less than a 1% change in quantity demanded. The value of the coefficient of income elasticity is between 0 and 1. For a luxury good, the income elasticity is greater than 1. The larger the proportion of income spent on the good, the greater the elasticity. Elasticity varies with the type of fish and the form in which it is presented. The sign of the coefficient is important. A negative sign indicates an inferior good because an increase in income results in decreasing quantities demanded. However, a positive sign indicates a direct relationship and a normal good. As incomes increase, quantities demanded also increase.

Price elasticity and total revenue

There is an important relationship between price elasticity of demand and total revenue. With an elastic demand, a decrease in price will result in a proportionately greater increase in quantity demanded. This is because the demand curve is relatively flat when demand is elastic. Thus, a decrease in price will increase total revenue $(TR = P \times Q$; where TR equals Total revenue, $P =$ Price, and $Q =$ Quantity). A price increase will result in a proportionately greater decrease in quantity demanded that will decrease total revenue if demand is elastic.

Conversely, with an inelastic demand, a decrease in price will result in a smaller proportionate increase in quantity demanded. Thus, a decrease in price will result in lower total revenue. An increase in price will result in a smaller proportionate decrease in quantity demanded. Thus, an increase in prices will result in increased total revenue.

Elasticity of supply

Elasticity of supply is a similar concept to that of demand elasticity. It measures the degree of responsiveness of the quantity supplied to changes in the price of the good.

The price elasticity of supply expresses the percentage change in quantity supplied in response to a 1% change in price, *ceteris paribus* and is calculated as follows:

$$ E_s = \frac{Percentage\ change\ in\ quantity\ supplied}{Percentage\ change\ in\ price} = \frac{\%\Delta Q}{\%\Delta P} $$

where
$E_s =$ elasticity of supply;
$Q =$ quantity supplied of good;
$P =$ price of the good;
$\Delta =$ small change.
A value of 0 means that supply is perfectly inelastic, or fixed. In other words, quantity supplied will not change irrespective of price changes. An elasticity value greater than 1 indicates that supply is elastic. In other words, a 1% change in price will result in a percentage change in the quantity supplied that is greater than 1.

Supply becomes more elastic as farmers have greater flexibility to respond to prices by holding crops for a better price or by switching to marketing a higher-priced product. Farmers who raise more than one type of crop have more flexibility when prices change. In such cases, supply can be more elastic. For example, some shrimp growers in Ecuador have diversified their farm production by co-culturing tilapia with shrimp in ponds. These species occupy different niches in the pond, offer more market opportunity, and can be used to reduce market risk.

Market structures and implications for competition and pricing

Economists use the term "market structure" to describe the factors that determine the competitiveness of the industry (Carlton and Perloff 2000). The degree of competitiveness is determined by (1) the number of firms (businesses), (2) the type of product (homogeneous, differentiated, or unique), (3) whether there is control over the price, and (4) the degree of freedom of entry and exit. The resulting classifications of market structures and differences among market structures with reference to seafood and aquaculture product processing are discussed in greater detail in Chapter 6. The market conduct of businesses within the industry affects market performance.

Fish farmers often discuss issues related to market power, or whether a particular level of the marketing chain has greater or less control over the price. Processors or other middlemen are often thought to exercise "unfair" control over prices. Market performance measures whether or not market power occurs in an industry. Metrics used to measure market performance include the rate of return, the price-cost margin, and Tobin's q (value of the market value of a firm to its replacement cost) (Carlton and Perloff 2000).

Special demand and supply conditions

The increasing production of farmed fish relative to wild-caught seafood implies that the productivity for farmed fish production has been increasing faster than that of wild-caught seafood (Asche 2001). If the two products are close substitutes, farmed fish can then win market share from wild-caught fish. Moreover, if demand is not perfectly elastic, the price will decline as will the income of the producers of wild-caught fish. However, if the goods are not substitutes there are no market effects and the increase in the supply of the farmed fish will only lead to a move down the demand schedule for farmed fish. Hence, for producers of farmed fish it is easier to expand when farmed fish has substitutes with established markets.

This situation changes if the potential substitute is fish from a fishery located on the backward-bending part of the supply schedule (Anderson 1985). Overfishing can result in a backward-bending supply schedule for captured fish (Fig. 2.6) (Anderson 1985). A backward-bending supply curve indicates that, as price continues to increase, the quantity supplied begins to decrease. This can happen when over-exploitation of fish stocks results in scarcity that causes price to increase. When the scarcity of fish stocks results in decreasing spawning and recruitment to the fishery, the decline in stocks leads to decreased quantity supplied, and the supply curve bends in a backward fashion. Since many of the world's fish stocks are reported to be fully or over-exploited, it is likely that the market equilibrium for them is on the backward-bending part of their supply schedule (Asche 2001). The increased supply of farmed fish can then lead to a greater supply of wild-caught fish in the short term that leads to sharp competition among suppliers. Price will decline and fishermen's revenues may increase or decrease depending on the slope of the backward-bending supply schedule. However, stock size will increase as one moves down the backward-bending

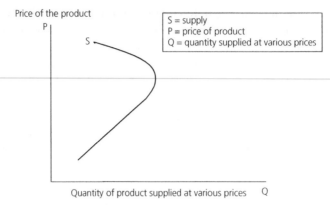

Price of the product

P

S

S = supply
P = price of product
Q = quantity supplied at various prices

Quantity of product supplied at various prices   Q

Fig. 2.6 Backward-bending supply curve.

supply curve. If a fishery is on the "normal" part of the supply schedule, the effects will be as for conventionally produced goods, and the reduction of supply that is caused by decreased price will also tend to enhance stock size.

Aquaculture market synopsis: salmon

The global supply of salmon has generally increased rapidly over time (Fig. 2.7). However, supplies of wild-caught salmon leveled off in the late 1980s, and the increases in world production since then have come from farmed salmon production. By 2012, farmed salmon production composed 71% of all salmon production worldwide.

The largest supplier of farmed salmon worldwide is Norway, with 60% of total farmed production in 2012 (Fig. 2.8). Chile is the second-largest producer, with 19% of farmed production in 2012. The Russian Federation, the U.S., and Japan are the major suppliers of wild-caught salmon. To provide perspective, Norway's production alone of farmed salmon exceeded that of the total wild-caught supply of salmon worldwide in 2012.

There are a number of different salmon species sold on the world market and several of these are cultured on farms. However, 92% of the aquaculture production of farm-raised salmon is the Atlantic salmon (*Salmo salar*) with some (8%) additional production of coho salmon (*Oncorhynchus kisutch*) and a very small amount of chinook salmon (*Oncorhynchus tshawytscha*). Wild catches of salmon are primarily based on pink salmon (*Oncorhynchus gorbuscha*), chum salmon (*Oncorhynchus keta*), and sockeye salmon (*Oncorhynchus nerka*).

Salmon culture technologies were originally developed for enhancement of wild stocks of salmon. Some of the earliest reports are of a U.S. Fish Commission salmon hatchery in California in 1872 (Thorpe 1980). This was soon followed by hatcheries in Japan and Alaska (U.S.) to produce salmon for re-stocking natural

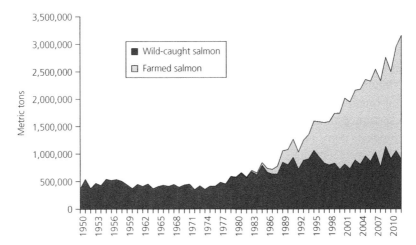

Fig. 2.7 Volume of wild-caught and farmed salmon, 1950–2012.

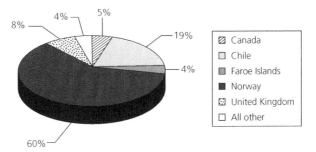

Fig. 2.8 Major suppliers of salmon, 2012. Source: FAO (2014).

populations. Culture technologies began to be adopted successfully on a commercial scale in the 1960s (Heen et al. 1993; Avault 1996; Anderson 2003).

Advances in foodfish production technologies led to rapid growth of the salmon industry in the 1990s. Asche and Bjorndal (2011) characterize these advancements collectively as developing control over production that provided a basis for more systematic and predictable supplies of salmon that led to stabilized processing and marketing strategies.

Such technological developments have reduced costs of production through improvements in efficiency (Asche and Roll 2013) and in quality. The development of pelletized feeds in the early years reduced wastage, and more recent developments have reduced feed costs by reducing the amount of more expensive fish meals in the diet. Nutritional advances further contributed to reductions in feed conversion ratios. Productivity growth of production of inputs such as smolts has also contributed to reduced costs (Sandvold and Tveteras 2014). Additional cost reductions were achieved through the development of vaccines that have reduced mortalities due to disease. Such improvements in health management of salmon resulted in a decrease in use of antibiotics and other chemicals

that further reduced costs and triggered rapid industrial growth. Additional technological advancements leading to industry growth included adaptive feeding
systems using cameras and infrared technology to reduce waste feed, synthetic
astaxanthin, and improved genetic lines through selective breeding (Thodesen
et al. 1999).

Wild-caught salmon has been sold since the earliest years of salmon processing as a canned product. However, farmed salmon is sold primarily as a fresh
fillet product. New packaging technologies, such as leak-proof Styrofoam™
packaging, were developed in the 1980s for farmed salmon and provided a
means to increase air shipment of fresh salmon (Anderson 2003). Improved
packaging and logistics infrastructure have made the distribution by air freight
feasible from Norway and Chile to major markets in Japan and the U.S., particularly for a high-valued product like salmon (Asche and Bjorndal 2011). The
major salmon products sold in recent years have been fresh and frozen fillets.

As supplies have grown and increased, the real price for salmon has dropped
over the long term (Fig. 2.9). Overall, price fell by nearly 70% between 1980
and 2007. Salmon has apparently moved from being a high-priced luxury product to more of a staple product.

International markets have developed over the last decade, mostly due to
aquaculture production (Bjorndal et al. 2003). The largest markets for salmon
globally are the European Union, Japan, and the U.S. However, new salmon
markets are developing in Central and Eastern Europe, Southeast Asia, China,
and South America. Most salmon is sold fresh to the European Union.

The salmon industry has turned to development of a wide range of pre-
packed and value-added products. In the European Union, various smoked
salmon products are popular product forms that account for sizeable proportions
of sales. The development of pin-bone-out salmon fillets in Chile resulted in
expansion of the U.S. salmon market into non-traditional market segments.

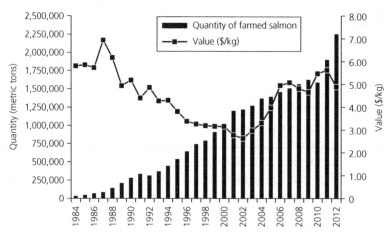

Fig. 2.9 Value and quantity of salmon supply, 1984–2012.

In the U.S., salmon is now sold in a wide variety of market outlets, including restaurants, cafeterias, and grocery stores (Bjorndal et al. 2003). Salmon is estimated to be on the menu of 39% of all restaurant menus in the U.S., including 71% of fine dining, 71% of hotel/motel, and 49% of casual/theme restaurant establishments.

Much of the growth in Chilean imports into the U.S. has been in the form of value-added fillets (Bjorndal et al. 2003). Salmon imported from Canada into the U.S. is mainly round (headed and gutted) product. While Denmark appears in statistics for salmon exports, these result from the re-export of fresh or smoked product produced from fresh salmon imported from Norway (Anderson 2003).

The farmed salmon industry has become increasingly concentrated over time. The share of salmon produced by the 10 largest companies increased from 44% to 54% in 2008 (Asche and Bjorndal 2011). The increasing scale of salmon farms has also contributed to reduced costs of production (Asche et al. 2013). However, advantages to scale may be more related to marketing and acquisition of services and in regulation compliance than in production. While the largest salmon company, Marine Harvest, accounted for more than 20% of global production in 2010 (Nilsen and Grindheim 2011), the overall levels of concentration in the two major producing countries, Norway and Chile, are moderate. Ownership structures have continued to become more international with Norwegian interests in both the Chilean and Scottish industries, although several of the largest multinational companies have had headquarters in The Netherlands and Japan as well as in Norway and Chile (Asche and Bjorndal 2011). Salmon farms have integrated vertically into processing facilities with sales offices in several countries. In Chile, the four largest firms accounted for 35% of exports in 2001 while the 10 largest accounted for 60% of exports (Bjorndal et al. 2003).

The growth of the salmon industry and market has been accompanied by a series of international trade conflicts. Asche and Bjorndal (2011) provide details of these, dating back to 1989. These have included the following:
- Scottish farmers vs. the European Union;
- U.S. commercial salmon fishermen vs. Norwegian salmon farmers;
- U.S. commercial salmon fishermen vs. Chilean salmon farmers;
- European Commission vs. Norway and Faroe Islands;
- Ireland and UK vs. Norwegian, Faeroese and Icelandic salmon;
- European Union farmers vs. Norway.

Many of the trade disputes involve allegations of dumping (selling below fair market price) or non-tariff phytosanitary issues related to food safety. Such disputes frequently are litigated over many years with ongoing adjustments and modifications.

The salmon industry has been at the center of numerous accusations from environmentalist non-governmental organizations (NGOs). Salmon production has been labeled as unsustainable and environmentally unsound for the following alleged reasons: (1) use of Atlantic salmon in Pacific waters has potential for escaped fish to weaken the genetic pool in the Pacific Ocean; (2) discharge of waste products from the net pens where salmon are raised pollutes surrounding waters; (3) mercury and polychlorinated biphenyl (PCB) concentrations are higher in farm-raised than in wild-caught salmon; (4) the use of astaxanthin in salmon feeds is unnatural and should be labeled as an additive; and (5) the use of fish meal and fish oil in salmon will lead to over-fishing of pelagic species upon which other species and fisheries depend. Many of these claims by NGOs have been exaggerated and information has been used incorrectly out of context. Moreover, technological developments have reduced discharges to the environment through new feeding and monitoring systems and dramatically reduced use of fishmeal and fish oil in salmon diets. Vaccines have resulted in similarly dramatic reductions in use of antibiotics and chemicals. Nevertheless, the very active opposition of some environmental NGOs to farm-raised salmon production has constrained sales and dampened market growth to some degree.

The salmon industry, as many other segments of aquaculture, is regulated by a variety of international, national, state or provincial, and local policies and regulations. The regulatory environment varies greatly by country. In the U.S., for example, resistance to offshore aquaculture and the prohibition of commercial foodfish production of salmon by the state of Alaska have effectively precluded industrial development in spite of abundant resources (Engle and Stone 2013; Kite-Powell et al. 2013). In Norway, salmon production is regulated primarily with a license that governs entry into the business, farm location, farm size, and ownership (Asche and Bjorndal 2011). The licensing system has been used to influence the rate of growth with respect to market development, support services (i.e., fish health and research and extension support), and industry structure. The regulatory environment in Chile is more complex with more than 30 types of compliance policies (Asche and Bjorndal 2011). Nevertheless, the Chilean government has actively promoted development of salmon production in Chile. Devastating disease outbreaks in 2009 led to a dramatic increase in staff and budget in the Chilean Servicio Nacional de Pesca to implement policies designed to manage and prevent additional disease outbreaks (Engle and Stone 2013).

Summary

Economics is the study of how resources are allocated to meet the unlimited wants and needs of consumers. The forces of demand and supply interact to determine the equilibrium price and quantity. Demand and supply curves can change as the determinants of either demand or supply change. Elasticities provide further insight into the demand and supply relationships.

Study and discussion questions

1 Draw the graphs in Figure 2.2. Now draw graphs showing an increase in demand and then a decrease in demand and discuss what would happen to price and quantity with each. Provide an aquaculture example (different from those in the text) for each.
2 Draw the graphs in Figure 2.3. Now draw graphs showing an increase in supply and then a decrease in supply and discuss what would happen to price and quantity with each. Provide an aquaculture example (different from those in the text) for each.
3 Draw graphs of what you think the following would look like:
   (a) Perfectly inelastic demand
   (b) Perfectly elastic demand
   (c) Elastic demand
   (d) Perfectly inelastic supply
   (e) Perfectly elastic supply
   (f) Inelastic supply.
4 Draw a backward-bending supply curve and explain how the biological growth curve and exploitation levels can result in this type of supply curve.
5 Choose five aquaculture species and list the types of changes in the factors that affect demand and supply that would cause the volume sold of those species to increase and those that would cause it to decrease.
6 Describe why it is important to understand the elasticity for the seafood and aquaculture products sold in terms of making pricing decisions. Provide an aquaculture example (other than those described in this chapter) of this importance.
7 Describe two examples of substitute seafood products and how a price change in one will affect sales of the other.
8 Describe two examples of complementary seafood products and how a price change in one will affect sales of the other.
9 Describe a seafood product that is income elastic and another that is income inelastic. Use examples other than those provided in this chapter. Explain how an effective marketing strategy would be different for each.
10 Describe some of the advances that have led to the rapid growth and development of the salmon industry.

References

Anderson, J.L. 1985. Market interactions between aquaculture and the common-property commercial fishery. *Marine Resource Economics* 2:1–24.
Anderson, J.L. 2003. *The International Trade in Seafood*. Woodhead Publishing Limited, Cambridge, UK.
Asche, F. 2001. Testing the effect of an anti-dumping duty: the US salmon market. *Empirical Economics* 26:343–355.

Asche, F. and T. Bjorndal. 2011. *The Economics of Salmon Aquaculture*. 2nd edition. John Wiley and Sons, Oxford, UK.

Asche, R. and K.H. Roll. 2013. Determinants of inefficiency in Norwegian salmon aquaculture. *Aquaculture Economics & Management* 17:300–321.

Asche, F., K.R. Roll, H.N. Sandvold, A. Sørvig, and D. Zhang. 2013. Salmon aquaculture: larger companies and increased production. *Aquaculture Economics & Management* 17:322–339.

Avault, J.W. 1996. *Fundamentals of Aquaculture*. AVA Publishing Company, Baton Rouge, Louisiana.

Bjorndal, T., G.A. Knapp, and A. Lem. 2003. Salmon—a study of global supply and demand. FAO/Globefish Research Programme Volume 73. FAO Fisheries Department, Rome.

Carlton, D.W. and J. M. Perloff. 2000. *Modern Industrial Organization*. Addison-Wesley Longman, Reading, Massachusetts.

Csavas, I. 1994. Important factors in the success of shrimp farming. *World Aquaculture* 25:34–55.

Dey, M.M., Y.T. Garcia, with P. Kumar, S. Piumsombun, M.S. Haque, L. Li, A. Radam, A. Senaratne, N.T. Khien, and S. Koeshendrajana. 2008. Demand for fish in Asia: a cross-country analysis. *Australian Journal of Agricultural and Resource Economics* 52:321–338.

Engle, C.R. and N.M. Stone. 2013. Competitiveness of U.S. aquaculture within the current U.S. regulatory framework. *Aquaculture Economics & Management* 17:251–280.

Heen, K., R.L. Monahan, and F. Utter. 1993. *Salmon Aquaculture*. Fishing News Books, Oxford, UK.

Kite-Powell, H.L., M.C. Rubino, and B. Morehead. 2013. The future of U.S. seafood supply. *Aquaculture Economics & Management* 17:228–250.

Kouka, P.J. and C.R. Engle. 1998. An estimation of supply in the catfish industry. *Journal of Applied Aquaculture* 8:1–15.

Neira, I., C.R. Engle, and K. Quagrainie. 2003. Potential restaurant markets for farm-raised tilapia in Nicaragua. *Aquaculture Economics & Management* 7:1–17.

Nilsen, G.B. and J. Grindheim. 2011. *The World's 30 Biggest Salmon Producers*. Intrafish Media, Bergen, Norway.

Sandvold, H. and R. Tveteras. 2014. Innovation and productivity growth in Norwegian production of juvenile salmonids. *Aquaculture Economics & Management* 18:149– 168.

Thodeson J., B. Grisdale-Helland, S.J. Helland, and B. Gjerde. 1999. Feed intake, growth and feed utilization of offspring from wild and selected Atlantic salmon (*Salmo salar*). *Aquaculture* 180:237–246.

Thorpe, J.E. 1980. *Salmon Ranching*. Academic Press, New York.

World Bank. 2013. Fish to 2030: prospects for fisheries and aquaculture. Agriculture and environmental services discussion paper; no. 3. World Bank Group, Washington, D.C.

CHAPTER 3

Seafood and aquaculture marketing concepts

The purpose of this chapter is to help readers understand some of the key marketing concepts used throughout the rest of the book. This chapter is particularly useful as a quick review of these concepts with detailed explanations and illustrations provided in the following chapters. The market synopsis on shrimp at the end of the chapter outlines the development of an industry that has grown to have major impacts on the world supply of that species and has had a major effect on world trade.

What is marketing?

There are various definitions of marketing. According to the American Marketing Association (2013), "Marketing is the activity, set of institutions, and processes for creating, communicating, delivering, and exchanging offerings that have value for customers, clients, partners, and society at large." In simple terms, seafood and aquaculture marketing covers all the processes that occur between the moment the product leaves the farm or fishing boat and when it is consumed by the end user. Seafood and aquaculture products must be harvested, transported, and assembled in adequate volume for re-sale. Many products are processed in some fashion before re-sale and consolidated by product form to provide volumes that are large enough to be traded and negotiated. Advertising programs are designed to increase demand for the product by communicating the attributes of the product and are included in the marketing process. Sales are also a part of marketing but, especially in today's complex economy, sales will not occur in the absence of the other marketing processes.

Seafood and Aquaculture Marketing Handbook, Second Edition. Carole R. Engle,
Kwamena K. Quagrainie and Madan M. Dey.
© 2017 John Wiley & Sons, Ltd. Published 2017 by John Wiley & Sons, Ltd.

Marketing plan

Every seafood and aquaculture business should have a well-defined marketing strategy defined in a written marketing plan. A comprehensive marketing plan includes an assessment of the current market situation, identification of opportunities and threats to the business, and a clearly defined marketing strategy. The marketing strategy developed should include a market summary, description of market demographics, market trends, market growth, analysis of strengths and weaknesses of the company, product offerings, specific objectives, financial analysis of the relative costs associated with these market objectives, and a monitoring and control plan. The plan should define the product, identify buyers and sellers, and articulate the market rules. Chapter 9 includes more detail on developing marketing plans.

Market products

Selection of the specific product or products to be marketed is a key decision for the business. The company needs to effectively articulate what is new or different about the product and understand how consumers view their product(s) when compared to the competition. When products available on the market are virtually identical, the products are said to be homogeneous; if products are significantly differentiated, they are said to be heterogeneous. For example, if the only tilapia product available on the market were a frozen fillet, then tilapia would be considered a homogeneous product. However, if there are diverse tilapia products with different characteristics, such as battered, breaded, stuffed, dried, marinated, or canned, it is possible to distinguish or differentiate each specific product from the other. Any product or service offered to the market must be defined clearly and compared to competing products available on the market.

Supply chain and value chain

The terms "supply chain" and "value chain" are often used interchangeably. Although there is not one standardized definition for either term, these concepts differ in terms of their emphasis (Yu et al. 2008; Bjorndal et al. 2014; Dey et al. 2015). A supply chain is a network of product-related business enterprises used to produce and deliver a product to its final consumers. A value chain analysis puts a dollar value on each step in a supply chain. Value chains add incremental value to the products in the nodes of a supply chain either by value addition or value creation (Dey et al. 2015). A typical seafood value chain consists of harvesting (either through aquaculture or fishing or a combination of both), processing, distribution, and finally consumption. Jacinto and Pomeroy (2011) provide a description of value chain analysis as it relates to seafood and aquaculture.

Processors

Processors add what is called "form utility" to raw farm products by processing live fish into more convenient product forms such as fillets, steaks, or nuggets. Processors may also add value to aquaculture products by providing transportation from farms to processing plants, and storing processed product in coolers and freezers until it is sold. Some processing companies even extend credit to farmers to help them finance their production operations. Chapter 6 presents additional detail on seafood and aquaculture product processing.

Some food processing industries have a dominant core of a few large firms that produce well-known brands, advertise, and have a strong influence on product price. National processors concentrate their selling efforts on innovation, quality, and other forms of non-price competition. Others consist of a large number of smaller firms that process products under wholesaler and retailer private labels. These competitive fringe food processors rely largely on price competition for their success. Brands allow processors to differentiate products and certify product quality. Food processing firms are among the nation's leading advertisers of food products, and food products are the most heavily advertised consumer products.

The trend in food processing has been to consolidate into fewer, but larger processing companies. This concentration is typically expressed as the share of the market controlled by the top food processing firms. As an example of recent trends, the top 20 food processing firms in the U.S. increased market share from 36% in 1987 to 51% in 1997 (ERS 2002–2014).

Processing involves significant investment in facilities and equipment, and thorough planning is critical to select the most efficient size levels of processing plants. However, it is often difficult to determine the optimal number and size of plants. Plants will run efficiently when running close to full capacity because a high proportion of the fixed costs are in building and equipment infrastructure. Supplies from fish farms may fluctuate due to the time of the year, changing feed prices, or availability of fingerlings or seedstock. Market demand for seafood affects the price, as do prices of other similar types of fish that may be caught from the wild or imported at lower prices. Fluctuations in supply will affect the plant's ability to operate near its capacity. Replacing a single, large plant with several smaller ones reduces some assembly and transportation costs but may require sacrificing the operational efficiencies of large-scale centralized plants.

The major market channel for U.S. catfish farmers has been through processing plants. In 2005, 73% of all foodfish produced in the United States were sold to processing plants (U.S. Department of Agriculture 2005). U.S. catfish processing plants are specialized in processing catfish and have well-automated systems designed to handle large quantities of that species. While higher prices can be obtained in other marketing channels, processing plants are the only market channel option that can absorb the production volume of the majority of catfish farms (Kinnucan et al. 1986).

Market or distribution channels

Market channel decisions are some of the most important decisions made by a company. Marketing channels can be thought of as customer value delivery systems in which each channel member adds value for the customer. Examples of companies that have successfully identified and utilized a market and distribution channel as a key component of their overall business strategy include: (1) FedEx in small package delivery; (2) Dell Computer sales directly to consumers; (3) Charles Schwab delivering financial services on the Internet; and (4) Caterpillar's network, powerful support, and partnership with dealers. Their respective market channel strategies have led these companies to become dominant in their respective product categories. Additional information on market channels can be found in Chapter 5.

Most producers use some type of intermediary to get their product to market, thus forging a distribution channel. The use of intermediaries is most common when growers or fishermen do not have the resources to develop marketing capabilities in transportation and storage. The importance of intermediaries lies in enhancing efficiency of the distribution system and thereby reducing marketing costs. Intermediaries also help producers reach otherwise unreachable customers. An efficient organization of the distribution system requires the performance of some key distribution functions. These include:

1 gathering information and conducting market research and intelligence important for market planning;
2 promotion and advertising products;
3 search and contact by finding and communicating with prospective buyers;
4 matching products to buyer needs through grading, assembling, and packaging;
5 marketplace negotiations on price and other contractual arrangements;
6 physical distribution of products through transportation and storage;
7 financing the costs of distribution;
8 assuming some commercial risks by holding stocks.

All of the above functions need to be undertaken in any marketplace. Intermediaries perform these functions to create a supply chain and a total distribution system that serves customers. In some cases the food processor sells directly to the consuming public. Many food companies market directly to consumers during holiday seasons such as Christmas and Thanksgiving in

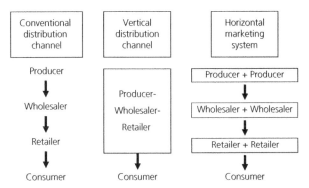

Fig. 3.1 Various forms of distribution channels.

addition to marketing through traditional intermediaries. While the use of inter-mediaries requires the grower to give up some control, the contacts, experience, specialization, and scale of operation often allow intermediaries to offer the firm more than it can achieve on its own (Armstrong and Kotler 2003).

There are three basic forms of marketing channels for delivering products from the producer to the consumer (Fig. 3.1): conventional distribution channels, vertical distribution channels, and horizontal marketing systems. Conventional distribution channels typically consist of one or more independent producers, wholesalers, and retailers. Each is a separate business that seeks to maximize its own profits, not those of the system as a whole.

Vertical distribution channels are composed of producers, wholesalers, and retailers that are part of one marketing system. Typically, one member is strong enough to either own the others or wield enough power to insist on cooperation (Armstrong and Kotler 2003). Vertically integrated corporations are single business entities that control all phases of the distribution channel. Some vertically integrated systems are based upon contracts among independent firms at different levels of production and distribution. Franchises, retailer cooperatives, and wholesalers that organize voluntary chains of independent retailers integrate vertically through contracts. Members of a top brand, such as Walmart, can administer or exert strong influence on suppliers through the company's size and power.

Horizontal marketing systems are those in which two or more companies join together at one level to pursue a new marketing opportunity by combining capital, production capabilities, and marketing resources. Kentucky Fried Chicken franchises in a Shell gas station create joint benefits from co-location of retail outlets. Companies may use different types of distribution channels to target different market segments.

E-commerce and other technological advances have led to a degree of disintermediation (bypassing intermediaries to sell directly to final buyers). Web-based sales allow sellers to capture the profit margins of the entire marketing chain. However, to be successful with Internet marketing, the seller must assume all the customer service, shipping, and advertising functions that are critical for success. Moreover, some types of products are less amenable to Internet sales than some others. Examples of marketing channel systems for specific commodities are presented in Chapter 5.

Transportation

Transportation is a marketing function that provides what is called "place" utility. In order to sell a product, the buyer and seller need to be able to physically make the exchange. The availability of adequate transportation alternatives can affect the type of product that can be sold, the quality of the product, the timeliness of deliveries, and the volume of product that can be moved. Lack of transportation can result in lack of access to specific markets, reduce competitiveness in preferred markets, and restrict the growth and development of the business. The types of transportation available can affect storage requirements, inventory costs, and the location of processing plants. Transportation costs will affect marketing margins and food prices.

Improvements in transportation technologies have allowed for the emergence of complex, global markets for a wide variety of food products. Much of the fish and seafood consumed around the world is transported to other countries, other continents, or the other side of the world while maintaining freshness and quality.

Transportation is especially critical to marketing fish and seafood because of its perishable nature. Moreover, the diversity of types of aquaculture products and markets has resulted in a wide variety of methods of transportation used throughout the world for fish and seafood markets. Even today, while shrimp, salmon, and tilapia are air freighted around the world, fish and seafood are still transported by bicycle (Kada 1997; Jagger and Pender 2001), pickup trucks (Jagger and Pender 2001), or boats with live holds (Phan et al. 2009).

Many fish and seafood products continue to be transported live. Live fish often sell for a higher price in many countries. Within the U.S., Europe, and China, farmed fish are hauled live in large hauling trucks equipped with liquid oxygen to processing plants or pay lakes. Baitfish species are hauled to wholesale distributors and re-loaded onto smaller trucks for transportation to retail bait shops. Fingerlings of many species are hauled by truck from hatcheries to grow-out facilities that may be only a short distance away or may be a several-hour drive. Boats with wells and mechanical oxygen are used in the salmon and other industries to transport live fish to processing plants (Schoemaker 1991). Boats with hulls that have holes in the sides to allow for water exchange have been used in China (FAO 1978). In the Mekong Delta region of Vietnam, fish are transported in net-like enclosures suspended beneath boats (Phan et al. 2009).

Fish fry are often shipped live in plastic bags with oxygenated water. The bags are typically packed in Styrofoam™ boxes as insulation for temperature control so that fry can be shipped around the world in good condition.

Shellfish must be kept shaded and cool in a humid environment (Schoemaker 1991). While shipping practices vary by species, typically they undergo some type of conditioning process prior to packing. Packaging frequently involves molded Styrofoam™ boxes with plastic liners and ventilation holes (Wingenter et al. 2013). Modified atmospheric packaging can extend shelf life by 48–72 hours (Pastoriza et al. 2004). An oxygen-saturated atmosphere has been shown to increase survival rates of scallops (Christophersen et al. 2007), abalones (Bubner et al. 2009), and mussels (Pastoriza et al. 2004).

Processed fish may be transported to local grocery stores or restaurants or shipped across the world. Fresh fillets are chilled before packaging and arranged to avoid direct contact between fillets and ice. Refrigerated transport is required for shipping frozen fish products with temperatures maintained at −30°C to −25°C for most products.

Fresh tilapia fillets are flown from Central America to markets in the U.S., while most frozen tilapia fillets are imported into the U.S. from China, Taiwan, and Indonesia. Careful coordination of harvesting, processing, and shipping times has been a key to the success of tilapia companies that export to the U.S.

In developing countries, access to markets can be a critical problem, especially if there are few transportation alternatives. Poor road conditions, unreliable vehicles, and lack of ice can prevent aquaculture products from reaching those markets with the greatest demand for their product. Leyva et al. (2006) developed an analysis of optimal markets for different sizes of tilapia farms located in different locations in Honduras. The mixed-integer trans-shipment mathematical programming model explicitly accounted for varying costs associated with different truck sizes, varying distances to various markets, and seasonality of demand. The models were used to suggest recommendations for farmers on the most profitable cities and outlets to target.

Wholesaling

The wholesaling process includes all functions associated with selling products to companies that then re-sell the products to other buyers. Typically, wholesalers buy from producers and re-sell to retailers or other wholesalers. To do so, they operate buying offices, warehouses, trucking and delivery services. Businesses will choose to sell to a wholesaler often because the wholesaling company may be more efficient at selling and carrying out other marketing functions. Wholesalers promote products, build variety to meet customer demand, break bulk quantities into smaller lots for customers at lower prices, warehouse product to offer adequate inventory, and also transport the product. Some wholesalers finance customers and suppliers with credit and bear risk related to title and theft, damage, spoilage, and obsolescence. A wholesaler may provide information to suppliers and customers on competitors, new products, and price development.

They may help retailers train sales clerks, improve store layouts and displays, and set up accounting and inventory control systems. Since retailers are focused on servicing customers, they often find it difficult to search out suppliers to source all the types of products offered in the store. Similarly, processors often are not in the best position to fully meet needs and demands of retailers. Successful wholesalers step into this gap to facilitate coordination between processors and retailers. Additional information on wholesaler marketing can be found in Chapter 5.

Recent market trends have made it more difficult to distinguish between retailers and wholesalers. Wholesaler clubs and hypermarkets may be operated by retailers but perform wholesale functions. Some large wholesalers such as SuperValu may perform retail functions. Rising costs combined with demand for increased services squeeze wholesale profit margins and require wholesalers to find ways to deliver more value to customers. Many large wholesalers are now expanding to operate on a global level.

Brokers

Brokers and agents primarily buy and sell products and earn a commission on the selling price. They often specialize in a particular product line, but do not take title to the goods (Armstrong and Kotler 2003). Brokers are paid by the party that hired them and do not carry inventory. Agents, on the other hand, represent either buyers or sellers on a more permanent basis. Some agents represent two or more manufacturers of complementary product lines. Selling agents have contractual authority to sell the entire output of a manufacturer while purchasing agents have a long-term relationship with buyers and make purchases for them. Commission merchants take a load of commodities to a market, sell it for the best price, deduct a commission, and send the balance back to the growers. Commission merchants are most often used by farmers who do not belong to a growers' cooperative.

Retailing

Retailing includes everything involved in selling products to the end consumers. Thus, retailing includes grocery stores, restaurants, and direct sales to consumers. Food retailing is one of the most expensive parts of the food marketing chain, and retailers have considerable market power in the food industry. Retail businesses take many different forms and aquaculture marketers should carefully understand the differences to identify potentially profitable marketing alternatives. It is important to understand that retail shopping patterns and consumer demographics change rapidly. Additional detail on retail market trends can be found in Chapter 4.

There are a number of examples of successful retail fish marketing concepts. For example, fish and chip shops in the United Kingdom (UK) face severe competition from other fast food retailers but continue to be popular.

Supermarkets have become the primary form of food grocer in the U.S. and in many other countries. In France, 73% of seafood consumption occurs at home (Food Export Association of the Midwest USA 2012). Supermarkets tend to be fairly large grocery stores that sell high volumes at low cost and are organized as self-service businesses. A supermarket can be described as a full-line, departmentalized, cash-and-carry, self-service food store. Supermarkets were products of growth in suburban areas and became an American symbol of innovation, affluence, abundance, efficiency, and the good life. Chain stores represent both a horizontal affiliation of retail stores and a vertical affiliation of food retailing, wholesaling, and sometimes processing businesses. Chain stores developed to take advantage of the efficiencies to be gained through large-scale buying and selling. The food chain store movement triggered competitive reactions on the part of independent retailers and service wholesalers, who developed their own joint activities (retailer-owned cooperative, wholesaler and wholesaler-sponsored voluntary retail chains).

Food grocers

Supermarkets have experienced slow sales growth in recent years with slower population growth and increased competition from convenience stores, discount food stores, superstores, and increased consumption away from home. Fresh seafood departments have been used to attract customers away from competing outlets. Market basket pricing gives the retailer latitude in pricing any one food. Loss-leaders can attract business without each individual item being priced based on wholesale prices.

The growing market share of multiple retail stores (supermarkets and hyper-markets) in food distribution has also changed patterns of production, supply, and distribution. Hypermarkets are stores with more than 200,000 square feet of selling space in groceries, sporting goods, auto supplies, etc. A warehouse food store eliminates some services and frills to reduce retail costs and prices. Superstores are larger supermarkets (up to 60,000 sq. ft.) that seek to supply all the products, food and non-food, that consumers want. Superstores grew at the rate of 25%/yr while supermarkets grew at only 1%/yr in the early 2000s. However, while grocery sales in superstores and hypermarkets continue to com-pose substantial volume, growth is slowing. New growth areas will be based on online sales and convenience retailing sales. Walmart, with more than 10,000 stores in 27 countries, has begun to add an increasing number of smaller stores, to take advantage of the trend towards more convenience. Warehouse or whole-sale clubs, such as Sam's Club and Costco, sell annual membership fees, often $45 to $100, and then sell a variety of grocery and non-food items at deeply discounted prices. Warehouse club sales have increased through the 2000s while traditional supermarket sales have slowed (Retail Leader 2014). Supermarkets, superstores, and wholesale clubs all handle a variety of aquaculture products. Convenience stores (small stores located near residential areas) and specialty stores (stores that sell a narrow product line) rarely handle aquaculture products. Other common types of retailers such as department stores, discount stores, and off-price retailers other than warehouse clubs do not generally sell food products.

There has been an increase in cooperative organizations among retailers. Voluntary chains such as the Independent Grocers Alliance (IGA), for example, are retailers that have formed an association to purchase in bulk and to merchandise jointly. Chain stores are companies with two or more retail outlets. Larger in size than independent grocers, they can purchase in bulk to benefit from lower prices. Associated Grocers is a retailer cooperative that has established a central buying organization to conduct joint promotion efforts. Other, non-grocer, examples of retail organizations include corporate chain stores (Pottery Barn), franchises (Subway, 7-Eleven), and merchandising conglomerates such as Dayton Hudson.

Livehaulers

Livehaulers buy live fish from producers and function as middlemen. Livehaulers market fish to a variety of outlets including processing plants, fee fishing operations, community fishing ponds, retailers, or other outlets. In the U.S., 12.7% of foodfish sales in 2005 were to livehaulers (U.S. Department of Agriculture 2005).

Restaurants

Restaurants are also retail outlets that operate in an extremely competitive environment. The away-from-home food consumption market is very different from the home food preparation market. Food service managers are often more concerned with standardization, portion control, and labor-saving foods than are grocery store managers. Prices typically are more stable in the restaurant trade, and prices cover a higher cost ratio of marketing services to food. In restaurants, for example, 45–65% of the price charged is in non-food costs as compared to supermarkets in which only 20% of the costs are non-food costs.

Direct sales

The complexity of market channels can be avoided by moving smaller quantities directly to the end consumer without any intermediaries (Palfreman 1999). However, for direct sales to be feasible, the grower must develop the capacity to transport and possibly store the product.

In the U.S., 1.2% of foodfish sales were direct sales in 2002 (U.S. Department of Agriculture 2005). Much of these direct sales occur through fee fishing operations. Fee fishing operations charge either a fee for customers to fish in their ponds or charge by the unit weight of fish caught. Most successful fee fishing businesses provide picnic areas, concession stands, bait, on-site dressing, fishing piers, and ice (Cichra et al. 1994). Locations close to a large customer base and constant restocking of large fish are important to the success of fee fishing businesses (Engle 1997). Inhabitants of local and nearby towns accounted for 88% of customers in a Kentucky study (Cremer et al. 1984). Direct marketing to outlets other than processing plants results in higher prices but retail outlets are relatively limited in size and not a feasible option for the entire crops of large-scale farms (Wiese and Quagrainie 2004).

Local food sales tripled in value from 1992 to 2007, from $404 million to $1.2 billion (Tropp 2014), growing at twice the rate of overall agricultural sales. In 2008, local food sales were estimated to be $4.8 billion. Local food sales tend to be greater in metropolitan areas and have been concentrated in the Northeast and on the West Coast. While local food sales are dominated by vegetable, fruit, and nut farms, the growing demand presents opportunities for seafood. The states of Rhode Island, Massachusetts, and New Hampshire dominate local food sales. More than half of local food sales were from farms selling exclusively through intermediated marketing channels such as grocers, restaurants, and regional distributors.

Producers may form cooperatives to assemble and sell produce directly to consumers. There have been a number of attempts to develop aquaculture cooperatives. Like many other forms of business, the failure rate typically is high. A successful cooperative must have a strong, skilled manager who is viewed as fair to all members and who has the marketing and business acumen to position the cooperative's products competitively.

Profit margins

As a product moves through the various levels of the market channel, the price increases at each stage in accord with the value added to the product. The amount added, or the marketing margin, is affected by the time of sale and the price paid for the raw product. Government price controls, producer organizations, types of products, and level of market concentration will affect the amount of the marketing margin.

While intermediaries, or middlemen, of the market channels are frequently called such names as "coyotes" and viewed as abusive of growers, legitimate costs are incurred as value is added to the product by intermediaries. In addition to storage, packaging, and transportation costs, time spent by the intermediary to identify buyers and coordinate with suppliers also has a value. Fish farmers and fishermen often forget to budget a cost for their time spent on marketing functions. This cost is referred to as an opportunity cost. Opportunity costs are defined as the cost of an input in its next best alternative use. Owners should value their time spent in marketing activities at their true opportunity cost (what they could earn working for someone else or spending the time on producing fish rather than splitting it between fish production and marketing) to ensure that prices charged reflect all costs.

Retailers typically seek either high markups or high volume but rarely both. Specialty stores typically select high markup on low volumes, while supermarkets have lower markups on higher volumes. That said, retail grocer markups for seafood tend to be higher than for other store products and have been even higher over the last several years. Retail margins for seafood can range from 25% to over 30% (Seafood Business 2001–2003). Such margins are greater than for many other food sales categories and demonstrate that seafood is often used by grocery stores to compensate for lower margins in other food sales categories.

Economies of scale in marketing

Economies of scale refer to decreasing costs with increasing size of the business. This is particularly true with the growth of large supermarket and other chains that take advantage of the large economies of scale in food distribution (Asche 2001). These economies of scale allow for productivity growth to occur throughout the value chain for fish (Zidack et al. 1992).

Economies of scale in marketing seafood are one of the reasons for consolidation among seafood suppliers (M&A 2013). Mergers and acquisitions that began to occur frequently through the late 1990s and into the 2000s have continued to increase the degree of consolidation in seafood value chains. Seafood companies are driven to distribute product more quickly and at lower cost, and acquisitions and mergers are one way to increase the company's control over the marketing chain and its costs. Seafood supply companies such as Tri Marine International (Chicken of the Sea, StarKist, and other brands) and Trident Seafood (Louis Kemp and other brands) have annual sales exceeding $1 billion. Companies such as Bumble Bee, Thai Union International, and Nippon Suisan USA have sales that range from $710 to $950 million. Aquaculture supply companies must compete with these large conglomerates in the seafood marketplace and with the marketing economies of scale that come from the ability to supply a wide variety of seafood products.

Supply chain management

Supply chain management is a term that has emerged to refer to the complexity of efficiently managing the flow of goods and information from suppliers to resellers and final users. Improved logistics associated with tracking inventories and moving product efficiently through market channels have provided a mechanism for managing the entire supply chain.

Supply chain management involves far more than just the marketing logistics, or physical distribution, of product to consumers. Supply chain management is more of a customer-centered approach that works backwards from end consumers in the market to the producer and back to the resources that are used as inputs. Efficient supply chain management can result in better service to customers or lower prices that may offer a competitive advantage to the business. It may also result in cost savings to both the business and its customers. Moreover, retail trends towards increased product diversification have made supplying large customers more complex. Information technology provides tools to manage supply in ways previously unknown, such as with point-of-sale scanners, uniform product codes, satellite tracking, web-based systems, and electronic orders and payments.

While the early stages in the value chain tend to receive more attention (Asche 2001) in modern retail markets, buyers increasingly demand that products can be traced to determine origin and history. Hazard analysis of critical control point (HACCP) plans are expected and required. HACCP regulations require the development of a plan that identifies potential food safety hazards in processing and develops procedures to minimize food safety problems (National Fisheries Institute 2015).

The seafood market has demanded greater traceability over time. The European Union has mandatory traceability requirements for seafood products, while Japan and Canada have more limited requirements for traceability. Traceability programs can range from paper-based systems (the least expensive to operate) to electronic bar-coding systems to the more recently developed radio frequency identification (RFID) systems. Paper systems can be adequate for smaller companies, but can quickly become too cumbersome for larger volumes of seafood.

Pricing systems

Price determination

In a purely competitive market situation, prices typically are determined by the interaction between supply and demand in the market (see Chapter 2 for more details on equilibrium prices). Some of the best examples of purely competitive pricing in seafood markets are the fish auctions that continue to operate in various countries around the world (United Kingdom, Denmark, The Netherlands, Norway, Germany, Kenya, Tanzania, New Zealand, Faroe Islands, U.S., and others). In auction markets, potential buyers bid for various lots of fish, and prices are bid either upwards or downwards depending on the particular auction's guidelines, until buyer and seller agree on a price (Palfreman 1999).

In corporate settings, however, pricing decisions are made based on other processes. Administered pricing describes all pricing in which a seller or buyer announces a non-negotiable selling (buying) price (Breimyer 1976). Prices paid by catfish processing plants to catfish farmers are examples of administered prices. In the case of catfish processing plants, these prices are based on whole-sale prices received by the plants from brokers and food service distributors.

Other companies may use cost-plus pricing in which a set margin is added to costs of production to determine selling price. In cost-plus pricing, an arbitrary amount of profit is added to the production costs. This pricing mechanism may be effective for highly-valued products for which few substitutes exist. Some companies base their pricing on competition-oriented pricing in which pricing is based on prices for similar and competing goods. Competition-oriented pricing is more common in markets with one price leader with a dominant market share. Other companies then set prices that are relative to the price leader. Other com-panies use demand-oriented pricing. This is especially true for customers with different quality standards. Sales of higher-priced species need to be supported by advertising the quality attributes to those population segments willing and able to pay higher prices for a high-quality product. Lower-cost species are sold by emphasizing the corresponding lower price to market segments that seek out more inexpensive types of seafood. Regardless of the pricing mechanism, the price for a particular product should be established based on in-depth under-standing of the targeted consumers, their attitudes and preferences, and where the product is to be positioned within the price-quality matrix.

Psychological pricing involves establishing prices that either look better or convey a certain message to the buyer. An example would be to charge $6.58/kg ($2.99/lb) instead of $6.60/kg ($3.00/lb) to make the product appear to be more of a bargain. Perceived-value pricing promotes the product based on non-price factors such as quality, healthfulness, environmental sustainability, or prestige.

Some temporary pricing strategies are used to increase sales and market share. Skimming involves introducing the product at a relatively higher price for more affluent, quality-conscious consumers, and then lowering the price as the market becomes saturated. Discount pricing offers customers a reduction from advertised prices for specific reasons. Discount coupons in the newspaper or radio ads may attract new customers. With loss-leader pricing, a portion of the product is offered at a reduced price (below cost) to attract customers. This is used to attract new customers to farmers' markets or supermarkets. Market-penetration pricing is a strategy in which a low price is charged to gain increased market share.

Marketing margins, marketing bill, and farm-retail price spreads

The marketing margin is that portion of the consumer's food dollar that goes to businesses engaged in marketing (Armstrong and Kotler 2003). Another way to view the marketing margin is that it represents the difference between what the consumer pays for food and what the farmer receives. It must be remembered that this difference includes costs associated with all marketing functions performed. Thus, the price that the consumer faces includes both the farm price and the marketing price of food. These two prices may not always move in the same direction.

The size of the marketing margin cannot be used to measure efficiency. Shorter marketing chains may have smaller margins but are not always the most efficient. A fish farmer who sells directly to the public may have a small marketing margin, but it may not be efficient for the farmer to make a large number of deliveries to satisfy his or her customers.

The size of the marketing margin reflects the marketing costs involved, not the number of intermediaries. Marketing costs include profit to each intermediary, but while middlemen may be eliminated, the costs of the required marketing functions minus the intermediary profit will still exist. Eliminating middlemen will not decrease the marketing margin if the farmer cannot perform the marketing functions as efficiently as the middlemen. Increased marketing margins also increase the retail value and price of food.

The food marketing bill is the difference between total consumer expenditures for all domestically produced food products and what farmers receive for equivalent farm products (Kohls and Uhl 1985). The marketing bill includes all transportation, processing, and distribution of foods as well as foods consumed both at home and away from home. It provides an aggregate view of the division of consumer food expenditures between farmers and food marketing businesses. The increasing share of the marketing bill reflects market trends towards more complex processing and distribution systems related to increasing food expenditures away from home and the growth in convenience foods for at-home consumption. Figure 3.2 shows that, of the marketing bill, labor is the largest portion, followed by packaging, profits, transportation, rent, advertising, depreciation, business taxes, energy, interest, and repairs.

The farm value share of the food dollar has declined continuously over time while the marketing bill share has increased. The decline in the farm value share does not necessarily mean that farmers' welfare has declined. Examination of the relative costs of production and returns is necessary to evaluate the economic and financial health of the farm sector. Changes in the farmers' share of the marketing bill can occur due to changes in supply and demand at the farm or retail levels or changes in marketing costs. Commodities for which the marketing agencies provide a relatively large share of utilities are products that require a lot of processing, are highly perishable, are seasonal, have high transportation costs, and are bulky in relation to product value.

Assembly market functions tend to be a small portion of the marketing bill. Products such as seafood that are frequently marketed fresh tend to have larger farm values, with the reverse being true for highly processed products. Transportation and wholesaling costs tend to be higher for more perishable goods.

Farm value
$123.3 billion   Marketing bill $537.8 billion

Consumer expenditures, 2000
$661.1 billion

Labor   Packaging   Transportation   Energy   Profits   Advertising   Depreciation   Rent   Interest   Repairs   Business taxes   Misc. costs

Fig. 3.2 Food marketing bill, 2000. Source: ERS (2004).

While the marketing bill is concerned with expenditure margins, the farm-retail price spread is concerned with price margins for individual foods. It measures the gross return per unit to food marketing, or the profits and costs of all marketing functions. The spread is the difference between the retail price per unit and the farm value of an equivalent amount of food sold by farmers. There is wide variation by different food crops. Figure 3.3 shows that the farm-retail price spread (calculated as the farm value share of the retail price) for beef is 50%; that of flour is 26%, while for cereals it is only 6%. Farm-raised aquaculture products would fall somewhere in between with values that would range somewhere between 15 and 20%.

Pricing at different market levels

Elasticities can vary for different market levels (Chapters 2 and 10 present information on how elasticities are calculated and additional detail on interpretation). Kinnucan et al. (1988) estimated demand to be elastic at the processor level but inelastic at the farm level for U.S. farm-raised catfish. However, more recent studies have shown retail catfish demand in the U.S. to be closer to

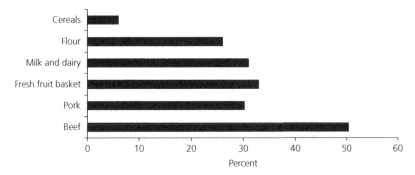

Fig. 3.3 Farm-retail price spreads (calculated as the farm value share of the retail price, 2013). Source: USDA (2014).

unitary elasticity (Singh et al. 2014). Other researchers (Kouka 1995; Kinnucan and Miao 1999; Norman-López and Asche 2008) found similar elasticities nationally for U.S. catfish. Elasticities have important implications for pricing strategies. If the demand for a product is inelastic, increasing price will result in greater total revenue. However, if demand is elastic, increasing price will result in lower total revenue.

Most estimates show elastic demand for seafood products (Anderson 2003). For example, cod (Brooks and Anderson 1991), flounder (Brooks and Anderson 1991; Wessells and Wilen 1994), salmon (Herrmann et al. 1993; Wessells and Wilen 1994; Kinnucan and Myrland 2002), and catfish (Lambregts et al. 1993) were shown to be price elastic at the retail level. Only imported shrimp (Keithly et al. 1993) and tuna at the retail level (Wessells and Wilen 1994) were shown to be price inelastic. More recent work (Singh et al. 2014) confirms the generally elastic supermarket demand for salmon, tilapia, whiting, cod, flounder, pollock, mahi-mahi, swordfish, and perch. On the other hand, halibut, orange roughy, and tuna, were shown to be price inelastic by Singh et al. (2014). More importantly, Singh et al. (2014) demonstrated that own-price elasticities varied according to location. Salmon demand was inelastic in the West South Central, unitary elastic in the South Atlantic and Pacific regions, but elastic elsewhere in the U.S. whereas catfish demand was highly elastic in the East South Central, but inelastic in the South Atlantic region. Additional detail on seafood demand elasticities is presented in Chapter 11.

Price behavior, trends, and fluctuations

Prices react to a variety of different forces, shocks, and events that can occur over the long term or the short term. Fish and seafood prices tend to be more volatile and exhibit greater fluctuations than do those of some other, less perishable, types of products. Shortages of certain species of fish, whether in the off-season

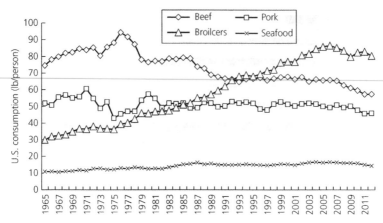

Fig. 3.4 Per capita consumption of protein in the U.S., 1965–2012.

or due to over-fishing, will tend to drive prices up, while increased supplies from aquaculture production or during the peak fishing season for that species will tend to drive prices down. Weather disasters may affect prices of products raised in affected areas. Food scares or reports of contamination in aquaculture grow-ing areas may cause demand to decrease and prices to drop as a result.

Longer-term trends in prices will be affected by changing trends in demand as well as in supply. Figure 3.4 shows the trends in per capita consumption of sources of animal protein in the U.S. from 1965 to 2012. The increased interest in fish and seafood consumption is driven, in part, by recommendations to eat more fish and seafood as part of a healthy diet. Increasing per capita consump-tion, combined with a world population that continues to grow, will increase overall demand and put upward pressure on price. However the increased supplies of shrimp and salmon that have come from aquaculture production have tended to drive prices of those products downwards over time.

Geographic markets

Consumer preferences vary by geographic region and careful consideration of regional variation is important in developing marketing strategies. Singh et al. (2014), for example, showed that own-price elasticities of many seafood species in U.S. supermarkets varied more by region of the country than by season. In Europe, consumer preferences varied dramatically across countries (Asche and Bjorndal 2011). For example, Spain and France are the largest seafood markets in Europe and prefer salmon primarily fresh. French consumers particularly enjoy smoked salmon whereas the United Kingdom is the most important market for canned salmon. Other countries such as Poland have not tradition-ally been major consumers of salmon, but there has been substantial growth in volume of imported salmon into Poland since 2002.

Product storage

The timing of the production of food commodities does not always coincide perfectly with demand for those commodities. Meeting consumer demand for products typically will require some type of storage.

Working inventory is necessary for an efficient marketing chain because, without stocks stored, there will be disruptions in supply. Storage is necessary for products that are harvested in a short time but are consumed throughout the year. Carryover stocks are those that are left from one marketing year to the next. Some farmers will store stocks of a product to wait for a higher price; these stocks are considered speculative stocks.

Processing plants often provide an important storage function in the supply chain of a product. With fish processing, unlike some other commodities, processors typically store processed volumes, while farmers "store" raw material (i.e., live fish), in ponds until it can be sold.

There are a number of costs associated with storing products. Direct storage costs include such items as repairs to storage facilities, depreciation, insurance, and utilities to maintain optimal temperatures. There is also a cost that represents the interest on the financial investment in the product while it is in storage. If the product deteriorates in quality during the storage period, another cost is incurred. If consumers prefer, and will pay higher prices, for fresh product than frozen, then storage incurs the costs of the price differences. There is also the risk that the price of the product may decline while in storage and that shrinkage will increase the costs of the product.

Market power

Market power is the ability to affect the behavior and performance of exchanges in the marketplace to the advantage of the particular firm. It is often expressed as the ability to affect prices, but can represent influence over marketing functions, product flows, quality, or other factors.

Some growers choose to form marketing cooperatives in order to increase market power. Cooperatives are businesses that are owned and controlled by those working in them (Palfreman 1999). Membership is open to all employees and each member has one vote, irrespective of shareholding. Profits typically are shared according to agreed upon rules. In most countries, cooperatives must be registered. Palfreman (1999) points out that cooperatives may have lost ground in the UK because they resist efficiency-enhancing change such as computerized buying and selling. In Europe, Fish Producers Organizations (FPOs) are cooperatives but are also companies limited by guarantee. They are backed by a group of subscribers who guarantee their debts.

Few studies have been conducted on market power in aquaculture. Early research indicated that in certain areas the U.S. catfish industry may exhibit monopsonistic (one buyer) control (Kinnucan et al. 1986). In West Alabama, for example, when only one processor existed, an imbalance in market power between catfish producers and processors could have resulted in lower prices paid to producers. Kouka (1995) also found evidence for market power at the processor level due to its degree of concentration. However, more recent studies (Hudson 1998; Wiese and Quagrainie 2004; Bouras and Engle 2007) provide evidence for competitive behavior of the catfish processing and farming sector in spite of the degree of concentration at the processor level. Results were attributed to the relatively small size of the catfish industry as compared to dominant food service companies (such as Sysco) and large retailers (such as Walmart). Such a structure can make it difficult to pass cost increases through to the end consumers. In the salmon industry, Marine Harvest has emerged as a major multinational company. Through mergers and acquisitions, Marine Harvest grew to supply 23% of global product in 2008. However, the top four salmon-producing countries accounted for about half of total world production of salmon in 2008. Thus, in spite of one major supplier, overall concentration in the salmon industry may not be as high as expected.

Forming a cooperative is a form of collective action that can be taken by farmers to improve their marketing outcomes. Historically, farmers have had little market power, or the ability to exert some degree of control over the price received for their product. They were strictly price takers. There are several federal statutes in the U.S. that provide legal protection for farmers when they seek to improve marketing outcomes through joint, collective action. The primary form of protection is the Capper–Volstead Act of 1922 that provides a foundation from which farmers can organize cooperatives or bargaining associations, or form marketing orders. Chapter 8 provides additional detail on options for collective action available to aquaculture growers.

Advertising and promotion

Promotion is a way of communicating a product's attributes to prospective consumers. Ultimately, it is the product itself that communicates with the consumer and the consumer decides whether or not the product meets his or her expectations. New products typically need to draw upon consumer perceptions through brand image, quality marks, labels, reputation of suppliers, and other point-of-sale information and support to introduce themselves to the consumer. For a consumer to reach a decision to purchase a new product, he or she must pass through the stages of awareness, interest, favorable perception, and evaluation (Marshall 1996).

Paid promotions are referred to as advertising. Advertising employs various media such as print media, radio, television, and other forms. Advertising reaches large numbers of possible consumers and subjects consumers to repeat messages. The type of advertising selected depends upon the stage of the life cycle. A new product that is being introduced will require informative advertising. Once competition increases, persuasive advertising becomes more important, to convince consumers of the benefits of one particular brand. Mature products require reminder advertising so that consumers do not forget about the product.

Advertisements should include a headline, picture, text, and information on where to buy.

Not all advertising programs need be costly, and there are low-cost ways to advertise products. Bold, funny, and striking graphics on tee shirts or in stores can be very effective. Large quantities of the product can be donated for people to try and spread news by word of mouth. Social media also provide opportunities to advertise effectively at a relatively low cost. Using creative company and product names such as Ben and Jerry's Ice Cream can be effective advertising. Some businesses even invite tourists to visit and promote their business as a tourist attraction.

Sales refer to building customer relationships with the expressed purpose of making sales. Personal selling is most effective to create preferences and purchase actions. In sales, there are two fundamental rules of thumb: (1) never promise more than you can guarantee, and (2) never deliver less than you guaranteed. Obtaining favorable publicity for a product to build a favorable corporate image is called public relations and often will take the form of press releases and special events.

Sales promotions refer to short-term incentives to encourage purchase of a product. Point-of-purchase displays, premiums, discount coupons, specialty advertising, and demonstrations can all be used to promote sales. Sales promotions are short lived but attract consumer attention and may be used to boost lackluster sales.

An increasingly common problem in marketing communications is that various groups within companies are not well organized or coordinated. Paid advertisements may send a message that is not well-supported by a price promotion or by the label. Strong brand identification only comes from seamless coordination and reinforcement of images and messages.

Product grades, quality, and marketing implications

Product attributes such as color, taste, aroma, texture, size, and shape can be combined in an infinite number of combinations. Product quality is a subjective evaluation of the value of the particular combination of attributes possessed by a specific product. Consumers perceive quality not only in terms of the sensory attributes such as taste, but also in terms of appearance, nutritional value, and safety of the product.

Quality standards can be used to sort varied mixes of product categories into uniform categories. This grading process can result in homogeneous product categories, with a pricing structure that conforms to buyer and seller preferences for the various bundles of attributes represented by each product grade.

The establishment of standards (commonly agreed upon yardsticks of measurement) to sort agricultural products into grades can simplify marketing and reduce marketing costs. Producers can charge price premiums for higher-quality products. Product grades provide clear product information to consumers, and consumers benefit as they select products more closely aligned with their needs. Although grading can be done at any stage of the marketing chain, food grades tend to operate mostly at the wholesale, not the retail, level.

One of the most critical steps in developing a food grading system is to select the criteria to be used to judge the adequacy of standards. Standards should be based on those characteristics considered most important by consumers. For standards to be successful, they should be those that can be measured and interpreted accurately. Individual grades that exhibit a great deal of internal variation in quality will reduce the usefulness of grading. The terminology used to identify grades must be understood clearly by consumers. Standards need to also capture a significant portion of the average production, and grading costs must be reasonable. The ultimate test is adoption in the marketplace.

There are a variety of problems associated with establishing grades, including types of tolerances and what terms should be used to identify grades. Positive terms are typically selected rather than those suggesting an inferior product.

Research has shown that consumers may not readily discriminate among different grades and may not be willing to pay price premiums for higher grades. This can be a problem particularly if grades were viewed as convenient for traders but were not consumer-oriented. Confusion can occur between grades and with federally required inspections related to food safety.

Farmers that produce the highest-quality product gain the most with grading systems, sometimes at the expense of farmers that produce lower-quality products. Typically, the larger, more specialized producers are the most receptive to developing grades. Programs that establish product grades can result in raising standards and quality across an industry, as producers seek to gain higher prices.

Large chain retailers benefit from grading because it simplifies their procurement decisions. Smaller processors also benefit because grades allow them to supply larger market outlets. Larger plants may be opposed to grades because federal grades may compete with their own brands. Grading may also result in decreased market concentration because it allows smaller packers to compete in the market.

There are a number of examples of product grading in the seafood market. In the U.S., the U.S. Department of Commerce has established standards for grades of fishery products that range from whole-dressed to frozen minced blocks to fillets to breaded products (U.S. Department of Commerce 2016). Individual tuna fish, for example, are assigned a grade that accompanies that fish through the market chain to its final sale (Bartram et al. 1996). Different grades of tuna are sold to different market niches. Tuna grading is done subjectively by visually inspecting the appearance and directly sampling a small section of fish muscle. The price of fresh tuna can range from $1.10/kg ($0.50/lb) to $121/kg ($55/lb) depending upon the grade assigned. There are four basic grade distinctions for the Japanese and U.S. markets: Grade #1 has bright red muscle, firm texture, clear flesh, and little fat; Grade #2 is red, firm, with some translucency, and no fat; Grade #3 has some red but some brown muscle, is firm and opaque, with no fat; while Grade #4 is brown and gray, soft, and opaque (Ledafish 1996). The top grade (#1) is used for high-end Japanese sashimi, Grade #2 in lower-end Japanese and Hawaiian sashimi, Grade #3 in lower-end restaurants in the U.S., and Grade #4 either canned or frozen. European markets use Grades #2 and #3.

Another example of product grades in seafood is for frozen raw breaded shrimp (U.S. Department of Commerce 2016). "U.S. Grade A" is a product that when cooked possesses good flavor and is rated over 85 points and above. "U.S. Grade B" is rated at 70–84 points and "Substandard" product fails to meet the standard of "U.S. Grade B".

International trade

Factors that affect demand and supply in a given country will play a role and interact in the international market. In addition, national regulations in each country and of international organizations will also affect the international flow of goods and services.

One of the basic economic principles underlying international trade is that of comparative advantage. Comparative advantage indicates that, if free trade conditions apply, some countries will specialize in production of the commodities that can be produced relatively most efficiently in that country. Other countries generally will be better off importing a commodity that is produced elsewhere at a lower cost than the commodity can be produced domestically, that is, it is cheaper to import the commodity than produce it at home. Thus, price ratios developed for both countries guide the flows of trade between them.

However, trade policies frequently are developed to "protect" domestic industries from competition from similar products imported into that country. Decisions to pursue a more free trade or a more protectionist policy should be based on the associated expected benefits and costs. Protectionist policies may be based on raising tax revenue, supporting producers' income, reducing consumers' food costs, attaining self-sufficiency, or countering interventions of other trading partners. Protectionist policy instruments are numerous and include those restricting quantities that can be imported (quotas), increasing the price of the imported product (tariffs), encouraging export (export subsidies), controlling exchange rates, or supporting domestic prices through the use of price premiums, marketing boards, supply quotas, commodity programs, etc. Export subsidies increase the share of the exporter in the world market at the cost of others; they tend to depress world market prices and may make them more unstable because decisions on export subsidy levels can be changed unpredictably (Pearson and Sharma 2003).

International trade research generally shows that the volume of international trade will be greater if trade policies are reduced or eliminated. This is referred to as trade liberalization. Free trade can raise aggregate economic efficiency (Suranovic 1997–2004). This increase in economic efficiency can include benefits from increased production efficiencies that result in producing more with the same amount of resources and providing more different types of goods and services that satisfy more consumer needs. Differences in how individuals seek profits along with differences in price will result in efficient trade under free trade conditions.

However, free trade will result in losses for some people, and protection from international competition may benefit some countries. Some groups may lose because it is difficult to quickly change investments from one industry to another in the short run, or in some cases, even in the long run.

While free trade appears to offer many economic benefits, people who are not trained in international economics tend not to favor free trade policies that relate to imports to their own nation. No major economic nation allows complete free trade. Companies that seek to export must learn the details of quotas, tariffs, subsidies, inspections, and certifications required by each country. Indeed, substantial trade barriers exist in many countries. Arguments often heard for supporting trade barriers are that they: (1) prevent dumping (i.e., selling at unfairly low prices); (2) protect farm programs; (3) enhance food self-sufficiency; (4) help to manage the national economy; (5) maintain employment; (6) stabilize the industry; (7) protect an infant industry; (8) combat the presence of international monopolies; (9) engage with international politics; and (10) protect national security (Rhodes 1993; Suranovic 1997–2004). Additional detail on the international market for seafood and resulting trade conflicts is found in Chapter 7.

Aquaculture market synopsis: shrimp and prawns

The total world supply of shrimp and prawns has grown rapidly over the past decade, primarily due to increased production of farmed shrimp (Fig. 3.5). By 2012, farmed shrimp production composed 64% of the total world supply of shrimp and prawns. Given the substantial differences in use of the common names for shrimp and prawn species, the values presented in this synopsis include all species labeled as "shrimp" and "prawns" in the FAO FishStatJ database.

Figure 3.5 also demonstrates that the world supply of wild-caught shrimp and prawns was level from 2008 to 2012.

At least 25 species of shrimp enter world trade, although there are literally hundreds of species of saltwater and freshwater shrimp in the world. The industry divides them into two categories: (1) coldwater/northern (family Pandalidae), and (2) warmwater/tropical (family Penaeidae). However, only eight of the saltwater species dominate the market in the U.S. In addition, a number of freshwater species of prawns, of the genus *Macrobrachium*, are sold commercially in many parts of the world.

The leading species of shrimp and prawns farmed worldwide is the whitelegged shrimp (*Litopenaeus vannamei*), followed by the giant tiger prawn (*Penaeus monodon*), oriental river prawn (*Macrobrachium nipponense*), and the giant river prawn (*Macrobrachium rosenbergii*) (Fig. 3.6). Production of the whitelegged shrimp composes 66% of the total world production of farmed shrimp and prawns, with production of the giant tiger prawn composing 18% of the total farmed supply of shrimp and prawns.

The Asian region dominates the supply of shrimp worldwide. For the major species raised (whitelegged, or Pacific white shrimp), China produces 46% of the world's production, followed by Thailand, Ecuador, Indonesia, and India in

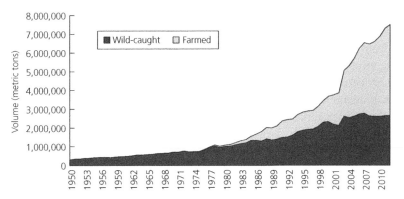

Fig. 3.5 Global production of farmed and wild-caught shrimp, 1950–2012. Source: FAO (2014).

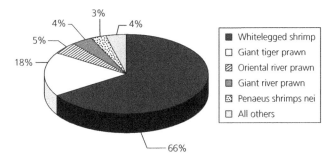

Fig. 3.6 Most important species of farmed shrimp and prawns. Source: FAO (2014).

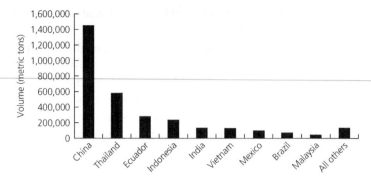

Fig. 3.7 Global production of *L. vannamei* by country, 2012. Source: FAO (2014).

descending order (Fig. 3.7). Vietnam leads world production of *P. monodon*, with 41% of production in 2012, followed by India, Indonesia, China, and Bangladesh.

The early development of the shrimp market was the result of Japanese trading companies and American importers urging developing nations to develop culture techniques in the 1950s and 1960s. There have been major shifts in the global supply of shrimp over the years. Taiwan had pioneered the development of many of the technologies for farming marine shrimp and led the world in shrimp production through the 1980s (Csavas 1994). However, disease problems resulted in a near-collapse of Taiwan's industry in 1988. China moved very rapidly into shrimp production in the early 1990s and, within just a few years, went from minimal production to dominating the global market for shrimp. Disease problems resulted in serious decline of the Chinese industry, too. Thailand learned from the Taiwanese and Chinese experiences and has been able to implement improved management practices to avoid the catastrophic losses that characterized the shrimp industries in other countries. By 2005, Thailand had become the world leader in farmed shrimp production.

In the past, *L. vannamei* was cultured only in the Western Hemisphere, primarily South and Central America and Mexico, but in recent years it has become widely cultured in Asia as well. While *L. vannamei* attracts a lower price due to its smaller size, Asian growers have found it to be more resistant to diseases and have begun to culture it for this trait.

The major species raised in Asia for many years was the black tiger prawn, or the giant tiger prawn (*Penaeus monodon*). The production of *P. monodon* dominated the total shrimp market from 1979 to 2002. However, while 2012 production of *P. monodon* was the highest ever recorded for that species, it was substantially less than the 2012 production of *L. vannamei*. Production of *L. vannamei* in Asia has increased rapidly over the past decade due to its greater resistance to disease as compared with *P. monodon*.

Shrimp is a globally traded commodity with primary markets in the U.S., Japan, and Europe. It is not considered exclusively a commodity because it is marketed by brand and packaged at the source of production in the final container in which it will be sold. The number of colors, sizes, and species also sets it apart from other commodities. Marketing is complicated because of the number of countries involved, the range of sizes sold, the number of species sold, the number of product forms, and the types of markets.

The development of new production technologies has been a key driving force behind the rapid expansion of global farmed shrimp production. In the early years, development of nutritionally complete feeds led to increased production. As the industry grew, supplies of post-larval seed from the wild were reduced, spurring development of new hatchery technologies. With the ability to supply seed on a regular basis, the industry has been able to grow rapidly. Periodic disease outbreaks have led to development of specific pathogen free (SPF) broodstock and also to the expansion of production of *L. vannamei* from the Western Hemisphere to Asia, to take advantage of its increased disease resistance, particularly to the white spot syndrome virus.

The market for shrimp is complex and highly differentiated. Shrimp products are differentiated and priced by size, by species to some degree, by product form, by quality, and by source. Shrimp are sold in units of counts per kg (lb). For example, 16–20 means that there are 16–20 shrimp per pound. These sizes range from under 10 (giant) per lb to over 300–500 (canned).

Product forms and packs are generally the same from species to species. Most shrimp are sold raw with the head off (green headless) and the shell on. Raw, without shells are referred to as "peeled." Heads-on, cephalothorax included, appear as the entire shrimp (known as "enteros" in Spanish). Peeled, undeveined (PUD) shrimp have the vein, or digestive tract, intact, varying in color from dark to light. Other product forms include peeled and deveined (P&D) and peeled, deveined, and individually packed (PDI). Tail-on peeled refers to a product form in which the tail fin and an adjacent shell segment are left on. Tail-on round refers to undeveined shrimp with the tail on. Butterfly shrimp, also referred to as split or fan-tail shrimp, have been cut along the vein. "Western-style" shrimp refers to splitting the shrimp through the first four segments.

Cooked shrimp are usually sold individually quick frozen (IQF), often as P&D tail-on or P&D tail-off and shell on. Other forms include minced, canned, dried, and value-added (marinated, flavored, or breaded). The U.S. Department of Commerce has established standards for green headless and breaded shrimp.

Breading consists of two components: wet, adhesive batter and a dry, crunchy breading. The percentage of breading by weight is critical and is regulated by the U.S. Food and Drug Administration. Labeling standards require that breaded shrimp be more than 50% shrimp, lightly breaded more than 65% shrimp, and imitation breaded products must be more than 50% shrimp.

There are a variety of forms of breaded shrimp. Whole breaded can be tail-on or tail-off, usually headless (although called "whole"), and deveined if less than 70 count. Butterfly breaded are split partway on the vein side (dorsal) and spread open. Split breaded is completely bisected ("Western" or "cowboy" style). Hand-breaded is labor intensive and expensive, but more attractive, and usually prepared tail-on. Machine-breaded is done either tail-on or tail-off. If the tail remains, it may or may not be breaded. Not breaded is referred to as "pinched."

Green headless shrimp are usually packed in 2.3 kg (5 lb) blocks (net weight). With ice, the total weight of the box is often 2.7–3.2 kg (6–7 lb). Blocks are packed in two styles: (1) layer or finger packed, and (2) random jumble or shovel pack. Individually quick frozen shrimp are usually in bags (1–30 lb), labeled to net weight, without glaze. Breaded shrimp are packed in boxes with a moisture-resistant barrier and are completely sealed.

Some shrimp are dipped in solutions as a preservative. Sodium tripolyphosphate (STP) is added on peeled and breaded shrimp to reduce drip loss (maintain weight). The label must advise of the use of STP. Sodium bisulfite is used primarily for shell-on shrimp to prevent melanosis ("black spot"). The limit in the U.S. is 100 ppm, but is higher in Europe. "Ever-Fresh" (4-hydroxyresorcinol) is a naturally occurring, generally regarded as safe (GRAS) compound, but is expensive.

The shrimp industry worldwide includes many different buyers and sellers. Nearly 40 different countries produce farm-raised shrimp that are exported to countries around the world. Thus, the overall structure of the shrimp industry worldwide is quite competitive.

Shrimp processors are key intermediaries between producers or shrimpers and the market. Shrimp processing has two stages: (1) turning the shrimp into a form in which it can be traded as a commodity, and (2) changing it from a commodity into a value-added product (e.g., peeled, cooked, IQF). Packaging is improving in developing nations where most of the production occurs and refrigerated vessels are more readily available. Air shipments of fresh and live product are becoming more common.

The nature of shrimp markets varies widely from country to country. In countries such as the U.S., most shrimp production is consumed domestically, often outside the home. In the U.S., for example, the majority (75%) of shrimp is consumed outside the home. However, in many of the world's leading shrimp-producing countries, the majority of the production is destined for export, with smaller sizes of shrimp sold in local domestic markets.

Most of the international trade in shrimp flows from developing nations to industrialized countries. Financing for international shrimp trades often is provided by the importer who typically opens an irrevocable letter of credit (LC) in favor of the exporter. Importers are marketers themselves and usually sell to wholesalers, distributors, re-processors, restaurant chains, and supermarket chains. Financing within the producing country is often provided by exporters who finance the processor, who in turn finances the farmer. An exporter may be a processor, farmer, or an independent third party that takes financial responsibility and communicates with the importer. Many governments require that prices be set before shipment. Others set minimum sales prices and quality parameters.

Importers may purchase shrimp outright from foreign traders, paying for the purchase at full invoice value either at the time of shipment or upon passing through customs inspections. Alternatively, the importer may work on a consignment arrangement whereby an advance is made to the exporter by means of an LC. In some cases, the importer acts as a sales agent (broker) for the exporter and collects a commission. Importers can also make pre-season advances to producers (therefore tying up their production). This is how Japanese importers typically maintain a strong grip on Asian sources. The availability of supply determines the direction shrimp markets will take, and the level of supply determines prices. Prices, as they relate to competing products, determine quantity demanded.

Japanese and U.S. shrimp markets are interdependent; prices prevailing in one market tend to affect the other. Fluctuations in each country's rates of exchange can cause a reaction in both markets and elsewhere. If the U.S. dollar is strong, exporters will target sales to the U.S., while a weak U.S. dollar and a weak yen favor sales to Europe. Increased flow to the major markets of the U.S., Japan, and Europe often results in decreased supplies to minor markets, resulting in firm prices in the minor markets. U.S. importers, however, are not as concerned with foreign exchange markets as are Japanese importers because the U.S. dollar is a major medium of currency exchange.

Over $5.3 billion of shrimp were imported into the U.S. in 2013 (ERS 2014). The frozen shrimp category showed the greatest increase in sales, nearly 90% of the increase. The primary countries exporting shrimp to the U.S., in decreasing order of importance in 2013, were India, Indonesia, Thailand, Vietnam, and Ecuador.

Prices are cyclical and subject to a variety of influences. Shrimp price cycles tend to match price levels in prosperous and recession years and appear to follow consumer discretionary income. Price breaks tend to follow sizeable accumulations of secondary and substandard quality product. While shrimp imports into the U.S. have been increasing, particularly since the mid-1990s, the average value (calculated by dividing the global value of farmed shrimp by total global production; FAO 2014) of farmed shrimp imported has been mostly level from 2003 to 2012 (Fig. 3.8). However, the value of farmed shrimp has remained at a level that is 30% lower than its average value from 1984 to 2002.

The shrimp industry continues to face challenges from environmentalist groups. Some groups allege that shrimp farms have had negative environmental and social impacts. It is unclear to what extent the actions of the environmentalist groups have had an effect on the overall market for shrimp. Nevertheless, it is likely that there will be continued pressure for the industry to continue to adopt environmentally friendly production practices.

The shrimp industry will also need to learn to adapt to a market position in which shrimp is regarded less and less as a high-value luxury good and more of a lower-priced good to be consumed more frequently. The challenge will be for growers to improve efficiencies to maintain profitability with higher volumes and lower prices.

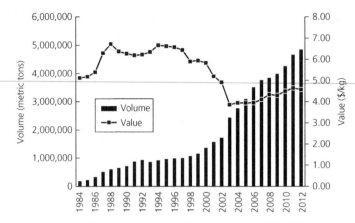

Fig. 3.8 Value and volume of farmed shrimp, 1984–2012.

As with many species of seafood, there have been some international trade conflicts. This is not surprising given the volume of shrimp traded globally. In 2003, a consortium of shrimpers from the Gulf of Mexico filed an anti-dumping lawsuit against farm-raised shrimp imported from a variety of countries. The International Trade Commission ruled in their favor in 2004 and imposed countervailing duties on Brazil, Ecuador, India, Thailand, China, and Vietnam. New countervailing duties were petitioned in 2013 and the countries of Indonesia and Malaysia were added to the petition by the Coalition of Gulf Shrimp Industries. However, the petition was denied by the International Trade Commission, who found that the U.S. shrimp industry was not injured by foreign imports.

Summary

Marketing is a broad term that encompasses all the interactions involved from the point of production to the end consumer. Marketing functions have grown in scope and complexity as consumer income levels, sophistication of consumer demand, and technology have grown. This chapter introduced terminology and fundamental concepts of marketing. It lays the groundwork of terminology and conceptual understanding for the discussion that follows in subsequent chapters.

Study and discussion questions

1 Compare and contrast vertical and horizontal marketing systems.
2 Explain the marketing functions provided by wholesalers and why these are important.
3 Explain the differences between brokers and food service distributors.

4 List the different types of retail outlets.

5 How have economies of scale affected seafood marketing?

6 Describe supply chain management.

7 Explain the differences between the terms marketing margin, marketing bill, and farm-retail price spread.

8 What is market power? What does it mean for the aquaculture industry?

9 What types of collective action can be used by growers to enhance their marketing outcomes?

10 Outline a marketing channel for an aquaculture product sold in your home town. Calculate the marketing margins.

References

American Marketing Association. 2013. Definition of marketing. Available at www.ama.org/AboutAMA/Pages/Definition-of-Marketing.aspx. Accessed March 25, 2016.

Anderson, J.L. 2003. *The International Seafood Trade*. Woodhead Publishing Limited, CRC Press, Boca Raton, Florida.

Armstrong, B. and P. Kotler. 2003. *Marketing: An Introduction*. Pearson Education, Upper Saddle River, New Jersey.

Asche, F. 2001. Testing the effect of an anti-dumping duty: the US salmon market. *Empirical Economics* 26:343–355.

Asche, F. and T. Bjorndal. 2011. *The Economics of Salmon Aquaculture*. 2nd edition. John Wiley and Sons, Oxford, UK.

Bartram, P., P. Garrod, P., and J. Kaneko. 1996. Quality and product determination as price determinants in the marketing of fresh Pacific tuna and marlin. SOEST 96-06, JIMAR Contributions 96-304. www.soest.hawaii.edu.

Bjorndal, T., A. Child, and A. Lem (eds). 2014. Value chain dynamics and the small-scale sector: policy recommendations for small-scale fisheries and aquaculture trade. FAO Fisheries and Aquaculture Technical Paper No. 581. FAO, Rome. 112 pp.

Bouras, D. and C.R. Engle. 2007. Assessing oligopoly and oligopsony power in the U.S. catfish industry. *Journal of Agribusiness* 25:47–57.

Breimyer, H. 1976. *Economics of the Product Markets of Agriculture*. The Iowa State Press, Ames, Iowa.

Brooks, P.M. and J.L. Anderson. 1991. Effects of retail pricing, seasonality and advertising on fresh seafood sales. *The Journal of Business and Economic Studies* 1:77–90.

Bubner, E.J., J.O. Harris, and T.F. Bolton. 2009. Supplementary oxygen and temperature management during live transportation of greenlip abalone, *Haliotis laevigata* (Donovan, 1808). *Aquaculture Research* 40:810–817.

Christophersen, G., G. Roman, J. Gallagher, and T. Magnesen. 2007. Post-transport recovery of cultured scallop (*Pecten maximus*) spat, juveniles, and adults. *Aquaculture International* 16:171–185.

Cichra, C.E., M.P. Masser, and R.J. Gilbert. 1994. Fee-fishing: an introduction. Southern Regional Aquaculture Center, Publication No. 479, November.

Cremer, M.C., S.D. Mims, and G. Sullivan. 1984. Pay lakes as a marketing alternative for Kentucky fish producers. Kentucky State University Community Research Service Program, Report Number 8, Frankfort, Kentucky.

Csavas, I. 1994. Important factors in the success of shrimp farming. *World Aquaculture* 25:34–56.

Dey, M. M.T. Bjorndal, and A. Lem. 2015. Guest editors' introduction: value chain dynamics in aquaculture and fisheries. *Aquaculture Economics & Management* 19:3–7.

Engle, C.R. 1997. Marketing of the tilapias. In: Costa-Pierce, B.A. and J.E. Rakocy, eds. *Tilapia Aquaculture in the Americas*, vol. 1. World Aquaculture Society, Baton Rouge, Louisiana.

Economics Research Service (ERS). 2002–2014. United States Department of Agriculture, Washington, D.C. Can be found at www.ers.usda.gov.

Food and Agriculture Organization of the United Nations (FAO). 1978. Marketing. FAO, Rome. Can be found at www.fao.org.

Food and Agriculture Organization of the United Nations (FAO). 2014. FishstatJ. FAO, Rome. Can be found at www.fao.org.

Food Export Association of the Midwest USA. 2012. European seafood market profile. Food Export Association of the Midwest USA. Available at www.foodexport.org. Accessed July 4, 2012.

Herrmann, M., R. Mittelhammer, and B.H. Lin. 1993. An international econometric model for wild and pen-reared salmon. In: Hatch, U. and H. Kinnucan, eds. *Aquaculture: Models and Economics*. Westview Press, Colorado.

Hudson, D. 1998. Intra-processor price-spread behavior: is the U.S. catfish processing industry competitive? *Journal of Food Distribution Research* 29:59–65.

Jacinto, E.R. and R.S. Pomeroy. 2011. Developing markets for small-scale fisheries: utilizing the value chain approach. In: Pomeroy, R.S. and N.L. Andrew. *Small-scale Fisheries Management*, pp. 160–177. CABI Publishing, Wallingford, UK.

Jagger, P. and J. Pender. 2001. Markets, marketing and production issues for aquaculture in East Africa: the case of Uganda. *Naga*, the ICLARM Quarterly 24:42–51.

Kada, Y. 1997. Transporting fish by car. Can be found at www2.lbm.go.jp/comparePhoto/E_104.htm.

Keithly, W., K.J. Roberts, and J.M. Ward. 1993. Effects of shrimp aquaculture on the US market: an econometric analysis. *In:* Hatch, U. and H. Kinnucan, eds. *Aquaculture: Models and Economics*. Westview Press, Colorado.

Kinnucan, H. and Y. Miao. 1999. Media-specific returns to generic advertising: the case of catfish. *Agribusiness* 15:81–93.

Kinnucan, H. and O. Myrland. 2002. Optimal seasonal allocation of generic advertising expenditures with product substitution: salmon in France. *Marine Resource Economics* 7:103–120.

Kinnucan, H.W., O.J. Cacho, and G.D. Hanson. 1986. Effects of selected tax policies on management and growth of a catfish enterprise. *Southern Journal of Agricultural Economics* 18:215–225.

Kinnucan, H., S. Sindelar, D. Wineholt, and U. Hatch. 1988. Processor demand and price-markup functions for catfish: a disaggregated analysis with implications for the off-flavor problem. *Southern Journal of Agricultural Economics* 20:81–91.

Kohls, R.L. and J.N. Uhl. 1985. *Marketing of Agricultural Products*. Macmillan, New York.

Kouka, P.J. 1995. An empirical model of pricing in the catfish industry. *Marine Resource Economics* 10:161–169.

Lambregts, J., O. Caps, and W. Griffin. 1993. Seasonal demand characteristics for U.S. farm-raised catfish. In: Hatch, U. and H. Kinnucan, eds. *Aquaculture: Models and Economics*. Westview Press, Colorado.

Ledafish. 1996. Can be found at www.ledafish.com/tuna.htm.

Leyva, C.M., C.R. Engle, and Y.S. Wui. 2006. A mixed-integer transshipment model for tilapia (*Oreochromis* sp.) marketing in Honduras. *Aquaculture Economics & Management* 10:245–264.

M&A. 2013. The seafood industry: a sea of buyers fishing for M&A opportunities across the entire value chain. M&A International, Inc.

Marshall, D.W. 1996. *Food Choice and the Consumer*. Blackie Academic Professional, Glasgow.

National Fisheries Institute. 2015. Canned tuna HACCP guide. Available at www.seafood.com.

Norman-López, A. and F. Asche. 2008. Competition between imported tilapia and U.S. catfish in the U.S. market. *Marine Resource Economics* 23:199–214.

Palfreman, A. 1999. *Fish Business Management*. Fishing News Books, Blackwell Science, Oxford, UK.

Pastoriza, L., M. Bernardez, G. Sampedro, M.L. Cabo, and J. Herrera. 2004. Elevated concentrations of oxygen on the stability of live mussel stored refrigerated. *European Food Research and Technology* 218:415–419.

Pearson, R. and P. Sharma. 2003. Export subsidies. Agreement on Agriculture. www.fao.org.

Phan, L.T., T.M. Bui, T.T.T. Nguyen, G.J. Gooley, B.A. Ingram, H.V. Nguyen, P.T. Nguyen, and S.S. DeSilva. 2009. Current status of farming practices of striped catfish, *Pangasianodon hypophthalmus* in the Mekong Delta, Vietnam. *Aquaculture* 296:227–236.

Retail Leader. 2014. Warehouse clubs expand amid tepid economic growth. Stagnito Business Information. www.retailleader.com. Accessed June 6, 2014.

Rhodes, V.J. 1993. *The Agricultural Marketing System*. 4th edition. Gorsuch Scarisbrick Publishers, Scottsdale, Arizona.

Schoemaker, R. 1991. *Transportation of Live and Processed Seafood*. INFOFISH Technical Handbook 3. INFOFISH, Kuala Lumpur, Malaysia.

Seafood Business. 2001–2003. Diversified Business Communications, Portland, Maine, USA.

Singh, K., M.M. Dey, and P. Surathkal. 2014. Seasonal and spatial variations in demand for and elasticities of fish products in the United States: An analysis based on market-level scanner data. *Canadian Journal of Agricultural Economics* 62:343–363.

Suranovic, S. 1997–2004. International trade theory and policy. The International Economics Study center. internationalecon.com/v1.0.

Tropp, D. 2014. Why local food matters: the rising importance of locally-grown food in the U.S. food system. Agricultural Marketing Service, USDA, Washington, D.C.

Unites States Department of Agriculture (USDA). 2005. Census of Aquaculture (2005). National Agricultural Statistics Service, United States Department of Agriculture, Washington, D.C.

Unites States Department of Commerce (USDC). 2016. Standards for grades of fishery products. Seafood Inspection Program, U.S. Department of Commerce. www.seafood.nmfs.noaa.gov/standards/23_standards1.html.

Wessells, C.R. and J.E. Wilen. 1994. Seasonal patterns and regional preferences in Japanese household demand for seafood. *Canadian Journal of Agricultural Economics* 42:87–103.

Wiese, N. and K. Quagrainie. 2004. Market characteristics of farm-raised catfish. Master's thesis. Department of Aquaculture and Fisheries, University of Arkansas at Pine Bluff, Pine Bluff, Arkansas.

Wingenter, K., B. Jujica, and T. Miller-Morgan. 2013. Development of live shellfish export capacity in Oregon. ORESU-T-13-001. Oregon Sea Grant, Oregon State University, Corvallis, Oregon.

Yu, R., L.E. Kam, and P. Leung. 2008. The seafood supply chain and poverty reduction. *In:* Briones, R.M. and A.G. Garcia, eds. *Poverty Reduction Through Sustainable Fisheries: Emerging Policy and Governance Issues in Southeast Asia*, pp. 145–187. Institute of Southeast Asian Studies, Singapore.

Zidack, W., H. Kinnucan, and U. Hatch. 1992. Wholesale- and farm-level impacts of generic advertising: the case of catfish. *Applied Economics* 24:959–968.

CHAPTER 4

Market trends

The role of imports in U.S. seafood markets

The literature on trade economics suggests that free trade among nations is better than restricted trade because free trade promotes competitive markets. A competitive market is efficient and results in better resource allocation and utilization. It also improves social welfare. Trade among nations occurs when a nation has a comparative advantage in producing a particular good or service which another nation wants and is willing to pay for. The advantage comes from the country's ability to produce that particular good or service at a relatively lower opportunity cost than other nations. A country can have a comparative advantage because of resource use including labor, transportation, productivity, and natural resources. Seafood is a major category of products for which the U.S. seems to have a comparative disadvantage. Therefore, the U.S. imports a significant part of its seafood needs (Fig. 4.1).

The U.S. depends on imported seafood products to fulfill its consumption needs. The total value of imported edible seafood products in 2012 was $16.7 billion compared to $10 billion in 2002. The total value of imported seafood has grown by about 67% between 2002 and 2012. The major species of seafood imported are crustaceans including shrimp, lobster, and crab. Shrimp accounted for about 26.7% of total imported seafood in 2012, crab 8.1%, and lobster 5.4% (Fig. 4.2). The major sources of seafood imports by volume in 2012 were China, which accounted for 23% of imports; Thailand, 12%; Canada, 12%; and Vietnam, 8%. Other top seafood exporting countries to the U.S. in 2012 included Indonesia, Chile, Ecuador, India, and Mexico. In 2012, the U.S. imported seafood from an estimated 188 countries (NOAA, NMFS 2013).

In terms of exports from the U.S., the major seafood products by volume in 2012 were ground fish, salmon, squid, and surimi. Crustaceans such as crabs, lobsters, and shrimp also contributed significantly in terms of value (NOAA,

Seafood and Aquaculture Marketing Handbook, Second Edition. Carole R. Engle, Kwamena K. Quagrainie and Madan M. Dey.

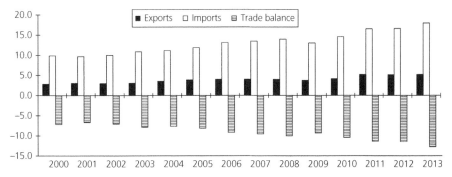

Fig. 4.1 Value of US seafood exports, imports, and trade balance, 2000–2013. Source: USDC-NOAA (2014).

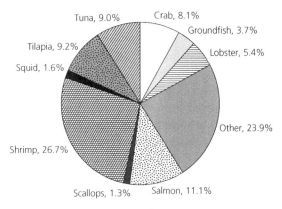

Fig. 4.2 Major species of seafood imported into the U.S., 2012. Source: USDC-NOAA (2012).

NMFS 2013). Major importers of U.S. exports in 2012 included China, Japan, Canada, South Korea, and Germany. However, the value of U.S. seafood exports has consistently been far below imports (Fig. 4.1). The increasing dependence on imported seafood has resulted in a seafood trade deficit which keeps growing. In 2013, the U.S. trade balance in edible seafood products was about $12.8 billion compared to $7 billion in 2002 (Fig. 4.1). Over 90% of seafood consumed in the U.S. was imported in 2012 (NOAA, NMFS 2013).

The persistent U.S. seafood trade deficit is of concern to government, trade policy analysts, and stakeholders in the seafood industry. The seafood trade deficit could be attributed to a number of factors such as the rapid growth in aquaculture for the production of seafood products around the world, particularly in Asia. Many Asian countries have seen significant growth in aquaculture production because of comparative advantage especially from low labor costs. For example, shrimp production from aquaculture in Asia has experienced significant growth, and most of it is exported to the U.S. In 2012, Asia accounted for 59% of U.S. seafood imports by volume from major geographic areas (NOAA, NMFS 2013); Canada and Mexico (the other North American countries) accounted for 17% of U.S. seafood imports by volume, probably due to their proximity to the U.S., and the consequently lower transportation costs. The U.S. seafood market is seen by seafood producing countries as a major market.

The relative strength of the U.S. economy and the value of the U.S. currency relative to other major currencies have also contributed to increased exports of seafood products into the U.S. However, in times when the U.S. economy is weak and/or the value of the U.S. dollar relative to other major currencies is also weak (as occurred in 2008 through 2010), seafood exporting nations tended to export to other nations instead of to the U.S. In 2008 for example, shrimp exports from Southeast Asian countries increased to the European Union (EU), diverting supplies that would otherwise have been exported to the U.S.

The seafood trade is thus competitive as seafood companies in the U.S. continually strive to develop a global network of sources for seafood. Most major seafood companies are global and report having representatives around the world for sourcing, marketing, and distributing seafood products. For example, East Coast Seafood, Inc., one of the largest distributors of live lobsters in North America, boasts of an integrated network of international subsidiaries that assist the company in worldwide sales, marketing, distribution, and customer service. East Coast Seafood established East Coast Europa, a seafood sales and distribution operation in Europe, with offices in Paris, Madrid, Milan, Frankfurt, Brussels, and London. Inland Seafood, another major seafood distributor in the nation, also has representatives across Europe and Asia who look for seafood products for the U.S. market.

Most importers continue to seek new sources of seafood and appear to be the major agents developing the seafood market in the U.S. In 1998, Seafood Connection, a seafood importer and distributor in Honolulu, Hawaii, was featured in *Pacific Business News* as the fastest growing independent seafood importer and distributor in mainland Hawaii. The company handled seafood products such as lobster, scallops, caviar, salmon, and crab and was known to import more exotic seafood products such as Russian caviar from the Caspian Sea, lobster tails from South Africa, and a variety of unique premium seafood items from Chile, Australia, and Africa for Hawaii's upscale restaurants and hotels (Zimmerman 1998).

U.S. seafood consumption

The general trend depicted in Figure 4.3 suggests that domestic consumption of seafood is not increasing in the U.S. Total consumption of seafood products appears to have peaked in 2004, remained fairly stable through 2010, and declined thereafter. In 2004, total consumption of seafood was almost 5 billion pounds but in 2011, it was 4.7 billion pounds, then 4.4 billion pounds in 2012. Data are not available relative to where consumers purchase their seafood, but it is traditionally known that most seafood products are consumed away from home. Miller (1985)

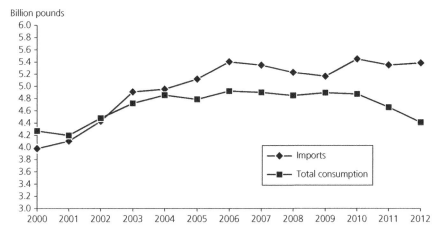

Fig. 4.3 Seafood imports and total seafood consumption in the U.S., 2000–2012.
Source: USDC-NOAA (2012).

suggests that the tourism and restaurant industries and other away-from-home outlets account for more than half of total U.S. seafood consumption. It also appears that there is a general lack of knowledge about how to prepare seafood at home for many U.S. seafood consumers. Zhang et al. (2004) reported that concerns over preparation time, lack of preparation knowledge, and product smell made consumers less likely to consume oysters and catfish at home.

Since Americans traditionally consume their seafood away from home, the status of the U.S. economy and consequently the incomes of households affect consumption. The health of the U.S. economy has not been very strong since the 2008 economic downturn. This appears to have affected seafood consumption as households reduced expenditures on away-from-home dining.

Total U.S. seafood consumption is calculated on the basis of disappearance of fishery products supplied on a round-weight (live, whole fish) equivalent basis. The total supply of fisheries products consists of both edible and non-edible (industrial) imports and domestic landings (in edible weight). The disappearance in supply consists of exports and industrial uses, which are deducted from the total supply of fisheries products to obtain total seafood consumption. This appears to be the standard for calculating consumption by the Food and Agriculture Organization of the United Nations (FAO) and many countries. Where beginning and ending stocks of the commodity are available, the calculation of total consumption accounts for these stocks.

Per capita consumption of seafood in the U.S. is low compared to that of other advanced countries in Europe and Asia. The long-term outlook of seafood consumption in the U.S. suggests a potential increase with population growth, increased awareness of the health benefits of consuming seafood, and low seafood prices. Since their discovery in the 1970s, omega-3 essential fatty acids have been the subject of several studies and clinical trials. The acids have been shown to aid in the treatment of asthma symptoms, obesity, Alzheimer's disease, bipolar disorder, and especially overall heart health and brain function (Nettleton 1995). They also benefit the heart of healthy people, and those who are at high risk of or have cardiovascular disease (Kris-Etherton et al. 2002). The American Heart Association recommends eating fish (particularly fatty fish) at least two times a week because fish is a good source of protein and does not have the high saturated fat of fatty meat products. Fatty fish such as mackerel, lake trout, herring, sardines, albacore tuna, and salmon are high in two kinds of omega-3 fatty acids, eicosapentaenoic acid (EPA) and docosahexaenoic acid (DHA). Some wild game, grass-fed meat, and some enhanced eggs have levels of EPA and DHA.

In spite of the benefits of eating fish, seafood continues to face some negative publicity. The aquaculture industry is often alleged to be using antibiotics, pesticides, and other chemicals in raising farmed fish, dissuading consumers from eating farmed fish. Fish that is captured in the wild is alleged to contain mercury, toxins, and other contaminants from the aquatic environment that accumulate in the bodies of fish. These concerns are food safety issues. Despite these challenges, it is expected that the increasing concern of American consumers about health issues and the benefits of consuming seafood will drive consumers to eat more fish. Because of the importance of a healthier heart to consumers, the benefits of eating fish would usually outweigh the risks associated with eating it.

Food consumption away from home

The share of household food dollars allocated to away-from-home meals and snacks has been increasing for more than a century. Total away-from-home expenditures include all food dispensed for immediate consumption outside the consumer's home. In 2012, total away-from-home expenditure was $680 billion. An average of $1,668 was spent per person on food in 1970 of which away-from-home meals and snacks captured 36% (USDA-ERS 2014). By 1990, the average food expenditure per person for away-from-home meals had increased by about 20% with snacks capturing 45% of the food dollar (Fig. 4.4). In 2012, about 50% of the food dollar was spent on away-from-home meals indicating that American consumers now spend half of their food dollars on meals and snacks at food service facilities such as restaurants, hotels, and schools. It is anticipated that households will continue to increase spending on food service meals and snacks at an annual rate of about 1.2% in real (inflation-adjusted) terms (Blisard et al. 2003).

Rising incomes, increasing participation of women in the labor force, the growing incidence of non-traditional households, and other demographic developments such as smaller household sizes and more affordable and convenient fast food outlets have enhanced the growth in away-from-home food

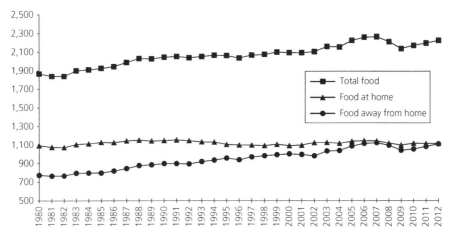

Fig. 4.4 Food expenditures at home and away from home in the U.S., 1980–2012.
Source: USDA-ERS (2014).

expenditures in the U.S. (Stewart et al. 2004; USDA-ERS 2014). There has also been a significant increase in advertising and promotion by large food service chains on away-from-home meals. Stewart et al. (2004) forecasted that consumer spending at full-service and fast food restaurants would continue to grow between 2000 and 2020, and that a modest growth in household income plus expected demographic developments would result in per capita spending rising by 18% at full-service restaurants and by 6% for fast food between 2000 and 2020. However, the aging of the population will decrease spending on fast food by about 2% per capita.

Convenience in food preparation and consumption

The need for convenience in food preparation and consumption continues to grow among American consumers as people are overwhelmed by product choices and starved for time. Convenience in home-prepared foods comes in a number of ways including ready-to-eat, heat-and-eat, quick preparation, easy-to-cook, and packaged complete meals for on-the-go consumption. Even the restaurant industry's off-premises market has outpaced growth in the dine-in option. "To-go" meal sales were approximately $7.4 billion at casual dining restaurants in the year ending August 2001, which is about 12% of total dollars spent in the casual dining segment.

Sales of convenient "dinner solution" meals continue to grow. It is estimated that sales of "dinner solutions" meals have grown by an 8-year compound annual growth rate of 7.5%, and "breakfast solutions" meals by about 6.6% (IRI 2002). Information Resources, Inc. (IRI), a global market research firm that has clients that include Anheuser-Busch, ConAgra, Johnson & Johnson, Philip

Morris, Procter & Gamble, PepsiCo, Unilever HPCE, and top retailers, reported that sales of "dinner solutions" meals added an average of over 385 million meals sold over 7 years (IRI 2002). In 2001, frozen entrées and meals reached retail sales of $9.3 billion, up 5.8% in supermarkets, unprepared frozen meat increased by 12.8%, ground beef by 16.8%, frozen unbreaded fish by 10.2%, frozen unbreaded shrimp by 36.9%, and other unbreaded seafood by 34.4% (Heller 2002). Other popular meals among consumers include ready-to-cook, pre-seasoned and prepared fresh meats, poultry, and fish/seafood, which collectively with precooked seafood accounted for 25% of supermarket seafood counter sales in 2001 (Bavota 2002).

It is anticipated that the success of ready-to-eat and ready-to-cook items will spawn new issues and opportunities (Sloan 2003). For example, there is the tendency to use only one appliance to prepare a meal, and it is anticipated that consumers will soon demand side dishes that can be cooked simultaneously in the microwave or oven in about the same length of time as the precooked entrée. Whether cooking is gourmet or everyday, any product that eliminates work or cleanup will likely have enormous appeal (Sloan 2003). Food products designed for easy home entertaining such as frozen pizza have also seen increased interest from consumers.

Demand for healthy and wholesome foods

The 1990s was a period of increased health awareness that resulted in consumer demand for foods and beverages that provided nourishment, health benefits, and good taste, at the right price. Consumers, in seeking to lead healthy life-styles, have consequently recognized the appeal of fresh and particularly natural and wholesome products, with their implied benefits of safety and wellness. Increasingly, nutritionists and food manufacturers are publicizing foods as healthy and making consumers aware of content of saturated fats and trans fatty acids. Consequently, since 2006, the U.S. Food and Drug Administration (FDA) regulations have required all food marketers to disclose the level of trans fats in their packaged products. Several large food companies such as Frito-Lay, Nabisco, and Tyson Foods have eliminated or reduced the level of trans fats in their products. The food-away-from-home sector is generally exempt from the mandatory nutrition labeling regulations, which public health advocates find unacceptable. They have called for the inclusion of the sector in the law to inform consumers about the nutritional content of these foods. Many fast food companies voluntarily follow the FDA guidelines.

Trends in healthy and wholesome food have focused on nutritional platforms such as fiber enriched, vitamin fortified, high protein, gluten-free, and omega-3 enriched. They have also focused on some general food attributes such as low fat, organic, low carbohydrate as well as the health benefits of food such as natural foods and functional foods. In 2002, total sales of natural products were estimated to be $36.4 billion with about 77% of total sales realized at the retailing and mass-market channels. Natural product retailers sold the most natural and organic foods, valued at $10.4 billion in 2002 (Spencer and Rea 2003). Food constituted 60% of total sales for natural products retailers, and 44% of the category was organic. In 2010, total natural product sales were valued at $81 billion, out of which all retailers accounted for $65 billion or 80% of the sales (Soref 2011). Of the total retail sales, natural product retailers accounted for $36 billion (44%) while sales by conventional retailers totaled $29.19 billion (36%). The non-retail sector comprising practitioners, the Internet, mail order and multi-level marketing accounted for the remaining 19% in total sales.

There is growing concern over the purity, quality, and lack of chemicals in food products. The Natural Marketing Institute (NMI) maintains a Health and Wellness Trends Database (HWTD) based on an annual research study of over 2000 U.S. consumer households. NMI reported sales of $59 billion within the consumer packaged goods health and wellness industry in 2002, representing 7.3% growth over 2001 sales (NMI 2003). The study indicated that functional and fortified foods/beverages constituted 11% of sales, organics 17%, and natural/organic personal care 15%. Vitamins, minerals, and herbals continued to thrive as about 30% of consumers indicated they made an effort to regularly eat a meatless meal, while 19% considered themselves an occasional vegetarian (NMI 2003). The study projected a 10% compound annual growth rate in the consumer packaged goods health and wellness industry, with sales of $86 billion by 2006.

Sustainability and seafood

Sustainability has become a major issue in the seafood industry over the past decade. The application of the word "sustainable" to seafood is primarily based on environmental, biological, and social principles. Various criteria are used to assess sustainability in fisheries and aquaculture. For fisheries, the criteria for sustainability include fishing practices that do not overexploit fish stocks, management practices that have minimal effects on non-targeted species (bycatch) and the ecosystem, and the adoption of conservation practices. For aquaculture, sustainability has been applied to farming practices that reduce the environmental impact of fish production from feeding practices and pollution, avoiding use of chemicals and antibiotics, and animal welfare considerations from stocking densities.

Over the years, various groups and organizations have championed the course of sustainable seafood from a niche to a major feature in the seafood industry. This has been accomplished through activist strategies such as boycotts and demonstrations as well as through seafood buying guides and ecolabeling

(Roheim 2009). There are a number of international organizations that promote sustainable seafood guides with the primary objective of influencing consumer choices for seafood. The most well-known organization is perhaps the Monterey Bay Aquarium (MBA).

The MBA is a non-profit organization with a mission of ocean conservation. The organization plays a major role in seafood markets through its Seafood Watch program, launched in 2000. The program makes science-based recommendations on various seafood products to inform consumers, chefs, and businesses on their choice of seafood to purchase. They have developed a Seafood Watch National Guide that helps buyers to make informed choices based on sustainable seafood production from both commercial fisheries and fish farming. The guide classifies seafood into three categories: *Best Choices*, *Good Alternatives* (previously called *Proceed with Caution*), and *Avoid*. A sample of the guide is shown in Fig. 4.5. Seafood guides for consumers are also produced by several other international organizations such as the Blue Ocean Institute's Guide to Ocean Friendly Seafood,

Fig. 4.5 The Monterey Bay Aquarium's Seafood Watch National Consumer Guide.

the Environmental Defense Fund's Seafood Selector, the World Wildlife Fund's Sustainable Seafood Guides, the Marine Stewardship Council, the Aquaculture Stewardship Council, Fish Choice, the Conservation Alliance for Seafood Solutions, and the National Geographic Society. Most of the guides provided by these organizations rely on the recommendations in MBA's national guide.

Certification of sustainability

The concerns of seafood consumers and environmental advocates relating to overfishing of marine resources, fisheries management, and fish farming practices have led to the development of certification programs for fisheries and aquaculture around the world. The certification programs are important considerations, especially with the increasing level of international trade in seafood products. The certifications are mainly voluntary, and producers/suppliers and marketers of seafood products have adopted them, in some cases, to have a competitive edge in the seafood trade (FAO 2011). Stakeholders generally adopt the certifications to help improve the sustainability of wild fishery resources and aquaculture practices, and as a strategy to increase market share.

There are a number of international organizations involved in certification programs for fisheries and aquaculture around the world. The following organizations have programs that are internationally recognized and have been adopted by many fisheries stakeholders.

The Marine Stewardship Council (MSC)

The MSC operates a market-based certification program that recognizes and rewards sustainable fisheries. The MSC was created in 1997 and provides a mechanism for labeling seafood products from wild-caught fisheries that have met MSC's robust sustainable fishing standard. The three principles of MSC's sustainable fishing standard are:

1 Sustainable fish stocks: The fishing activity must be at a level which is sustainable for the fish population. Any certified fishery must operate so that fishing can continue indefinitely and is not overexploiting the resources.
2 Minimizing environmental impact: The management of fishing operations should maintain the structure, productivity, function, and diversity of the ecosystem on which the fishery depends.
3 Effective management: The fishery must meet all local, national, and international laws, possess the ability to respond to changing circumstances, and maintain sustainability.

The MSC program also includes a traceability component: any company seeking to sell MSC certified fish and display the ecolabel (Fig. 4.6) on products must obtain MSC Chain of Custody (CoC) certification. CoC certification ensures that products sold with the ecolabel are traceable back to an MSC certified fishery. Results of independent DNA testing on MSC labeled products show that over 99% are correctly labeled, confirming supply chain integrity for MSC certified sustainable seafood.

Fig. 4.6 The Marine Stewardship Council (MSC) logo.

Fishery improvement projects

Fishery Improvement Projects (FIPs) have been implemented to reward fisher-men and -women based on progress towards sustainability by improving market access (Sampson et al. 2015). Often funded by non-governmental organizations and the private sector, FIPs form part of a supply chain partnership oriented towards increasing the sustainability of seafood. FIPs constitute something of an intermediate step towards certification by the Marine Stewardship Council until sufficient data become available to achieve full certification status. In 2015, there were more than 130 fisheries in FIPs worldwide (Sampson et al. 2015). FIPs require continuous progressive improvements towards MSC certification stand-ards to benefit from having access to important markets.

The Global Aquaculture Alliance (GAA) Best Aquaculture Practices (BAP)

The GAA certifies aquaculture facilities based on a set of standards. The BAP standards are based on environmental and socially responsible practices, animal welfare, food safety, and traceability for the facilities outlined below (GAA 2014). Facilities must comply with the appropriate BAP standard. Current standards apply to seafood processing and repacking plants; finfish (currently tilapia, channel catfish, and *Pangasius* species), crustacean (shrimp), mussel, and salmon farms; fish and shrimp hatcheries; and feed mills. BAP certification is a process that involves site inspections and effluent sampling with sanitary controls, therapeutic controls, and traceability. Certified facilities can use the BAP certifica-tion mark (Fig. 4.7). The BAP certification program is subject to annual site audits.

The GAA also has a Registered Buyer Program (RBP) in which it acknowledges seafood marketers on its web site for demonstrating support for farmed seafood products from environmental and socially responsible aquaculture practices.

The Aquaculture Stewardship Council (ASC)

The ASC was formed in 2010 to manage the ongoing development of global standards for responsibly farmed seafood. There were a number of "Aquaculture Dialogues" – roundtable discussions on specific seafood coordinated by the World Wildlife Fund (WWF). The dialogues involved representatives from the global aquaculture industry, retail and food service sector, NGOs, government, and the scientific community, and resulted in the development of standards for responsible fish farming. The standards apply to species that include abalone, bivalves (oysters, mussels, clams, and scallops), freshwater trout, *Pangasius*, salmon, shrimp, tilapia, and seriola/cobia (ASC 2014).

Fig. 4.7 The Global Aquaculture Alliance (GAA) Best Aquaculture Practices (BAP) certification mark.

Fig. 4.8 The Aquaculture Stewardship Council (ASC) logo.

The operations of the Council are similar to those of MSC and GAA, with an aquaculture certification program and seafood label. ASC certifies aquaculture operations but also partners with seafood processing facilities and seafood retail and food service companies. A certified facility can bear the ASC logo (Fig. 4.8), assuring customers and consumers that the seafood is sourced from a farm that adheres to environmentally and socially responsible production practices.

GLOBAL Good Agricultural Practice (GLOBALG.A.P.)

GLOBALG.A.P. is a private sector body that has standards for the certification of agricultural and aquaculture products. The GLOBALG.A.P's aquaculture standards cover the entire chain to account for the origin of the farmed product through various stages of the food supply chain. Thus, activities associated with broodstock, hatchery, fingerlings, feed, farming, harvesting, and processing are part of the certification system (GLOBALG.A.P. 2014). The latest version of GLOBALG.A.P's aquaculture standard, Version 5, has 231 control points: 65 associated with food safety; 65 associated with the environment; 45 associated with animal welfare; 30 associated with workers' welfare; and 26 associated with traceability. The GLOBALG.A.P system is more of a business-to-business certification process and is also designed to assure consumers of food safety.

Traceability and labeling of seafood products

There are mandatory requirements on traceability and labeling in the seafood value chain in most countries. This has become necessary because of increasing food-related recalls and food-borne illnesses. Traceability systems are therefore essential to identifying the source of the problems and taking appropriate corrective measures, especially with increased international trade. That is why the sustainability certification processes of international organizations associated with seafood also include traceability.

In the U.S., food wholesalers and distributors have been required to maintain full traceability of food items they handled as they made their way throughout the supply chain. There are thousands of prepared, perishable, and packaged food products being offered to the consuming public, therefore handlers of food products in the supply chain are required to have control and sanitation procedures. One of these new requirements is the establishment of written hazard analysis and critical control point (HACCP) programs that are mandatory for all processors and handlers of meat, poultry, seafood, and fruit/vegetable juices.

There have been a number of initiatives to enhance traceability. One is the Public Health Security and Bioterrorism Preparedness and Response Act of 2002, a law that required all food companies to develop compliance plans and register with the FDA. There is also a U.S. Animal Identification Plan that identifies premises used for livestock operations. The primary goal of this program is rapid containment of animal disease when it occurs. There is also a mandatory country-of-origin labeling (COOL) law that directly affects seafood marketing, which is discussed below.

With the dependence on imports for over 90% of seafood needs in the U.S., the FDA has instituted a project known as Fish SCALE (Seafood Compliance and Labeling Enforcement) in collaboration with the Office of Food Safety (OFS), Office of Compliance (OC), and Office of Regulatory Affairs (ORA) to regulate inaccurate and false labeling of seafood products. The project involves the development and implementation of regulatory genetic methods for the proper identification of fish species on seafood labels. The project is also meant to assess potential risks associated with certain seafood products (i.e., processing related hazards, natural toxins, allergens, etc.) and assure consumers of seafood safety.

Country-Of-Origin Labeling (COOL)

In the 2002 Farm Act, the U.S. Congress amended the Agricultural Marketing Act of 1946 and required retailers to use country-of-origin labeling (COOL) for certain covered commodities. The implications of mandatory COOL included record-keeping and tracking systems to verify country of origin. However, the main purpose of COOL was to allow end consumers to make more informed decisions when making their purchases by telling them the country of origin of food products. The reasons for making traceability systems mandatory included facilitating and monitoring the ability to trace product back to its origin to enhance food safety, addressing consumer information about food safety and quality, and protecting consumers from fraud and producers from unfair competition (USDA-ERS 2002).

The COOL rules identified two broad categories of entities that have responsibilities under COOL: suppliers and retailers. The rules identified three classes of suppliers: (1) those initiating suppliers who have the responsibility of initiating a country-of-origin declaration; (2) the intermediary supplier, who is any supplier other than the initiating supplier; and (3), the catchall class, any person engaged in the business of supplying a covered commodity to a retailer, whether directly or indirectly. The rules require that each covered commodity offered for sale individually, in a bulk bin, carton, crate, barrel, cluster, or consumer package bear a legible declaration of the country of origin and, if applicable, the method of processing. The responsibility for such disclosures is on the retailer. The purpose of record keeping is to ensure that a proper audit trail exists to allow the government or other enforcement authority to track the covered commodities from origin to retailer or vice versa.

In 2005, the U.S. Department of Agriculture established the final rules for COOL. Fish and shellfish are covered commodities under the rules and are required to be labeled at the retail level to indicate country of origin and method of production (i.e., wild or farm raised). Under the final rule, processed seafood is exempted; therefore food service establishments, such as restaurants, lunchrooms, cafeterias, food stands, bars, lounges, and similar enterprises are exempted from the mandatory COOL requirements (USDA-AMS 2004). However, some southern states have passed laws that require COOL labeling by food service establishments, such as restaurants.

The European Union introduced labeling measures in 2002 that required labels to include information on the commercial and common name of fish, production method (capture or farmed), catch area (ocean, freshwater, or farmed), and country of origin for farmed fish. This information is required throughout the supply chain.

Ecolabeling of seafood products

Ecolabeling is generally voluntary but can be mandatory when backed by the government. An ecolabel is a market parameter used to create consumer demand for seafood products from sustainably-managed fisheries and aquaculture, thus providing incentives for sustainable production practices. Ecolabeling schemes use demand-side factors as mechanisms to influence production. The main assumption underlying ecolabeling is that consumer awareness of environmental issues results in the demand for ecolabeled seafood products rather than non-ecolabeled products. Certification programs and ecolabeling have thus become important business/investment decisions for producers/suppliers of seafood products.

Certification and ecolabeling are interrelated tools being adopted to support sustainable fisheries and aquaculture. Some certification programs come with an ecolabel as described previously. Ecolabeling programs disclose information and take the form of assurance schemes, certification, and seals of approval.

Different criteria or standards are adopted by different entities in ecolabeling programs. Ecolabeling schemes can be classified into three categories (Wessells et al. 2001; Viðarsson 2008): first-, second-, and third-party labeling schemes. A first-party scheme involves individual commercial companies that set their own eco-standards for products based on some environmental issues of interest to their customers. Whole Foods Market, a grocery store chain that sells only natural and organic food products, has a scheme that represents such first-party labeling. The company has its own "Responsibly Farmed" label for farmed fin-fish and shrimp products that meet its quality standards. This form of ecolabeling is self-declaration, which has drawbacks. For example, it could encounter credibility issues with consumers because it lacks verification from an independent party.

The second-party scheme applies to programs established within an industry, such as product labeling programs established by industry associations for their members. Verification of compliance is achieved internally and members/users pay the label owner a logo-licensing fee to use the label. In the U.S., an example is Massachusetts's "Commonwealth Quality," a brand used for a certification process which has criteria that include use of best management practices, sustainability, and environmental friendliness. For aquaculture, best practices for all fish and shellfish farms include not treating waters with pesticides, antibiotics, hormones, and growth stimulants that may be harmful to the native habitat. Shellfish farmers should utilize natural conditions of the water to feed and grow their products to minimize the environmental impact. Once a grower or producer satisfies these and other requirements, they can obtain the "Commonwealth Quality Program" (CQP) certification and use the brand logo.

Alaska also has a Responsible Fisheries Management (RFM) certification scheme that is based on the Food and Agriculture Organization's (FAO) code and guidelines on responsible fisheries management. The certification label is owned by the Alaska Seafood Marketing Institute (ASMI) and available to organizations or individuals that have a written agreement with ASMI. This type of labeling scheme could also face a lack of credibility due to lack of verification from an independent third party.

The third-party labeling scheme involves three parties: the public or private organization that owns the label, a certification entity, and the producer. The certification processes of the Marine Stewardship Council, the Global Aquaculture Alliance, and the Aquaculture Stewardship Council outlined earlier fall under this scheme. These organizations set standards and criteria and own the label, which the certification entity uses to evaluate production processes or practices. Once the standards or criteria are met, the label can be used as a seal of approval and assurance to customers about the seafood product. This labeling scheme is generally assumed to be credible because of the independent third-party certification.

Ecolabeling is being used as a market-based tool to achieve environment-friendly production practices in fisheries and aquaculture. However, concerns have been raised by some countries and industry groups that ecolabeling requirements are tools being adopted by seafood-importing countries to protect domestic industries and restrict market access. Many developing countries complain about the costs associated with meeting international labeling and certification standards (FAO 2011).

Seafood and the "local food" movement in the U.S.

There is increasing appreciation by consumers of their food sources, which has given rise to various movements supporting food production. The "local food" movement, in particular, places emphasis on local food systems and sources, and is being embraced by policy makers, food producers, marketers, and the consuming public. The "local" label often applies to political boundaries or geographical distance and sometimes on the food production process and distribution. Local foods are publicized to be fresh, healthy, and environmentally friendly because the foods do not have to be transported over long distances. Local food systems also benefit the local economy through various economic activities.

A consequence of the local food trend is an expansion in state-sponsored agricultural and food marketing programs, generally including product labeling and sometimes slogans, which are used as marketing strategies to differentiate the state's products from those of other states. These programs aim to project a perception or image of quality to increase demand for the state's (local) products. In certain cases, the programs have adopted grading and certification processes to support the slogan and label.

All states along the coastal regions of the U.S. have programs relating to seafood; some states include aquaculture. Massachusetts's CQP brand is used for products that are grown, harvested, and processed in Massachusetts. It has a certification process as outlined in the previous section. New Jersey's program is known as "Jersey Fresh." It is an advertising and promotional program for agricultural products grown in New Jersey. Based on the success of the "Jersey Fresh" program, other brands have been developed which include "Jersey Seafood," "Jersey Grown," and "Jersey Equine." In North Carolina (NC), the slogan is "When you want the best, it's Got to Be NC"! The slogan is used to promote North Carolina's agricultural products through various marketing channels. The program includes seafood and aquaculture products, which have a "Freshness from NC waters" logo. All these state labels are used on point-of-purchase materials to inform consumers about the availability of state-grown products.

Organic seafood

One of the major factors influencing consumer food choices is health and food safety. Concerns about chemical residue on foods and food-borne diseases are making more and more consumers turn to organic foods. Organic foods are perceived by consumers to be safer and healthier than foods produced with non-organic materials. Organic products account for only about 1% of food sales nationwide, but sales of organic foods have quadrupled since 1990. The Organic Trade Association reported that overall sales of organic products reached $35.1 billion in 2013, which includes $32.3 billion in organic foods (Organic Trade Association 2014). The fastest-growing categories are fruits and vegetables which accounted for $11.6 billion in sales in 2013. More than 10% of the fruits and vegetables sold in the U.S. are now organic (Organic Trade Association 2014). Organic meat, poultry, and fish sales amounted to $675 million in 2013. The organic food sector has moved from a niche to a mainstream industry in the U.S. (Organic Monitor 2003).

There are growing consumer concerns about conventional production methodologies in both terrestrial and aquatic farming, perceived health benefits of food raised without the use of synthetic chemicals or drugs, and desires for humane treatment of livestock. The healthy perceptions of seafood are helping to increase per capita seafood sales, especially the health benefits of omega-3 fatty acids. Unfortunately, there are no official organic certification standards for the U.S. aquaculture industry.

For farmed seafood in general, organic principles involve biological, environmental, social, and food processing factors. The criteria for farmed seafood include specific site selection issues for aquaculture farms, minimal environmental impact of farming practices, prohibition of chemical use, use of biological and natural disease control measures, use of fish feed ingredients from organic agriculture, prohibition of fish feed ingredients from genetically modified organisms (GMOs), and use of fishmeal and fish oil in feed sourced from sustainable fisheries. Seafood processing should also adhere to some strict organic standards.

There have been several years of efforts and activities to establish USDA organic standards for aquaculture but there has not yet been a final rule. However, organic seafood is available in the U.S. market: organic salmon, shrimp, mussels, tilapia and other seafood are certified under European Union, Canadian, and other third-party private standards such as that of Naturland and the Soil Association.

The lack of USDA standards for organic aquaculture products represents lost income opportunities for the U.S. aquaculture industry. However, the industry is ready to pursue organic aquaculture once official standards are put in place. While there has been disagreement on several issues, the underlying principles related to the designation of organic include breeding, feed, health-care, living conditions, and record-keeping standards (ISEES, University of Minnesota 2001).

Wholesale-retailer integration in the food system

Mergers and acquisitions continue to change the structure of food wholesaling. Many food wholesalers and distributors are acquiring retail food operations. This process diminishes the share of retail food distribution accounted for by traditional third party wholesalers. For example, SuperValu, a leading broad-line wholesaler, was ranked second among the top 10 national leaders in grocery wholesaling and ninth in grocery retailing based on 2005 sales (Martinez 2007).

The reason for an acquisition depends on the company's position in the value chain. Seafood producers may pursue acquisitions for operational synergies and reducing the cost of doing business, while seafood distributors and retailers may seek acquisitions for better control over supply. Achieving operational synergies and cost reductions helps to obtain higher margins. Seafood is a perishable product and requires efficient distribution systems.

Changing consumer demands are allowing the seafood industry to have demand-driven production processes that integrate seafood producers, processors, and retailers. Companies also want greater control for traceability purposes for their products because of the increasing emphasis and requirements on food safety. A company's involvement in harvesting, processing, and distribution activities allows it to address traceability and sustainability issues. Involvement in both processing and distribution activities also allows a company to market differentiated products, mostly under their own brands. For example, major U.S. food distributors such as Sysco and US Foods have partnerships that allow them to market their branded seafood products.

Another trend that has developed in the past two decades is food retailing by non-traditional food retailers selling food and non-food grocery products. Mass-merchandisers such as Walmart and Target, and warehouse clubs such as Costco, Sam's Club, and BJ's are now major food retailers. Most of these are self-distributing wholesalers who have established a growing presence in food retailing, positioning themselves within the food industry by creating new shopping formats that appeal to consumers and by lowering costs (Martinez 2007). Walmart is the top grocer in the U.S. and accounted for 15.2% of all supermarket sales in 2003 (Tarnowski and Heller 2004).

Electronic Data Interchange (EDI)

Food wholesalers and distributors handle thousands of commodities and operate complex distribution centers and delivery fleets. Therefore their operations require an optimized and synchronized system that involves labor, inventory, warehouse space utilization, tracker and trailer utilization, and customer/vendor accounts receivable. The wholesale industry is competitive and, given the current industry trends, food wholesalers and distributors must have a technology focused on the unique aspects of food distribution that reduces costs and increases profitability. The electronic data interchange (EDI) system allows businesses to order merchandise, streamline delivery, and reduce overall costs. Any EDI system requires that suppliers and retailers use compatible computer systems.

The Efficient Consumer Response (ECR)

In 1992 the food supply industry developed the efficient consumer response (ECR) system, which shares information between retailers and vendors. It allows for deliveries to be based on sales, lowering storage costs. Prior to ECR was the quick response (QR) system, which focused on shortening the retail order cycle: the total time elapsed from the point merchandise is recognized as needed to the time it arrives at the store. Goods that once took 8 weeks or more to be ordered and received were ordered and delivered on a weekly basis, hence "quick response." The advantage gained was that the shorter the order cycle, the lower the inventory levels required, which provided significant financial leverage for a business. Order cycles were shortened through the use of EDI and bar codes to automatically identify products.

ECR was built on QR techniques but addressed order cycle as well as a wide variety of business processes involving new product introductions, item assortments, and promotions. ECR uses technology to improve every step of the cycle (or business process), which results in making every step faster and more accurate (Food Marketing Institute 2004). ECR also uses collaborative relationships in which any combination of retailer, wholesaler, broker, and manufacturer works together to seek out inefficiencies and reduce costs by looking at the net benefits for all players in the relationship. The ultimate goal of ECR is to drive the order cycle and all the other business processes with point-of-sale data and other consumer-oriented data, giving an accurate read on consumer demand (Food Marketing Institute 2004). The data are passed by way of EDI to the manufacturer so products can be made in quantities based on actual consumer demand and then distributed to the end consumer in the most efficient manner, hence the term "efficient consumer response." The ECR system is intended for the grocery industry to focus on the efficiency of the total grocery supply system to maximize consumer satisfaction and minimize cost (Food Marketing Institute 2004).

In 1996, Walmart tested a new system of EDI called "Collaborative Planning, Forecasting, and Replenishment (CPFR)." The system involves sharing sales forecasts of the manufacturer and those of Walmart, and tailoring orders and deliveries accordingly (Kinsey 1999). A modified version of CPFR that is now commonly used in the food industry is scan-based trading (SBT). SBT is also known as "Pay-on-Scan (POS)."

The SBT system allows food manufacturers to bill retailers for their inventory only after the goods are scanned and sold (Kinsey 1999). Inventory is therefore on consignment basis from vendors. There is some lag time in billing, which could be up to 30 days. Some advantages of the system to the grocery retailer include savings in labor cost and improvements in cash flow since capital is not tied up in inventory. For the food manufacturer or wholesaler, the store's scanner data allow the company to monitor product movement and replenish products, thus increasing sales. The scan-based trading depends on mutual trust and accurate scanning.

Other leading food companies have proposed an Internet-based platform, called UCCnet, which operates on the World Wide Web. One element of UCCnet is CPFR, involving manufacturers and retailers separately forecasting future sales and sharing these forecasts to arrange orders and deliveries.

The Efficient Food Service Response (EFR)

A comparable system that has been initiated in the food service sector is the efficient food service response (EFR). The system helps improve efficiencies in the food service supply chain by linking manufacturing plants to distribution warehouses to operator's tables. A study conducted by Computer Sciences Corporation, Consulting and Systems Integration, and the Stanford Global Supply Chain Forum of Stanford University titled *Enabling Profitable Growth in the Food-Prepared-Away-From-Home Industries* was the blueprint for the project. The report documents $14.3 billion in annual supply chain savings that may be achieved across five strategies: (1) equitable alliances, (2) supply chain demand forecasting, (3) food service category management, (4) electronic commerce, and (5) logistics optimization. Savings to food service wholesalers, in particular, would amount to $4.7 billion (Harris et al. 2002).

A study conducted by the EFR project in 2003 suggested that, despite steady progress by the food service industry in using bar codes on cases and inner packs, the industry required more efforts in both the use and quality of bar codes to achieve real benefits of supply chain (EFR 2003). The study reported that case coding among food service manufacturers had increased since 1999. Case coding among respondents was 54% in 1999, 61% in 2000, 69% in 2001, and 77% in 2002. The EFR project has an industry-wide goal of 96% use of bar coding. While the use of bar codes had increased from previous years, the survey revealed that the quality of bar coding efforts had slipped. The 2003 data showed 74% of case codes were scanned accurately, compared to 82% in 2002 and 89% in 2001.

The 2003 survey also revealed significant variations in the use of case coding within different product categories. Equipment and supplies had the highest rate of case coding at 83%, followed by dry grocery at 80%, frozen and refrigerated foods at 73%, and produce at 23% compliance. The variations were consistent with 2002 data among the same categories. The survey also showed that 68% of cases were marked with bar codes on at least two sides, while 32% had a code printed on only one side. EFR recommends placing bar codes on two adjacent sides.

The 2003 survey recorded 29,579 cases in six different distribution facilities including three regional broad-liners in the Southwest, Southeast and Northeast regions of the U.S., two national broad-liners in the West Coast and Mid-Atlantic regions of the U.S., and a systems distributor. The survey recorded cases from 1719 different suppliers.

Food service distributors and operators have been advocating for food companies to use bar codes in order to increase supply chain efficiencies and ensure better product traceability. The EFR project believes that, as more companies use bar codes, there will be better tracking of products from manufacturer to end user, reduction in invoice discrepancies, more accurate communications, and effective electronic capture of company and product information.

E-commerce

Technology has significant impacts on the way businesses operate, and the food industry is no exception. The common use of the Internet and the need for speed has forced the food industry to reexamine how it does business. Food companies are looking to technology to decrease costs, increase service levels, and improve the bottom line; therefore e-commerce is becoming popular. Food companies engaged in e-commerce expect to reduce costs and improve efficiency in the supply chain by reducing fragmentation within it.

There are various forms of e-commerce in use in the food industry. Some systems simply serve as a registry of suppliers and buyers and provide a forum for business transactions. Other systems are market based, allowing for trading, including auctions. For example, the Uniform Code Council (UCC) operates a web-based system called UCCnet. It is a registry and synchronization service that helps to improve the accuracy of members' supply chain products and location information. Suppliers provide product, location, and trading partner information to the UCCnet Registry service and the system then validates the data with demand side partners, ensuring that all trading partners use identical UCC standards. Over 3000 companies have signed on to UCCnet, including several food industry companies. UCCnet facilitates the delivery of products to reduce out-of-stocks and excess store inventory.

Foodconnex is another example of an e-commerce platform that offers services including a catalog database for National Fisheries Institute members, marketing products, and customized business-to-consumer or business-to-business transactions. Clients of Foodconnex include Del Monte, Campbell's Foodservice, and the National Frozen Food Association.

In 2000, some food service leaders including McDonald's, Sysco, Tyson, and Cargill teamed up to form the electronic Foodservice (eFS) Network for their own purchases across all food categories, including seafood. The network also caters to other segments of the food service industry. The Internet site provides a public exchange and a private exchange for confidential customer–supplier trading. All procurements are online (*SeaFood Business* 2000).

Another example of an online marketplace serving suppliers and retailers is GlobalNetXchange. The exchange is designed to match retailers with suppliers and cut costs. Companies that utilize this exchange include Kroger, Sears Roebuck, Carrefour, Oracle, METRO AG, and J. Sainsbury. The Worldwide Retail Exchange is another business-to-business exchange for the retail industry including Albertson's, H.E. Butt, Wegman's, Kmart, and Target. Subway restaurants operate an extranet, called IPCnet, that links all its suppliers with distributors and store level operators. The system provides for tracking, invoicing, and auditing all supply chain activities and enables Subway operators to monitor the performance of distributors and manufacturers from different parts of the country.

In 2000, a number of seafood companies from Canada, U.S., and Iceland formed an Internet-based business called Seafood Alliance. The companies included Pacific Seafood Group, American Seafoods Inc., SIF Group, Pacific Trawlers/Crystal Seafoods Inc., Fishery Products International Limited, Clearwater Fine Foods Inc., Coldwater Seafoods, a subsidiary of Icelandic Freezing Plants Corporation, the Barry Group of Companies, and High Liner Foods Inc. The ultimate purpose of the alliance is to find an industry-specific solution to improving the financial performance of participating companies in the seafood industry. The alliance implements an independent platform that enhances business-to-business e-commerce in the interest of all seafood industry participants (Puget Sound Business Journal 2000).

At the 2000 Boston Seafood Show, several e-commerce systems were promoted. Among them were Gofish, an online seafood exchange with 400 subscribers, Fishmonger, Globalfoodexchange, Gotradeseafood, Gofrozen exchange with over 900 member companies, and Worldcatch. Each web site allows commodity buyers and sellers to exchange information and conduct product exchanges over the Internet (*SeaFood Business* 2000). However, since many major buyers and sellers of seafood are involved in other e-commerce systems, some of these seafood companies have struggled to get buyers to sign on. In 2001, Gofish eliminated its online seafood trading which was started in 1999, and Globalfoodexchange also ceased operations (*SeaFood Business* 2001).

Aquaculture market synopsis: *Pangasius* spp. (swai, basa, and tra)

The Asian catfish, commonly known as *Pangasius* spp., is native to Southeast Asia and is predominantly cultured in Vietnam. The species most commonly cultured are *Pangasius bocourti* (basa in Vietnamese) and *Pangasius hypophthalmus* (tra in Vietnamese), the latter accounting for over 95% of production (IDE-JETRO 2013). *Pangasius* has achieved remarkable worldwide market successes since its introduction as a commercially farmed fish in Vietnam in the early to mid-1990s. Other Asian countries such as Thailand, Cambodia, Lao People's Democratic Republic, Myanmar, Bangladesh, and China also farm *Pangasius*, which now competes with major freshwater farmed species such as tilapia on the world market and the channel catfish (*Ictalurus punctatus*) in the U.S.

The number of countries that import *Pangasius* from Vietnam has increased from 11 in 2001 to over 100 countries worldwide since 2007 (VASEP 2012). The major markets for *Pangasius* include the European Union, Eastern Europe, and the Americas.

Following the successful commercial farming of *Pangasius*, Vietnam began exporting to the U.S. in 1996, and by 2001 *Pangasius* had taken about 20% of the market for catfish frozen fillets, which otherwise was controlled by the domestically produced channel catfish. *Pangasius* products exported into the U.S. are mainly frozen, boneless fillets, which have similar appearance to fillets produced from U.S. farm-raised channel catfish; the price is relatively lower than that of channel catfish. Export of *Pangasius* frozen fillets from Vietnam to the U.S. increased from 0.05 million kg in 2001 to 7.76 million kg in 2012 (NOAA-NMFS 2014). The increasing competition from *Pangasius* contributed to the decline in U.S. channel catfish prices from the late 1990s to the mid-2000s (Quagrainie and Engle 2002). In 2009, Australia allowed the importation of *Pangasius* and it is believed to have affected the market for New Zealand's hoki, which experienced a 90% decline in market price (Globefish 2009).

Pangasius exports to the U.S. and other countries have increased significantly since 2003. Vietnam is the main supplier of frozen catfish fillets to the U.S. market and accounted for 94% of all imports in 2012. *Pangasius* frozen fillets accounted for 58% market share of the frozen catfish fillets market in 2012 (NOAA-NMFS 2014). The EU continues to be the main market for *Pangasius* from Vietnam with major markets in Spain, Germany, The Netherlands, Italy, and Poland. In 2008, about 30% of Vietnam's *Pangasius* exports went to the EU, 16% to the U.S., and 5% to Mexico. There have been increased exports to Latin America as well including Brazil, Mexico, Colombia, and Costa Rica. Mexico was the fifth largest importer of *Pangasius* in the world in 2009 (Globefish 2010).

The growth in the market for *Pangasius* has been attributed to a number of factors, the major factor being its relatively low price (Quagrainie and Engle 2002; Globefish 2009). The low prices helped with market penetration and expansion into Western and Eastern European markets, especially during the economic crises of 2008 and 2009. In 2009, for example, the unit value of Vietnamese *Pangasius* exports averaged US$ 2.20/kg but the processor price of U.S. channel catfish fillets, the main product competitor, averaged US$ 3.20/kg. In 2011, the average price of U.S. channel catfish fillets reached US$ 3.70/kg. The EU price of *Pangasius* averaged US$ 2.52/kg in 2009, which was lower than the price of other potential substitute products on the market such as Alaska pollock, cod, and hake fillets.

Consumers are also reported to prefer *Pangasius* because it is not so oily and has an attractive snowy white color (*SeaFood Business* 1999). These attributes have enabled *Pangasius* to establish an identity among buyers in the international market substituting it for other white fish such as tilapia, pollock, cod, and hake. In the EU, consumers consider *Pangasius* as the "tropical white fish," which has encouraged increased importation of *Pangasius* fillet.

The success of *Pangasius* on the U.S. market was not without challenges. The continuous decline in catfish market prices initiated some U.S. domestic policy changes to help the domestic catfish industry to compete. The U.S. Congress enacted Section 747 of the 2001 Agriculture, Rural Development, Food and Drug Administration and Related Agencies appropriations bill (Public Law 107-76), which prohibited the use of the label "catfish" on imported products other than fish from the family *Ictaluridae*. Vietnamese swai, basa and tra belong to the family Pangasiidae, but were marketed in the U.S. as catfish. This bill aimed at differentiating U.S. farm-raised channel catfish from other imports (especially from Vietnam) at the market level.

In 2002, an anti-dumping suit was filed against Vietnam for selling frozen *Pangasius* fillets in the U.S. at below production cost. The U.S. International Trade Commission in 2003 approved the anti-dumping suit, which resulted in the imposition of tariffs on Vietnamese *Pangasius* fillets that ranged from 37 to 64%.

The U.S. policy had a consequent effect of reduced exports and oversupply of *Pangasius* in Vietnam. Domestic prices plummeted as a result, forcing exporters to look for alternative export markets in Central America, Canada, Europe, Australia, and the Middle East (IDE-JETRO 2013). The diversification of export market opportunities helped domestic prices to recover in 2004, which in turn spurred significant growth in the Vietnamese *Pangasius* industry (IDE-JETRO 2013).

Export of *Pangasius* to other countries also encountered some challenges. Between 2008 and 2010 for example, the governments of Brazil, Mexico, Egypt, Russia, Italy, and Spain, among other countries, imposed temporary restrictions on imports of *Pangasius*, mostly because of concerns about quality resulting from poor sanitation and the use of antibiotics. These restrictions have since been reversed but were seen by many in the industry as protectionist schemes for the respective domestic industry (Globefish 2009; McGee 2010). *Pangasius* was seen to be in competition with the domestic channel catfish industry in the U.S., the carp industry in Brazil, the tilapia industry in Egypt and Latin American countries, and the hoki industry in New Zealand.

The world economic crises of 2008 and 2009 also contributed to some market challenges for *Pangasius*. Russia and Ukraine were major importers but reduced imports by about 66% and 49%, respectively, as a result of the economic recession (Globefish 2010). However, because of its relatively low price compared to other white fish seafood products, the EU continued to import *Pangasius* even during the recession.

Summary

A country can have a comparative advantage in the production of a product because of resource use. Seafood is a major category of product for which the U.S. appears to have a comparative disadvantage, probably because of the rapid growth in aquaculture for the production of seafood products around the world, particularly in Asia compared to the U.S. Therefore, the U.S. imports a significant part of its seafood needs. The increased dependence on seafood imports to meet domestic consumption needs in the U.S. has made the role of seafood importers very important in the seafood distribution system. The seafood trade is competitive. Thus, seafood companies continually strive to develop a global network of sources for seafood and appear to be the major agents developing the seafood market in the U.S. Many food wholesalers/distributors and non-traditional food retailers are becoming importers of seafood and offer seafood products along with food and other non-food grocery products to the consuming public.

Per capita consumption of seafood in the U.S. is low compared to that of other advanced countries in Europe and Asia. The long-term outlook in the U.S. suggests potential increase in seafood consumption with population growth, increased awareness of the health benefits of consuming seafood, and low seafood prices.

Consumers, in seeking to lead healthy lifestyles, have consequently recognized the appeal of fresh and particularly natural and healthy products, with their implied benefits of safety and wellness. It has been projected that, with a 10% compound annual growth rate in the consumer packaged goods health and wellness industry, sales will reach $86 billion by 2006. For example, the organic food sector has moved from a niche to mainstream industry with an average of 20–25% growth in sales over the past decade in the U.S. Sales of convenient "meal solutions" continue to grow. It is estimated that sales of "dinner solutions" meals have grown by an 8-year compound annual growth rate of 7.5%, and "breakfast solutions" meals by about 6.6%.

Sustainability has become a major aspect of seafood. For fisheries, the criteria for sustainability include fishing practices that do not overexploit fish stocks, management practices that have minimal effects on non-targeted species (bycatch) and the ecosystem, and the adoption of conservation practices. For aquaculture, sustainability has been applied to farming practices that reduce the environmental impact of fish production from feeding practices and pollution, not using chemicals and antibiotics, and animal welfare considerations from stocking densities. Certification and ecolabeling are interrelated tools being adopted to support sustainable fisheries and aquaculture. Some certification programs come with an ecolabel that takes the form of assurance schemes, certification, and seals of approval. Ecolabeling schemes aim to use demand-side factors as mechanisms to influence production. The main assumption underlying certification and ecolabeling is that consumer awareness of environmental issues results in the demand for ecolabeled seafood products rather than non-ecolabeled products. Certification programs and ecolabeling have thus become important business/investment decisions for producers/suppliers of seafood products. Similar assumptions also underlie the "local foods" movement as well as the various state agricultural marketing programs.

Food traceability has become an important public policy issue because of concerns about food-borne illness and diseases. Handlers of food products in the supply chain are being required to have control and sanitation procedures in place such as the hazard analysis and critical control point (HACCP) programs that are mandatory for all processors and handlers of meat, poultry, seafood, and fruit/vegetable juices. Other initiatives to enhance traceability include the Public Health Security and Bioterrorism Preparedness and Response Act of 2002 (a law that required all food companies develop compliance plans and register with the FDA); mandatory country-of-origin labeling (COOL); and the U.S. Animal Identification Plan that would require premises identification for livestock operations.

To optimize and synchronize the food supply chain system, some technology initiatives have been adopted in the supply chain that involve labor, inventory, warehouse space utilization, tracker and trailer utilization, and customer/vendor accounts receivable. Some of these initiatives include: electronic data interchange (EDI), a technology system that allows businesses to order merchandise, streamline delivery, and reduce overall costs; efficient consumer response (ECR), a collaborative relationship in which any combination of retailer, wholesaler, broker, and manufacturer works together to seek out a more efficient manner to distribute manufactured food products; scan-based trading (SBT), a technological system that provides food manufacturers instant information on their inventory in retailer outlets when the goods are scanned and sold; and efficient foodservice response (EFR), a technology system in the food service supply chain that links food manufacturers to distribution warehouses, and to restaurant outlets.

The Internet is becoming a common tool for food companies in the food marketing system to decrease costs, increase service levels, and improve efficiencies in their operations. E-commerce is becoming popular.

Study and discussion questions

1 Which countries were the main exporters of seafood to the U.S. in 2012? What was the dominant species exported to the U.S.?
2 What are some factors contributing to the increased import of seafood into the U.S.?
3 How is total seafood consumption calculated?
4 Discuss three reasons why consumers are turning to organic seafood.
5 Discuss three factors that are enhancing the growth in away-from-home food expenditures in the U.S.
6 Define sustainability and how it applies to fisheries and aquaculture.
7 Outline two initiatives that the federal government has implemented to enhance traceability in the food chain.
8 What is the major implication of COOL for consumers?
9 What are the types of ecolabeling schemes, and their drawbacks?
10 What role do certification and labeling schemes play in the demand for seafood?
11 Why are local food systems gaining much attention among policy makers, food markets, and consumers?
12 What is electronic data interchange (EDI)? How does this technology help reduce cost in the food supply chain?
13 Scan-based trading is becoming popular with retailers and food manufacturers. What has contributed to its popularity? Compare the advantages and disadvantages of SBT.
14 What are the various forms of e-commerce in use in the food industry? Describe two of them.

References

ASC (Aquaculture Stewardship Council). 2014. Standards, Certification and Accreditation. Available at www.asc-aqua.org/index.cfm?act=tekst.item&iid=6&iids=290&lng=1. Accessed August 6, 2014.
Bavota, M.F. 2002. Is ready-to-eat safe to eat? *Progressive Grocer*, 81(13): 58.
Blisard, N., J. Variyam, and J. Cromartie. 2003. Food expenditures by U.S. households: looking ahead to 2020. Agricultural Economic Report (AER) 821, U.S. Department of Agriculture, Economic Research Service.
Efficient Foodservice Response (EFR). 2003. Standard product ID and bar coding in the foodservice supply chain. The Efficient Foodservice Response annual benchmarking survey, 2003 Survey Results.
Food and Agriculture Organization (FAO). 2011. Private standards and certification in fisheries and aquaculture: Current practice and emerging issues. Technical Paper 553, FAO Fisheries and Aquaculture Department [online], Rome. www.fao.org/docrep/013/i1948e/i1948e.pdf.
Food Marketing Institute. Supply Chain, ECR-Efficient Consumer Response. www.fmi.org/about-us/careers-at-fmi/getting-to-know-fmi.

Global Aquaculture Alliance (GAA). 2014. Best Aquaculture Practices (BAP) standards. Available at bap.gaalliance.org/bap-standards/. Accessed August 6, 2014.

GLOBALG.A.P. 2014. The GLOBALG.A.P. Aquaculture Standard. Available at www.globalgap.org/uk_en/for-producers/aquaculture/. Accessed August 5, 2014.

Globefish. 2009. Market Reports, July 2009 – Pangasius. Available at: www.fao.org/in-action/globefish/market-reports/resource-detail/en/c/336932/. Accessed August 13, 2014.

Globefish. 2010. Market Reports, March 2010 – Pangasius. Available at: www.fao.org/in-action/globefish/market-reports/resource-detail/en/c/336928/. Accessed August 13, 2014.

Harris, J.M., P. Kaufman, S. Martinez, and C. Price. 2002. The U.S. Food Marketing System, 2002. Competition, coordination, and technological innovations into the 21st century. Agricultural Economic Report (AER) 811. USDA-ERS. www.ers.usda.gov/media/887670/aer811_002.pdf

Heller, W. 2002. 55th Annual Consumer Expenditures Study. *Progressive Grocer* 81(13).

Information Resources Inc. (IRI). 2002. Dollar Sales: IRI's combined supermarket/drug store/mass merchandiser, including Wal-Mart reviews database. "What Do Americans Really Eat?" A study report presented at the Consumer Connection 2003 Conference, "The Power of Knowing in an Era of Change," Rancho Mirage, CA. USA.

Institute of Developing Economies/Japan External Trade Organization (IDE-JETRO). 2013. Meeting Standards, Winning Markets. Regional Trade Standards Compliance Report East Asia 2013. Report prepared for United Nations Industrial Development Organization (UNIDO). Available at www.unido.org/tradestandardscompliance.html. Accessed August 12, 2014.

Institute for Social, Economic, and Ecological Sustainability (ISEES). 2001. National Organic Standards Board Aquatic Animal Task Force Final Recommendations. University of Minnesota, St. Paul, MN.

Kinsey, J.D. 1999. The Big Shift From a Food Supply to a Food Demand Chain. *Minnesota Agricultural Economist*, No. 698, Fall 1999. University of Minnesota, St. Paul, MN.

Kris-Etherton, P.M., W.S. Harris, and L.J. Appel. 2002. Fish consumption, fish oil, omega-3 fatty acids, and cardiovascular disease. *Circulation: Journal of the American Heart Association* 106:2747–2757.

Martinez, S.W. 2007. The U.S. Food Marketing System: Recent Developments, 1997–2006. Economic Research Report No. (ERR-42). USDA, ERS.

McGee, M. 2010. Pangasius for Western Aquaculture. *Global Aquaculture Advocate* 13:73–75.

Miller, M. 1985. Looking ahead at the U.S. seafood market. *National Food Review* 29: 18–20.

National Oceanic and Atmospheric Administration, National Marine Fisheries Service (NOAA-NMFS). 2013. Fisheries of the United States: 2012. Current Fishery Statistics No. 2012. www.st.nmfs.noaa.gov/Assets/commercial/fus/fus12/FUS2012.pdf.

National Oceanic and Atmospheric Administration, National Marine Fisheries Service (NOAA-NMFS). 2014. Annual trade data by product, country/association. www.st.nmfs.noaa.gov/commercial-fisheries/foreign-trade/.

Natural Marketing Institute (NMI). 2003. The 2003 Health and Wellness Trends Report (HWTR). Natural Marketing Institute, Harleyville, Pennsylvania.

Nettleton, J. 1995. *Omega-3 Fatty Acids and Health*. Chapman and Hall, New York.

Organic Monitor. 2003. The Global Market for Organic Food and Drink. www.organicmonitor.com/700140.htm.

Organic Trade Association. 2014. 2014 Press Releases: American appetite for organic products breaks through $35 billion mark. www.ota.com/news/press-releases/17165. Accessed March 28, 2016.

Puget Sound Business Journal 2000. Seafood companies join in e-commerce venture. October 27, 2000.

Quagrainie, K.K and C.R. Engle. 2002. Analysis of catfish pricing and market dynamics: the role of imported catfish. *Journal of the World Aquaculture Society* 33:389–397.

Roheim, C.A. 2009. An evaluation of sustainable seafood guides: implications for environmental groups and the seafood industry. *Marine Resource Economics* 24:301–310.

Sampson, G.S., J.N. Sanchirico, C.A. Roheim, S.R. Bush, J.E. Taylor, E.H. Allison, J.L. Anderson, N.C. Ban, R. Fujita, S. Jupiter, and J.R. Wilson. 2015. Secure sustainable seafood from developing countries. *Science* 348:504–506.

SeaFood Business. 1999. Vietnam catfish seeks an identity. *SeaFood Business* 18, #2:6.

SeaFood Business. 2000. The dotcom e-volution (electronic commerce and the seafood industry). *SeaFood Business* 19, #11.

SeaFood Business. 2001. Gofish axes online purchasing. *SeaFood Business* 20, #8.

Sloan, A.E. 2003. Top 10 trends to watch and work on: 2003. *Food Technology* 57:30–50.

Soref, A. 2011. Natural and organic products industry sales hit $81 billion. *The Natural Foods Merchandiser*, June 2011. Available at http://newhope360.com/news-amp-analysis/natural-and-organic-products-industry-sales-hit-81-billion. Accessed August 19, 2014.

Spencer, M.T. and P. Rea. 2003. Market overview: sales top $36 billion. *The Natural Foods Merchandiser*, June 2003. Pp.1, 20, 21, 22, 24.

Stewart, H., N. Blisard, S. Bhuyan, and R.M. Nayga, Jr. 2004 The demand for food away from home: full-service or fast food? Agricultural Economic Report No. 829. USDA-ERS.

Tarnowski, J. and W. Heller. 2004. The super 50. *Progressive Grocer*, May 2004.

Unites States Department of Agriculture, Agricultural Marketing Service (USDA-AMS). 2004. USDA issues regulatory action on mandatory country of origin labeling for fish and shellfish. AMS News Release No. 172-04, September 30, 2004.

Unites States Department of Agriculture, Economic Research Service (USDA-ERS). 2002. Traceability for food marketing and food safety: what's the next step? *Agricultural Outlook*, January–February 2002.

Unites States Department of Agriculture, Economic Research Service (USDA-ERS). 2014. Food Expenditures Series: Data tables: www.ers.usda.gov/data-products/food-expenditures.aspx

Unites States Department of Commerce, National Oceanic and Atmospheric Administration (USDC-NOAA), National Marine Fisheries Service. 2012. Fisheries of the United States, 2012. www.st.nmfs.noaa.gov/commercial-fisheries/fus/fus12/index

Unites States Department of Commerce, National Oceanic and Atmospheric Administration (USDC-NOAA), Fisheries Statistics Division. 2014. Imports and Exports of Fisheries Products – Annual Summary. 2014. www.st.nmfs.noaa.gov/commercial-fisheries/foreign-trade/

Vietnam Association of Seafood Exporters and Producers (VASEP). 2012. *Pangasius* 26 Q & A. Agricultural Publishing House, Hanoi, 2012.

Viðarsson, J.R. 2008. Environmental labeling in the seafood industry: Iceland's perspective. Available at www2.matis.is/media/utgafa/krokur/Jonas-Runar-Environmental_labelling.pdf. Accessed August 12, 2014.

Wessells, C.R., K. Cochrane, C. Deere, P. Wallis, and R. Willmann. 2001. Product certification and ecolabelling for fisheries sustainability. FAO Fisheries Technical Paper No. 422. Rome, FAO.

Zhang, X., L. House, S. Sureshwaran, and T. Hanson. 2004. At-home and away-from-home consumption of seafood in the United States. Selected Paper prepared for presentation at the Southern Agricultural Economics Association Annual. Tulsa, Oklahoma, February 18, 2004.

Zimmerman, M. 1998. Dealer goes to great depths for patrons. *Pacific Business News*, February 1. www.bizjournals.com/pacific/stories/1998/02/02/smallb2.html.

CHAPTER 5

Seafood market channels

Chapter 3 introduced some of the terms and concepts related to market channels. This chapter will go into more depth and review the dynamics of channel organization, ownership, and control in seafood marketing. Contrasts will be made with trends in agribusiness marketing.

Market channels for primary seafood products

A market channel (also called a channel of distribution) is a combination of interrelated intermediaries (individuals and organizations) who direct the physical flow of products from producers to the ultimate consumers. Market channels can be very simple and direct, as with direct sales, or can be complex and comprised of an array of brokers, sales agents, traders, distributors, wholesalers, food service operators, and importers.

Seafood distribution in developing economies

Fish marketing in developing countries often involves fish traders and middlemen such as brokers, wholesalers, wholesaler-retailers, and retailers. In Honduras, for example, fish traders buy and sell all kinds of freshwater, brackish water, and marine fish. However, the market channels for tilapia in Honduras are not complex (Leyva et al. 2006) (Fig. 5.1). As in many developing economies, small-scale fish farmers often keep some of their produce for home consumption, while medium and large fish farmers sell all of their harvest. Molnar et al. (1996) reported that the percentage of farmers keeping tilapia for home consumption decreased as pond area increased, indicating that increased pond area was associated with increased entry into the cash market economy.

There are large numbers of small-scale fishers throughout the world who supply fish primarily to local markets. However, this sector of the seafood supply

Seafood and Aquaculture Marketing Handbook, Second Edition. Carole R. Engle, Kwamena K. Quagrainie and Madan M. Dey.
© 2017 John Wiley & Sons, Ltd. Published 2017 by John Wiley & Sons, Ltd.

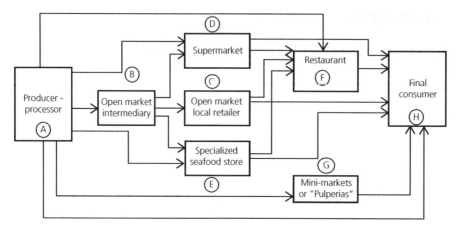

Fig. 5.1 Market channels for tilapia in Honduras. Source: Leyva (2004).

has been characterized as highly fragmented with small-scale fishers who tend to operate independently to capture and market seafood (Jacinto and Pomeroy 2011). The lack of organization of the sector can retard adoption of new technologies or efforts to improve sustainability of the relevant fisheries.

Seafood distribution in developed economies

In most developed economies, seafood market channels consist of a wide variety and a high number of actors including importers, agents, traders, wholesalers, processors, retailers, and restaurants. Large retail chains are strongly involved in the distribution of seafood products to their outlets. In Germany, the retail sector is highly concentrated such that the top five account for 63% of the total retail market value, putting a squeeze on food distribution (Lahidji et al. 1998). The German seafood market, as is the case with most seafood markets in developed economies, is heavily dependent on imports of seafood products to meet domestic demand. For example, the salmon supply in Germany is almost solely dependent on imports and, as Fig. 5.2 suggests, the retail and food service sectors are heavily involved in the flow of salmon in the seafood distribution system (Johnsen and Nilssen 2001).

In Northern Ireland, marketing of Dublin Bay prawns (also called Norwegian lobster or simply Nephrops) follows two main channels, depending on whether the prawns are tailed or whole. Figure 5.3 is a schematic flow of Dublin Bay prawn market channels from a sample of 44 seafood businesses surveyed in March 2000 (Rogers 2000). The prawns are the most important seafood species of Northern Ireland's fishing industry (Rogers 2000). The tails are mainly bought by the local prawn scampi processing sector which processes them into a range of breaded and peeled scampi products mainly for supermarkets and catering outlets in England and Scotland. Whole prawns are sold through wholesalers primarily to Spain, France, and Italy, with smaller quantities sold to England and

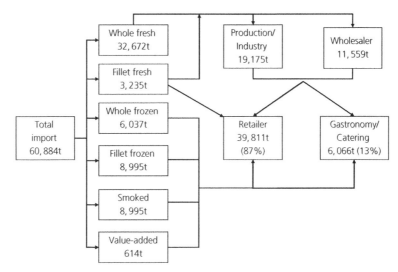

Fig. 5.2 Salmon flow in the German market channels. [Quantities in product weight in tons (t).]

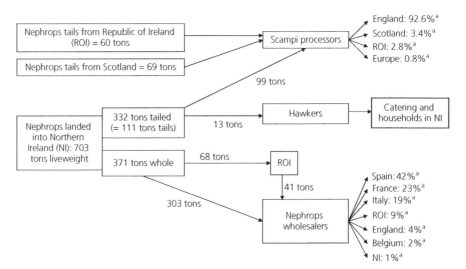

Fig. 5.3 Nephrops market channels in Northern Ireland, March 2000. Source: Rogers (2000).
[a]Based on turnover during the survey period (March 2000).

Belgium. Compared to prawns, the channels for whitefish, including cod, haddock, hake, dogfish and whiting, are different in Northern Ireland (Fig. 5.4) and involve hawkers (small businesses employing fewer than 10 full-time employees whose primary activity is filleting fish for catering and retail markets in Northern Ireland), inland merchants (businesses that process and wholesale marine fish situated more than 10 miles from a major fishing port), and port processors/wholesalers (businesses that process and wholesale sea fish, located within 10 miles of a major fishing port) (Rogers 2000).

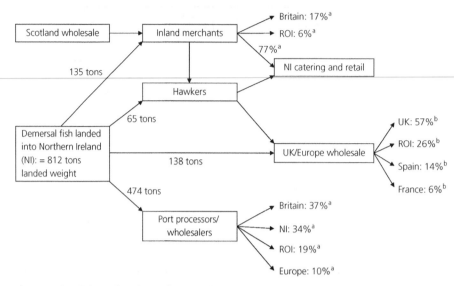

Fig. 5.4 Whitefish market channels in Northern Ireland, March 2000. Source: Rogers (2000). [a]Based on annual turnover from 1998 Sector Survey period. [b]Based on turnover during the survey period (March 2000).

Seafood distribution in the U.S.

The seafood distribution business in the U.S. is highly competitive and fragmented with several examples of flows in the distribution channel because of the wide variety of actors involved in seafood distribution (Fig. 5.5). While some seafood products flow directly to the consumer, others flow through processors, brokers, distributors, and retailers, with value added at any stage in the channel (Radtke and Davis 2000).

Supply chains used to supply local food products to local markets tend to be very short and may supply companies that source from a number of different suppliers. At the other extreme are very large companies, such as Applebee's, that source only from suppliers that can meet all their food product needs.

The U.S. imports more than 90% of the seafood it consumes. Approximately 89% of the total supply of shrimp, the top seafood product in the U.S., was imported in 2012, primarily from Southeast Asia (NMFS 2013). Domestic farmed shrimp production accounts for less than 5% of the total U.S. supply, with the remaining supply from the shrimp fishery. The flow of domestic processed shrimp begins with fishermen bringing harvested fresh shrimp to the dock. Some shrimp may be deheaded, sorted, and frozen. Processors buy fresh shrimp at the dock and in turn sell processed shrimp to distributors/wholesalers, brokers, or directly to retailer customers such as chain grocery and restaurant companies (USITC 2004). Most shrimp are imported by independent and family owned seafood companies engaged in general seafood import, distribution, and marketing. In 2003, the National Oceanic and Atmospheric Administration, U.S. Department

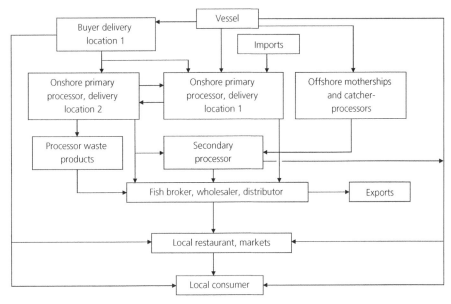

Fig. 5.5 U.S. seafood product distribution chain. Source: Radtke and Davis (2000).

of Commerce (NOAA-USDC) estimated that there were over 3500 seafood dealers operating in the U.S., and approximately 1000 were in the business of importing fish and shellfish. The major shrimp importers include companies such as Slate Gorton, Ocean Garden Products, Empress International, and Darden Restaurants. Both shrimp processors and importers serve national, regional, and multiple market areas. Some importers process shrimp into other value-added products such as marinated, sauced, or breaded shrimp that are then sold to retail chain grocery and restaurant companies and other customers.

Price discovery for primary commodities

A variety of pricing mechanisms are observed in U.S. seafood markets, including negotiation on a boat-by-boat basis at the time of landing, short-term marketing agreements, and sale on consignments (Anderson 2003). The specific provisions of transactions between buyers and sellers of seafood are generally proprietary with little available public information.

Contracting and vertical integration in U.S. seafood business

Some forms of contracting and marketing arrangements exist in U.S. seafood businesses. Some fish processing companies contract with fishermen by providing them with inputs such as nets, boats, motors, and gear, and fishermen in turn supply their catch to processors at a negotiated price (University of Alaska 2001). A study by the U.S. International Trade Commission (USITC) revealed that some shrimp contracts involved fixed prices or quantities and covered periods ranging from two months to a year. Long contracts for up to two years usually involved fixed prices and quantities and involved volume discounts (USITC 2004). A fair amount of catfish transactions between processors and farmers involve delivery rights. In Mississippi, some catfish processors sell delivery rights to fish farmers, which require delivery of a certain quantity and quality of catfish for a specified period at a negotiated price. Marketing arrangements between major seafood buyers such as large wholesalers, mass merchandisers, brokers, restaurant chains, and grocery chains involve quantity and price considerations, off-invoice marketing. Some forms of trade practices involve services, special packaging, and requirements for third-party food safety certification. Contracting assists large-scale buyers to guarantee supply and stabilize prices.

Vertical integration in the distribution chain helps seafood companies to reduce distribution costs and enhance control over product supply and price. The seafood sector is much less vertically integrated than the grain and livestock sectors. However, there are trends towards integration in the commercial fisheries sector due to the uncertain nature of commercial fisheries, sizes of fish runs, management, fishing regulations, subsistence fishing regulations, and quality standards. The competitive nature of the seafood business and the international scope of seafood trade have resulted in several processing companies in Alaska and the U.S. Pacific Northwest investing in vessels, processing plants, and distribution networks that allow them to offer their customers a wide variety of seafood products sourced from around the world.

The aquaculture sector has also become increasingly integrated with ownership from hatchery operations through processing and distribution to retail and food service customers. For example, Clear Springs Foods, Inc., is a vertically integrated company involved in trout farming, fish feed manufacturing, trout processing, and distribution. The farm operations include a broodstock facility that produces about 80 million rainbow trout eggs a year and farms that raise rainbow trout to market size. The company also owns a feed mill that produces feed formulations for its farming operations. The research and development center produces vaccines, monitors water quality, and provides an array of fish health services to its farms. The center is also engaged in research projects on nutrition, waste management, genetics, and fish culture. The research division provides a complement of quality assurance services to the other divisions of the company. Rainbow trout harvested at the farms are shipped live to the company's processing facility and then packaged under a hazard analysis and critical control point (HACCP) quality assurance program. In terms of distribution, Clear Springs Foods, Inc., operates its own fleet of refrigerated trucks to deliver products to customers across the U.S. It supports the sale of its products through a national network of regional sales managers and broker sales representatives.

The salmon industry also has exhibited a strong degree of vertical integration. Marine Harvest has emerged as the leading salmon producer worldwide, with operations in several different countries in addition to its strong base in Norway. Marine Harvest operations include feed manufacturing and hatcheries that supply its net pen farms located in various locations. It operates its own processing plants and produces fish meal and oil as byproducts from its processing operations. Given that the company processes in response to orders taken, this degree of control over the supply chain provides the company with a means to ensure the desired quality for its customers.

Other transaction types in U.S. seafood business

Pricing of landings from small commercial fisheries frequently occurs on an individual basis. Spot market prices offered by processors depend on marketing arrangements for processed fish, and ex-vessel prices offered often depend on their spread or margin (University of Alaska 2001). The Fulton Fish Market in New York City, New York, is the largest open spot market in the U.S. where many food retailers and restaurateurs come to buy seafood. Other commercial fish are sold through fish marketing cooperatives that negotiate prices with large processors and buyers. In the domestic shrimp market, shrimp are usually sold on the spot market with pricing negotiated by transaction, the spot market price, or prices reported in Urner Barry's industry price reports (USITC 2004).

There are some fish auctions in the U.S. The Portland Fish Exchange, the New England fish exchange auction, New Bedford whaling city seafood display auction, Gloucester seafood display auction, and fish auctions in Honolulu and Hilo provide venues for buyers to engage in competitive bidding for seafood products.

Futures markets have been used in the grain and livestock industries for many years but have developed only recently for seafood products. In broad terms, futures entail anticipated future prices of basic commodities based on current market and industry information. Futures are contractual agreements made between two parties through a regulated exchange in which the parties agree to buy or sell an asset (e.g., salmon) at a certain time in the future at a mutually agreed upon price. Each futures contract specifies the quantity and quality of the item, expiration month, time of delivery, and all details of the transaction except price, which the two parties negotiate based on current market conditions. Some futures contracts call for the actual, physical delivery of the commodity at contract termination, but others simply call for a cash settlement at contract termination.

A futures market for salmon was developed in Bergen, Norway, in 2005. Fish Pool ASA buys and sells salmon futures contracts and options. The contracts are in tons of fresh Atlantic salmon. Fish Pool ASA trades 3.64 billion Norwegian krone yearly. Japan has a futures contract for frozen black tiger shrimp on the Kansai Commodities Exchange in Osaka. Two futures contracts for shrimp introduced in the 1960s on the Chicago Mercantile Exchange were terminated after a brief period because of low trading volume. In 1994, the Minneapolis Grain

Exchange began a frozen white shrimp futures but that was also discontinued due to lack of interest.

Participation in food market channels

The market channels for food involve several players, including various types of distributors and wholesalers. Each channel has a role to play in the efficient movement of food from supply centers to the ultimate consumers and users. In general, distributors do not have the responsibility of selling products to delivery points, while wholesalers tend to own the merchandise and render services related to sales. A detailed discussion of the various roles played by these channel actors is presented next.

Distributors

A typical food distributor operates warehousing facilities and transportation services. The main function of a distributor is to receive, store, invoice, and deliver goods. Distributors usually handle a wide range of food products in addition to aquaculture and seafood products, but there are distributors who exclusively handle seafood and aquaculture products. Examples of distributors who specialize in seafood include H & M Bay, Inc., of Maryland and Preferred Freezer Services of New Jersey.

Major trucking fleet companies also provide logistics through warehousing, data management, shipping, distribution services, and invoicing. Such logistics providers handle dry, frozen, and refrigerated food products. Ocean Spray, a large agricultural cooperative of cranberry and citrus growers and processors in North America, uses Schneider Logistics, a major trucking company, for freight-related services for its processing plants, warehouses, and distribution centers in Canada and the U.S.

Distributors usually cover a multi-state region and are contracted by seafood and aquaculture processing companies to deliver to their customers. For example, Idaho Trout Processors Company contracts with distributors to deliver fresh and frozen dressed whole trout products to warehouses or distribution centers of major customers who are grocery wholesalers. The wholesaler then distributes the products further through the food system.

Wholesalers

Food wholesalers assemble, store, and transport goods to customers who include grocery retail food stores, food service companies, other wholesalers, government agencies, and other types of food businesses. A significant portion of wholesale sales target the food retail sector, accounting for over 40% of total wholesalers' grocery and related product sales (Fig. 5.6). Wholesalers play an important role in the timely delivery of assorted products from many different companies and

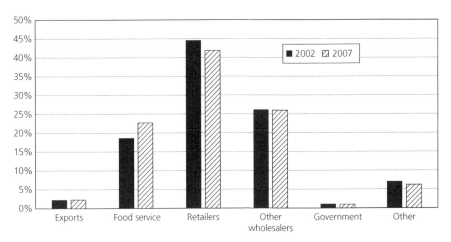

Fig. 5.6 Food wholesale sales by type of outlet. Source: USDA-ERS (2007).

sources to institutions and establishments. For many retailers and establishments, purchasing through wholesalers is a more convenient way of purchasing a diverse range of products. For such retailers, dealing with one major supplier, rather than several supplier company accounts, reduces administrative costs. Thus, wholesale marketing can improve efficiency in the distribution system and may reduce costs. In 1997 the food wholesaling business was estimated to be a $589 billion industry (Harris et al. 2002) increasing to about $980 billion in 2007 (USCB 2011).

There are a wide variety of types and sizes of wholesalers. The U.S. Census Bureau (USCB) classifies wholesalers into three major segments: (1) merchant wholesalers who buy and take title to the goods they sell; (2) manufacturers' sales branches and offices that sell products manufactured domestically by their own company; and (3) agents and brokers who collect a commission or fee for arranging the sale of merchandise owned by others. In 2011, merchant wholesalers accounted for about 65% of total wholesale sales, manufacturers' sales branches and offices accounted for 25%, and agents and brokers accounted for an additional 10% (USCB 2011).

Merchant wholesalers

Unlike distributors, merchant wholesalers own the products that they handle. Under the USCB classification, merchant wholesalers involved in food distribution primarily buy groceries and grocery products from processors or manufacturers, and resell to food retailers, institutions, and other businesses. Merchant wholesalers have distribution centers in strategic locations which serve thousands of independent grocery stores as well as their own stores. Merchant wholesale sales accounted for 57% of wholesale distribution to retail food stores in 2007 (USDA-ERS 2007). Profits are earned on the price spread and the services provided. Merchant wholesalers often repackage larger-sized loads of product into smaller units, or case sizes, for sale to clients. Merchant wholesaling requires demand/supply planning and collaboration, distribution that accounts for lead times and constraints, network optimization involving markets to serve and what products to serve, general planning to reduce costs, management, account-ing, evaluation, and reporting. Some large retail chains perform their own wholesaling functions. Some independent retailers have banded together in the form of a cooperative to provide their own wholesaling, or they may contract with a wholesaler.

Merchant wholesalers can also be categorized by the type of merchandise they handle (e.g., grocery food and non-food items) and the sector they serve (e.g., food service). Grocery wholesalers carry broad-line, specialty, and/or miscellaneous merchandise. Based on the setup and product handled, the merchant wholesale and distribution system can generally be separated into integrated and non-integrated grocery wholesalers.

Integrated grocery wholesalers

Integrated wholesalers serve the grocery and retail industry consisting of supermarkets, warehouse clubs, and convenience stores. These wholesalers own retail store chains and deliver most of the products they sell in their stores. They operate their own transportation and warehouse or distribution centers from where distribution is made to their retail stores. Seafood and aquaculture processors and food manufacturers usually deliver their products to the ware-houses or distribution centers of integrated grocery wholesalers using their own transportation networks or contracting the services of a distributor.

Large retail chains such as Kroger, Albertson's, Walmart, Safeway, Publix, and Ahold are examples of integrated grocery wholesalers that own their own distribution centers and are becoming a more significant part of the wholesaling sector. They are also known as self-distributing retailers. Figure 5.6 shows a decline in the percentage of total wholesale sales to the retail sector from 2002 to 2007. This suggests increasing integration by large retailers into food whole-saling and distribution, where they deal directly with food manufacturers. They buy directly from grocery and food manufacturers and producers, who then deliver products to the wholesale/distribution centers of these retailers.

Self-distributing food retailers account for about 34% of all food distribution (Kinsey 1999). In 1999, 47 out of the 50 largest food retailers in the U.S. were self-distributors (Harris et al. 2002). This type of wholesaling is beneficial to retailers because it reduces labor and general operating cost. The proportion of labor cost to sales at inventory for self-distributors is 0.9 percentage points lower than similar costs for other merchant wholesalers. Their non-labor costs are 1.3 percentage points lower (Kinsey 1999).

Warehouse clubs or cash-and-carry establishments such as Costco, BJ's, and Sam's Club are emerging as a significant segment of the wholesaling industry. Their activities are a blend of wholesaling and retailing of grocery food and other non-food items. These establishments require membership for shopping at the outlets. Members include both individuals and small businesses, including businesses in the hospitality industry. Although warehouse clubs are wholesalers, their prices are slightly above bulk wholesale prices.

Non-integrated grocery wholesalers

Non-integrated grocery wholesalers are also known as general-line grocery wholesalers. They normally do not own the retail or grocery stores that they serve, although some wholesalers own a percentage of the retail grocery outlets they serve. They normally procure grocery products, both food and non-food, for independent grocery and retail stores and smaller retail chains that do not own and operate buying offices, warehouses, trucking fleets, and store delivery services. The primary function of non-integrated grocery wholesalers is to serve independent grocery outlets.

General-line wholesalers are distributors and are sometimes referred to as broad-line or full-line distributors (e.g., SuperValu, Fleming, C & S Wholesale Grocers, and Nash Finch). They handle a broad line of dry groceries, perishable food products, health and beauty products, and household products. General-line wholesalers accounted for about 25% of grocery wholesale sales in 1997 (Kinsey 1999). As an example, SuperValu served as primary supplier to approximately 2460 stores, 29 Cub Foods franchised locations, and SuperValu's own regional banner store network of 267 stores in 2013, while serving as secondary supplier to approximately 1500 stores. In 2004, Supervalu owned 24 wholesale/ distribution facilities with approximately 14 million square feet of warehouse space, while Fleming owned 32 wholesale/distribution centers. Nash Finch is another food wholesale company that supplies products to independent supermarkets and military bases in approximately 30 states. The wholesale business accounts for about 75% of company sales. In addition to wholesaling, the company owns and operates approximately 85 retail supermarkets throughout the Midwest. The buying power of these merchants allows them to obtain volume discounts and leverage in food auctions.

Just as large retail chains such as Kroger, Albertsons, Walmart, and Safeway have integrated into food wholesaling and distribution, larger general-line wholesalers (such as SuperValu, Fleming, Giant Eagle, and Nash Finch) have also ventured into food retailing. In 2001, sales derived from retail operations accounted for 45% of SuperValu's $20.9 billion total sales, 64% of Giant Eagle's $4.5 billion total sales, 15% of Fleming's $15.6 billion total sales, and 25% of Nash Finch's $4.11 billion total sales value (Harris et al. 2002). However, not all wholesalers who have ventured into retailing have been successful. In 2003, Fleming and its operating subsidiaries filed voluntary petitions for reorganization under Chapter 11 of the U.S. Bankruptcy Code. Consequently, most of Fleming's retail stores were sold to competing retailers while C & S Wholesale Grocers acquired Fleming's wholesale grocery business. Spartan Stores, the seventh general-line grocery wholesaler in the U.S., with retail sales of 40% of its $3.5 billion total sales, divested a number of its retail stores in 2003 to focus on its core business of wholesaling.

Food service wholesalers

Food service wholesalers fall under the categories of general-line, specialty, or miscellaneous wholesalers. Some can also be categorized as integrated food service wholesalers that own self-distributing retail food service operations. They operate as merchant wholesalers and deliver a greater percentage of the products they offer consumers at their restaurant outlets. Food service wholesalers operate their own warehouses and transport centers from where distribution is made to their food service establishments. Major restaurant chains such as McDonalds and Shoney's are examples of integrated food service wholesalers. Non-integrated food service wholesalers serve hotels, restaurants, commercial cafeterias, hospitals, schools, and hotels and do not own any of the food service establishments that they serve. Examples of such wholesalers include Sysco, US Foods, and Alliant.

The food service sector has grown rapidly in recent years, with 5.5% annual growth in sales. The number of food service establishments has increased over the last decade in response to the growing trend of away-from-home food consumption (USCB 2004). In 1997, sales to food service institutions accounted for about 22% of all sales of groceries and related products by all wholesalers (Harris et al. 2002).

General-line or broad-line food service wholesalers typically purchase a wide range of food products from manufacturers and stock them at their distribution centers for distribution to their clients. They can carry up to 10,000 stock-keeping units (SKU) and price competitively using economies of scale as leverage (Friddle et al. 2001). Their prices may be negotiated or they may be cost-plus pricing. The major food service broad-line distributors include Sysco Corporation, US Foods, Alliant Foodservice, Performance Food Group, Gordon Food Service Incorporated, and Food Services of America. Sysco Corporation is the largest broad-line and seafood distributor in the U.S. (Foodservice.com 2014).

Most broad-line food service distributors offer more than just distribution services. Many also offer value-added services tailored to the needs of their customers. Sysco Corporation and U.S. Foodservice offer a variety of services and proprietary food product lines in addition to food manufacturer brands. Sysco Corporation owns a number of brands, including Buckhead Beef and Newport Pride (beef products) as well as Sysco Natural and FreshPoint (fresh produce).

Food service seafood wholesalers carry a full range of seafood products. They purchase seafood from processors and other wholesalers and sell primarily to restaurants. Cash- and-carry wholesalers typically supply small retail fish stores. Store owners travel to the cash-and-carry wholesalers to purchase fish, pay cash, and transport fish back to their store.

Specialized wholesalers

Specialty food distributors specialize in the distribution of a particular line of product items such as frozen foods, dairy products, poultry products, seafood, meat and meat products, fresh fruits and vegetables. Specialty wholesale distributors usually do not handle a wide range of products but focus on special products and niche markets. For example, a specialty wholesaler may handle Asian foods to service Asian markets or may specialize in servicing convenience stores. McLane Company is one of the nation's largest wholesale food distributors to convenience stores, drug stores, quick service restaurants, and movie theaters. Some of the specialty distributors among major seafood distributors include Inland Seafood, East Coast Seafood, Supreme Lobster and Seafood Company, Morey's Seafood International, and South Stream Seafoods. Inland Seafood, for example, handles over 1000 seafood products that include species such as salmon, lobster, shrimp, tilapia, tuna, red snapper, catfish, rainbow trout, scallops, crab, and clams. It has the largest inland holding facility for lobsters in the U.S. and sells about 35,000 kg of salmon a week (personal communications). East Coast Seafood specializes in fresh lobster (*Homarus americanus*), dogfish (*Squalus acanthias*), monkfish (*Lophius americanus*), skate (*Raja* spp.), scallops (*Placopecten magellanicus*), squid (*Doryteuthis pealeii*), and whiting (*Merluccius bilinearis*).

Jobbers are specialized versions of merchant wholesalers that have been important historically in delivering seafood from fishermen to restaurants or retail grocery stores. With the growing influence of large food service distributors such as Sysco and U.S. Foodservice, the role of jobbers in seafood marketing has diminished.

Engle (1997) showed that seafood wholesalers in Atlanta, Chicago, Los Angeles, New York, and San Francisco ranged in size from less than $20 million to over $100 million in annual sales. Seafood wholesale companies tended to specialize in either finfish or shellfish, but were equally likely to sell fresh and frozen product. Those that sold tilapia tended to be either in the smallest or largest size categories. Most of the tilapia was sold to retail grocers, primarily Asian and Hispanic, or to independent restaurants. A very few large wholesale companies had very high sales of tilapia, but of frozen, whole tilapia. Fresh tilapia products were purchased more frequently and in lower average purchase amounts than other types of seafood.

Miscellaneous wholesalers

Miscellaneous wholesalers are also known as system distributors. This category of food service wholesaler serves a customer base that includes chain restaurants with centralized purchasing and menu development. The leading wholesalers in this category are Sysco, U.S. Foodservice, Performance Food Group, Gordon Food Service, Food Services of America, Reinhart Foodservice, Inc., Shamrock Foods Co., Maines Paper and Food Service, Inc., Ben E. Keith Foods, and The IJ Company (Foodservice.com 2014). Miscellaneous distributors are primarily engaged in the wholesale distribution of a narrow range of dry groceries such as canned foods, coffee, bread, or soft drinks, accounting for 32% of grocery wholesale sales in 1997 (Harris et al. 2002).

Manufacturers' wholesaling

Manufacturers' sales branches and offices are mainly wholesale divisions and offices of grocery manufacturers and food processors that market the company's products. This type of wholesaling involves direct-store delivery by grocery manufacturers and food processing/manufacturing companies. Typical examples are Coca Cola Company and Frito-Lay. Direct-store deliveries account for 28% of distribution to retail food stores (Harris et al. 2002). Typically, the vendors deliver their products directly to individual retail stores and arrange products on display shelves for retailers. One of the ways in which grocery outlets and grocery/food manufacturers streamlined the supply chain and reduced inventory was the adoption of scan-based trading.

Several seafood processing companies also operate their own wholesale/ distribution divisions. Inland Seafood, for example, was the fourth largest seafood wholesaler/distributor in the U.S. in 2003, and purchased seafood from fishing ports and aquaculture farms to produce fresh, frozen, smoked, and specialty seafood products that included salmon, lobster, shrimp, tilapia, tuna, red snapper, catfish, rainbow trout, scallop, crab, and clams. With distribution facilities across five southern states, the company operates its own wholesaling and distribution functions. It has its own fleet of refrigerated trucks and utilizes air freight to deliver fresh and specialty seafood products to its restaurant, hotel, and grocery retail customers in the U.S.

Sales agents and brokers

Independent sales agents and brokers of seafood products function by locating buyers and negotiating a sale. They seek out information on the species, size, package, and price from seafood suppliers and then offer these products for sale at a certain price to prospective buyers. Sales can be negotiated between seafood processors and buyers such as wholesalers, retailers, exporters, or food service establishments. Alternatively, brokers seek out buyers and their specification needs and look for suppliers who can supply products according to those needs. The National Oceanic and Atmospheric Administration (NOAA), Department of Commerce listed about 571 major seafood brokers in 2003. However, some seafood and aquaculture product processing companies also maintain a staff of sales personnel who promote and sell only the company's products.

The services of independent sales agents and brokers are mostly compensated with commission fees when sale is completed. Transactions of brokers have traditionally been through phone contacts, but the Internet now plays a major role. Transactions usually do not involve contracts but consist of one-time purchases on a day-to-day or week-to-week basis. Typical transactions involve some specified quantity and price and shipping arrangements. The supplier usually delivers products to the buyer. There are exceptional cases where the broker pays some of the shipping costs if the demand is high but the supply is limited.

Food broker companies typically operate in regional market areas instead of nationally, but the global and competitive nature of the seafood business makes it necessary for them to have a worldwide sourcing network for the supply of quality products. Broker companies have a number of sales associates responsible for contacts with corporate headquarters of suppliers, warehouses for receiving samples, test kitchens, and conference areas for presentations by clients.

Homziak and Posadas (1992) interviewed 72 U.S. and Canadian tilapia brokers and reported that the brokers that handled tilapia had mean annual gross sales significantly greater than the average for all seafood companies. The brokers controlled nearly 10% of the seafood market and provided a diversity of seafood products. Approximately 52% bought fresh tilapia and 43% handled frozen fish, but these companies primarily purchased lower-priced, whole tilapia products (48%).

Food brokers can be classified into broad-line or specialty brokers, but the majority of food brokers fall into the broad-line category due to the number of products that they handle. For example, Asmussen Waxler Group LLC is a broad-line broker that handles a variety of products from different food manufacturers including Chicken of the Sea International (tuna products); Contessa Food Products (raw and cooked shrimp products); Country Select Catfish (farm-raised catfish products); Dean Foods/Land O Lakes Milk (lactose-free milk products); Fishking Processors, Inc. (value-added shrimp, scallops, oysters, salmon, surimi, and lobster products); Icelandic USA, Inc. (fresh and frozen fish and seafood products); Orca Bay Foods, Inc. (salmon, swordfish, tuna, halibut, mahi, and crab products); and Tyson Foods, Inc. (chicken products and branded concepts). The Asmussen Waxler Group operates in the Chicago area. Buzz Crown Enterprises, Inc., handles a variety of product lines similar to those of the Asmussen Waxler Group but the market area includes Washington DC, Baltimore, Richmond, Roanoke, Virginia Beach, and Charleston, South Carolina. ACH Food Service, Inc., operates in the Charlotte, Greensboro, and Raleigh areas in North Carolina as well as in Columbia, South Carolina. Food Sales West, Inc., is a major food broker that serves major cities in the West including Bakersfield, Fresno, Los Angeles, Sacramento, San Diego, and San Francisco in California, Las Vegas and Reno in Nevada, and Salt Lake City, Utah.

Channel ownership and control for secondary products

Seafood and aquaculture processors strive to achieve efficiency and reliability in terms of product supply because an efficient channel system for processors is important for customer loyalty and could greatly improve market share.

Consequently, the process of distribution of a company's product requires careful planning and execution to help determine the overall success of the marketing effort.

One of the fundamental issues that processors consider is the choice of intermediary to adopt for the distribution of their products. Important factors to consider include the type of customer, performance capabilities of the intermediary, and costs associated with the product's distribution. The choice of intermediary also depends on the company's overall management and sales strategy, how seafood consumers purchase seafood and fish products, and the extent to which processors wish to perform any of the many levels of channel functions in a cost-effective manner. Alternatively, a processing company may decide to perform distribution functions by itself. Whatever the choice of market channel, the seafood and processing company should tailor its choice to support the overall marketing strategy of the company. In certain instances, processing companies form distribution alliances with other processing companies. Such partnerships help to expand product distribution and allow more customers access to diverse products. The alliance also offers an opportunity for companies to grow through the distribution relationship.

Retail establishments have changed as a result of increased levels of mergers and acquisitions, expansion into food retailing by retail discount stores, and the use of information technology. In turn, food processors have become increasingly concerned about their ability to adapt to the changing needs of such large-volume buyers. Power struggles between consumer-product companies and retailers have emerged over issues such as merchandising standards, marketing control, pricing, and markdown management.

Bargaining power appears to have shifted to large retail buyers in which the buyer can potentially dictate terms of trade. For example, if a relatively large percentage of a processing company's products is sold through a wholesaler like Sysco, threatening the relationship would be disastrous for such a processor. Such a relationship can be key to the survival of the processor. Thus, the choice of distribution network can be critical to a successful marketing strategy.

Consolidation and channel control

Increased consolidation and control of market channels by a smaller number of actors leads to discussion of where bargaining power lies in the market channel. Various stages of a marketing chain can exhibit some degree of buyer and/or seller concentration that can create the potential for market power. Farmers generally face commodity markets in which they can sell their products at market price but have no individual bargaining power to negotiate transaction terms. Organizations of farmers such as cooperatives allow farmers to pool sales and input purchases, which provides them some degree of control and bargaining power with which to negotiate terms of trade with farm commodity buyers. In addition, farmers have increasingly engaged in contracts and vertical integration to seek greater stability in prices and markets for their products.

The food processing sector in particular has continued to consolidate vertically and horizontally through acquisitions to gain economies of size and scope and increase efficiencies through specialized production, more capital-intensive technology, and greater productivity (Harris et al. 2002). Increased economies of size can also increase market share, which in turn can increase bargaining power with respect to increasingly concentrated supply chain stages such as food wholesaling and retailing. For example, in red meat packing, market share of the four largest firms increased from 47% in 1987 to about 61–63% in 1993. Particularly in steer and heifer slaughter, the four largest firms controlled about 81% in 1999 compared to 70% in 1989; in hog slaughter, the four largest companies controlled 66% of the industry in 2005 compared to 70% in 1989. In pasta, the four largest processors had a 78% market share in 1992 and in malt beverages, the four largest firms controlled 90% in 1992. In 1998, companies with $800 million or more in sales accounted for 69% of U.S. dairy sales (Harris et al. 2002).

A great deal of consolidation has occurred in the food service sector. The top four distributors accounted for 23% of sales in 2000, compared with 14% in 1995 (Friddle et al. 2001). Acquisitions by broad-line and specialty distributors have been partly responsible for this consolidation and the subsequent growth. In 2004, a report by the Unison Capital Group concluded that there were over 6000 small to medium-sized independent distributors with sales between $10 million and $100 million, and that, since 1996, companies such as Sysco Corporation, JP Foodservice, U.S. Foodservice, Nash, Performance Food and others had acquired over 200 food distribution companies (Harris et al. 2002). Sysco, for example, has been active in acquiring other food service wholesalers and distributors, including specialty wholesalers, since about 1994. In 2001, the second leading seafood distributor, U.S. Foodservice, bought the third leading seafood distributor, Alliance Foodservice, further consolidating this sector of the marketing chain. By 2001, U.S. Foodservice accounted for 10% of total food service distribution sales as a result of mergers and acquisitions (Friddle et al. 2001). In 2014, Sysco and U.S. Foodservice merged to form one company (Wright 2014).

Merchants were the most concentrated in the retail food store wholesale sector. In 2007, the top four general-line merchant grocery wholesalers accounted for about 40% of sales, while the top eight accounted for 56% (USCB 2007). The leading general-line merchant wholesalers in 2008 were Supervalu, C & S Wholesale Grocers, Wakefern Food Corp., Associated Wholesale Grocers, and Nash Finch Co. A major reason for this concentration was that, besides serving thousands of independent grocery stores, the large wholesalers had also integrated vertically. For example, in 2003, Supervalu was the nation's tenth largest supermarket retailer and owned more than 1400 stores, including more than 800 licensed locations (Tarnowski and Heller 2004.). Company-owned grocery chains included Bigg's, Save-A-Lot, Cub Foods, Scott's Foods, Farm Fresh, Shop 'n Save, Hornbacher's, Shoppers Food Warehouse, and Deals.

In the retail sector, increasing concentration and consolidation of sales among large supermarket chains and supercenters have made retailer market power in the food industry a topical issue. As more and more products compete for space in supermarkets, retailers have gained increased power to determine what should be displayed on store shelves. There is a significant trend toward store brands that compete with national brands. Some food wholesalers also have their own brands. Thus, food manufacturers, wholesalers, and retailers aggressively compete with each other and with processors to achieve product differentiation.

Competition among diverse products has resulted in retailers demanding slotting fees as a means for signaling and screening new products and as a basis for achieving efficient cost sharing and risk shifting among manufacturers and retailers. Slotting fees are lump sum fees that suppliers pay to retailers for introducing new products to the supermarket shelves or for securing prime shelf areas. The fees have long been used in the supermarket industry for dry grocery items and have entered the fresh produce and other store departments. Slotting fees are also thought to lead to more efficient shelf space allocation and demand/supply apportionment. In contrast, opponents of slotting fees see the fees as an abuse of power by large retailers who use them to gain a competitive advantage over smaller rivals, as well as to discriminate among food manufacturers. When a greater proportion of processor sales is concentrated with a few distributors, wholesalers, retailers, or restaurateurs, processors could potentially lose their effective bargaining position with these customers. Major retailers such as Walmart, Albertson's, and Krogers capture more value through practices such as levying slotting fees for food processors to place products in prime shelf areas.

Channel coordination and leadership for secondary products

Coordination of the distribution channel is critical for effective management. In the absence of coordination mechanisms, various participants in a distribution system may conflict with each other if they are pursuing distinct objectives. Various mechanisms have been suggested to coordinate potential conflicting interests of channel members for mutual profit maximization. These include market-based mechanisms that coordinate the channels through short-term exchanges; administered channel coordination through non-market incentives such as promotions; contractual channel coordination through long-term contracts including franchising; and vertical integration that coordinates the channels through ownership and authority of members at various levels of the system.

In the food marketing system, coordination mechanisms take the form of specialized contracts between a food processor and a wholesaler or retailer. Such contracts often involve profit sharing or quantity discount arrangements, which allow risks and revenues to be shared by all members. Because of concentration, revenue sharing may not always be equitable. A contract would normally involve periodic or stochastic orders from the wholesaler, retailer, or restaurateur for specified quantities at some agreed price, with provisions to order additional quantities of products within the contract period.

Contract models differ depending on the product. Examples of contract models include quantity flexibility contracts, backup agreements, buy-back or return policies, incentive mechanisms, revenue-sharing contracts, allocation rules, and quantity discounts. Contracts would usually specify the rights, responsibilities, rewards, and sanctions for nonconformity for each member of the channel in the system. Food processors may use different market channels to reach diverse target markets with each channel involving a different set of intermediaries and contracts.

Although many market channels are organized by consensus among the members, some are organized and controlled by a single leader, called the channel leader. The channel leader may be a processor, wholesaler, or retailer. The channel leader normally possesses the greatest market power and ability to influence another channel member's goal achievement. Nevertheless, channel cooperation is vital if each member is to gain from the system and avoid conflicts with other supply chain members that can result in inefficient operations. There are several ways to improve channel cooperation. If a market channel is viewed as a unified supply chain that competes with other supply systems, then individual members will be less likely to take actions that create disadvantages for other members of the same supply chain. Channel members should agree to direct their efforts toward common objectives so that channel roles can be structured for maximum marketing effectiveness, which in turn can help members achieve individual objectives.

One of the mechanisms of coordination in the channel system is the electronic data interchange (EDI) that can be utilized by various members of the channel. EDI is a computer-to-computer exchange of business transactions in a standard format. The system allows a company to send information over communications links. The system can read information such as the net total, vendor name, or address from an invoice, and send it directly to the company's accounting application for payment preparation. The EDI system is used for inventory control, stock replenishment programs, warehouse management, customer management, pricing, and financial reporting. Some related software could be used to rank customers, products, and services by profitability, to optimize inventory and customer service levels as well as for business-to-business management.

Various channel stages may be combined either horizontally or vertically under the management of a channel leader. Vertical channel integration involves a combination of two or more stages of the channel under one management. An example is a situation in which one member of a market channel purchases the operations or simply performs the functions of another member, eliminating the need for that intermediary as a separate entity. Normally, members of a channel system work independently, but in vertical channel integration members coordinate efforts to reach a desired target market. The integration allows a single channel member to coordinate or manage channel activities to achieve an efficient, low-cost distribution system. Vertical marketing systems can take one of three forms: (1) a corporate system in which all stages of the market channel, from processor to consumer, are under a single owner; (2) an administered system in which channel members are independent, but with a high level of interorganizational management achieved through information coordination; and (3) a contractual system in which channel members are linked by legal agreements that spell out each member's rights and obligations. The last is the most popular type of vertical marketing system.

Combining channels at the same level of operation under one management constitutes horizontal channel integration, that is, merger between companies at the same level in a market channel. Although horizontal integration allows for increased efficiencies and economies of scale in purchasing, marketing research, advertising, and specialized personnel, it is not always the most effective method of improving distribution.

Channel agreements

Tying agreements

A tying agreement occurs when a processor or other supplier provides a product to a channel member with the stipulation that the channel member must purchase other products as well, such as the feed-for-fish program that once existed with Southern Farm Services. Related to this type of agreement is what is commonly known as "full-line forcing." In full-line forcing, a supplier requires that channel members purchase the supplier's entire line of products to obtain any of the supplier's products. Tying agreements are legal provided that: (1) the supplier alone can provide a line of products of a certain quality; (2) the intermediary is free to carry competing products; and (3) a supplier has just entered the market. Most other tying agreements are considered illegal.

Exclusive dealing

An agreement in which a processor or supplier forbids an intermediary to carry products of competing suppliers or processors is illegal: (1) if the agreement blocks competing suppliers from as much as 10% of the market; (2) if the sales revenue involved in the transaction is large; and (3) if the supplier is much larger and thus more intimidating than the intermediary. Exclusive dealing is legal if intermediaries have access to similar products from competitors or if the exclusive dealing contract strengthens an otherwise weak competitor.

Value chain analysis

Analysis of the value chain includes all activities related to bringing a product to final consumption and disposal rather than analyzing only one actor or participant in the supply chain (Jacinto and Pomeroy 2011). The goal of a value chain analysis is to seek to maximize profits across the entire chain, not just at one level. A value chain analysis is descriptive and includes the following: point of entry, mapping the value chain, product segments and critical success factors in final markets, how producers access final markets, production efficiency benchmarking, governance, upgrading, and distributional issues.

In developing countries, marketing relationships among actors in the market channel can influence how the market system operates. For example, as described by Jacinto and Pomeroy (2011), the *suki* relationship between supply chain actors in the Philippines provides a credit and marketing relationship. In a *suki* relationship, one actor provides credit to the other and that actor has to sell exclusively to that actor. Similar relationships can be found in Indonesia, Vietnam and other countries. While claims of unfair advantage have been made of the fish traders involved in a *suki* relationship, Pomeroy (1989) found that social and kinship ties inhibited exploitation of fishermen.

Market governance issues identified through a value chain analysis can be especially important. As described in Jacinto and Pomeroy (2011), governance issues such as how actors are governed among and between themselves, formal and informal roles that regulate the action of actors, who establishes the rules, who monitors the enforcement of the rules, what makes the rules effective, why the rules are needed, and the advantages and disadvantages of the existing role for each category need to be analyzed, evaluated, and factored in to develop effective marketing strategies. A value chain analysis identifies the distribution of benefits of economic agents and identifies who could benefit from organization, particularly in the case of poor communities and villages.

Channel conflict

Conflicts arise among channel members due to various issues such as self-interest, misunderstandings, disappointments, false expectations, communication difficulties, and disagreements. There appears to be no single method for resolving conflict among actors in the supply chain; nevertheless, partnerships can be maintained in which there is a clear understanding of the role of each channel member. Measures can be established for channel coordination that may require leadership and benevolent exercise of control. An important element in maintaining good relationships among channel members is ensuring that each member meets agreed-upon contract guidelines. Potential conflict areas include processor rebates, product promotion, billing payments, resellers with different brands, territorial issues, and direct sales.

Aquaculture market synopsis: trout

Rainbow trout (*Oncorhynchus mykiss*) have been cultured for over a century and have been introduced into countries across the world. Rainbow trout are prized as freshwater game fish as well as a preferred foodfish, and recreational angling for trout is popular around the world. Much of the early aquaculture of trout was developed in order to stock and re-stock natural waters to enhance fish populations to support recreational trout fishing. There are records of aquaculture production of rainbow trout from 1950 (FAO 2014a).

Capture fisheries for trout exist but in negligible quantities. Global production of trout has grown and expanded over the years. The growth has occurred especially in Europe and more lately in Chile. The Americas accounted for about 37% of global production in 2012 while the European region accounted for about 30% (Fig. 5.7). Production in France, Italy, Denmark, Germany, and Spain is mainly inland and meant for the domestic market while production in Chile and Norway is from mariculture in cages that targets the export market (FAO 2014a). Chile remained the largest producer of rainbow trout in the world in 2012 and accounted for 30% of total production (Fig. 5.8). Chile produced 254,353 metric tons of rainbow trout in 2012. Much of the Chilean trout is sold to markets in Japan, Iran, Turkey, and Italy. Denmark also produces significant amounts of trout. The largest markets for trout overall are Japan, the U.S., the European Union (mainly France, United Kingdom, and Italy), and the Russian Federation.

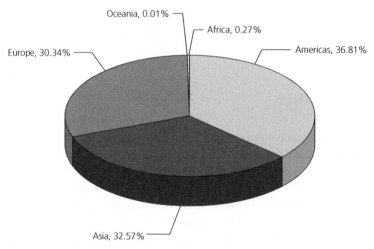

Fig. 5.7 Global trout production by region in 2012 (%). Source: FAO (2014b).

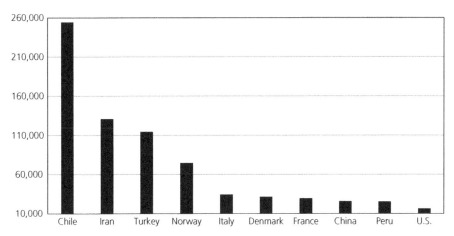

Fig. 5.8 Major trout-producing countries in 2012 (metric tons). Source: FAO (2014b).

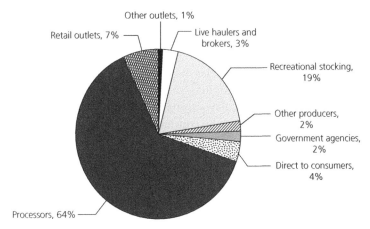

Fig. 5.9 Average percentage sales of foodfish trout (30.5 cm) by market channel in the U.S., 2000–2014. Source: USDA-NASS (2014).

In the U.S., trout are raised primarily in raceways located in areas with high volumes of high-quality surface water, typically springs. Trout are raised in cages in places such as Chile and Lake Titicaca in Bolivia and elsewhere. Trout are served traditionally as a whole fish with the head on. In some restaurants, butterfly fillets are served, but there are a variety of traditional preparations and forms served around the world.

Trout production in the U.S. continues to provide fish for angling as well as for the food market. Figures 5.9 and 5.10 show the distribution of food-sized (30.5 cm) trout and stocker (15–30 cm) trout in the U.S. On average, 64% of the food-sized trout produced are sold to processing plants and another 19% are sold to fee-fishing businesses to provide recreational fishing opportunities to anglers. Of the stocker trout sold, an average of 50% are sold to fee-fishing

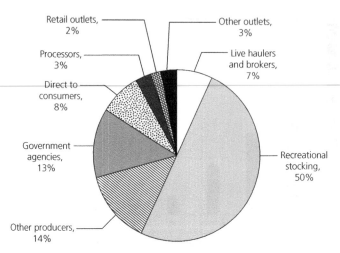

Fig. 5.10 Average percentage sales of stocker trout (15–30 cm) by market channel in the U.S., 2000–2014. Source: USDA-NASS (2014).

businesses and another 13% are sold to the government, primarily for stocking programs in natural waters.

The U.S. foodfish trout industry has moved into value-added product development in recent years with products such as breaded, stuffed, finger-food portions with a variety of recipes, flavors, and preparations. Many of the products are meant for minimal preparation by consumers.

The trout industry has had to cope with increasing regulations related to discharge of effluents into the environment. New treatment technologies, new feeds, and increased monitoring have been adopted as the industry has adapted to these changing demands.

Summary

A market channel (also called channel of distribution) is a combination of inter-related intermediaries (individuals and organizations) who direct the physical flow of products from producers to the ultimate consumers. A market channel can be very simple and direct, as with direct sales, or can be complex and comprise an array of brokers, sales agents, traders, distributors, wholesalers, food service operators, and importers. The complexity often depends on the type of seafood and the extent of development of a given nation. In most developed economies, seafood market channels consist of a wide variety and a high number of actors that include importers, agents, traders, wholesalers, processors, retailers, and restaurants. In the U.S., the physical flow of agricultural commodities through the market channel varies with commodity groups.

Evidence from the food distribution system in the U.S. indicates differences in the relative importance of specific commodity flows, how channel agents facilitate commerce, and price-discovery mechanisms. A variety of pricing mechanisms are used in the U.S. seafood market that include negotiation on a boat-by-boat basis at the time of landing, short-term marketing agreements, and sale by consignment.

One of the fundamental issues that processors consider is the choice of intermediary to adopt for the distribution of their products and the issue of control of the market channel that can lead to bargaining power. Some mechanisms to avoid conflict in coordination of supply chain actors include short-term exchanges, promotions, long-term contracts, franchising, and vertical integration. In the food marketing system, coordination mechanisms take the form of specialized contracts between a food processor and a wholesaler or retailer involving profit sharing or quantity discount arrangements, which allow risks and revenues to be shared by all members. Types of channel agreements include "tying agreements" and "exclusive dealings." One of the mechanisms of coordination in the channel system is the electronic data interchange (EDI) that is utilized for inventory control, stock replenishment programs, warehouse management, customer management, pricing, and financial reporting.

Wholesalers generally perform the functions of purchasing, transporting, assembling, storing, and distributing at reduced costs. They service food retailers, food service establishments including hotels and restaurants, hospitals, and government institutions such as schools, prisons, and other government catering operations. Wholesalers are classified into three major categories: merchant wholesalers, manufacturers' sales branches and offices, and agents and brokers. Merchant wholesalers mainly serve the grocery retail and food service sectors and can be classified into general-line, specialty, or miscellaneous wholesalers. There is an increasing trend toward integrating the wholesale business into other aspects of the food marketing system. Larger restaurants and retailers deal directly with food manufacturers and handle their own wholesaling functions. Large wholesalers are in turn operating food retail stores and therefore handle their own wholesaling functions.

Food agents and brokers also play a major role in wholesaling and distribution. They seek out information on the species, size, package, and price from seafood suppliers and then offer these products for sale at a certain price to prospective buyers. They also seek out buyers and their specification needs and look for suppliers who can supply products according to those needs.

Study and discussion questions

1 Describe the difference between commodities produced by farmers and the products demanded by consumers. Using a specific type of seafood/aquaculture commodity, suggest how the difference is bridged by the food system.
2 What factors determine the complexity of seafood market channels? Illustrate with an example.

3 What are futures and options? How can a futures market be used as a price-discovery mechanism?

4 What are the various criteria for classifying wholesalers? Describe the types of food wholesalers in the U.S. food system and the role each plays in the system.

5 Describe how integration operates in the wholesaling business. What are the advantages and disadvantages of integration in the wholesale business?

6 Suppose you are an independent fish processor who seeks a distributor for your products. What factors will you consider in your decision process?

7 Why is market power essential in the food distribution channel? What methods do businesses adopt to be able to increase market power?

8 Concentration has increased in the food wholesaling industry. How does that benefit the consumer?

9 What are the advantages and disadvantages of slotting fees?

10 Describe some specialized contracts between a food processor and a wholesaler or retailer that would involve non-market incentives. Give seafood and aquaculture examples.

11 Give two examples of channel agreements and indicate how they differ from each other.

References

Anderson, J.L. 2003. *The International Seafood Trade*. Woodhead Publishing Ltd., Abington Hall, Cambridge, UK.

Engle, C.R. 1997. Marketing of the tilapias. In: Costa-Pierce, B.A. and J.E. Rakocy, eds. *Tilapia Aquaculture in the Americas*, vol. 1. World Aquaculture Society, Baton Rouge, Louisiana.

Food and Agricultural Organization of the United Nations (FAO). 2014a. Cultured Aquatic Species Information Programme. *Oncorhynchus mykiss*. Text by Cowx, I.G. FAO Fisheries and Aquaculture Department, Rome. Updated June 15, 2005. Accessed September 5, 2014. www.fao.org/fishery/culturedspecies/Oncorhynchus_mykiss/en#tcNA00FE.

Food and Agricultural Organization (FAO). 2014b. Fisheries and Aquaculture Software. FishStatJ. FAO Fisheries and Aquaculture Department, Rome. Available at www.fao.org/fishery/statistics/software/fishstatj/en. Accessed September 9, 2014.

Foodservice.com. 2014. Foodservice Distributor's Directory. Available at www.foodservice.com/foodshow/restaurant-distributors.cfm. Accessed March 28, 2016.

Friddle, C.G., S. Mangaraj, and J.D. Kinsey. 2001. *The Food Service Industry: Trends and Changing Structure in the New Millennium*, Working Paper 01-02. The Retail Food Industry Center, University of Minnesota, St. Paul, MN, March 2001.

Harris, J.M., P. Kaufman, S. Martinez, and C. Price. 2002. *The U.S. Food Marketing System, 2002 – Competition, Coordination, and Technological Innovations Into the 21st Century*. USDA-ERS Agricultural Economic Report No. 811, June 2002.

Homziak, J. and B.C. Posadas. 1992. A preliminary survey of tilapia markets in North America. *Proceedings of the Annual Gulf and Caribbean Fisheries Institute* 42:83–102.

Jacinto, E.R. and R.S. Pomeroy. 2011. Developing markets for small-scale fisheries: utilizing the value chain approach. In: Pomeroy, R.S and N.L. Andrew. *Small-scale Fisheries Management*, pp. 160–177. CABI Publishing, Wallingford, UK.

Johnsen, O. and F. Nilssen. 2001. *The Future for Salmon in Germany*. Norwegian Institute of Fisheries and Aquaculture Research Report 10/2001, Tromsø, July 2001.

Kinsey, J.D. 1999. The big shift from a food supply to a food demand chain. *Minnesota Agricultural Economist*, No. 698. University of Minnesota, St. Paul, MN, Fall 1999.

Lahidji, R., W. Michalski, and B. Stevens. 1998. The future of food: an overview of trends and key issues. In: *The Future of Food: Long-term Prospects for the Agro-food Sector*. OECD, Paris.

Leyva, C.M. 2004. Optimizing tilapia (oreochromis sp.) marketing in Honduras (Order No. EP25751). Available from ProQuest Dissertations & Theses A&I; ProQuest Dissertations & Theses Global (305046334). Retrieved from http://search.proquest.com/docview/305046334?accountid=13360

Leyva, C., C. Engle, and Y.S. Wui. 2006. A mixed-integer transshipment model for tilapia (*Oreochromis* sp.) marketing in Honduras. *Aquaculture Economics & Management* 10:245–264.

Molnar, J.J., T. Hanson, and L. Lovshin. 1996. Social, economic, and institutional impacts of aquaculture research on tilapia. Research and Development Series, No. 40. International Center for Aquaculture and Aquatic Environments, Alabama Agricultural Experiment Station, Auburn University, Auburn, Alabama, USA.

National Marine Fisheries Service (NMFS). 2013. Fisheries of the United States, 2012. Office of Science and Technology Fisheries, Statistics Division, Silver Spring, Maryland, September 2013.

Pomeroy, R.S. 1989. The economics of production and marketing in a small-scale fishery: Matalom, Leyte, Philippines. PhD dissertation, Cornell University, New York.

Radtke, H. and S. Davis. 2000. Description of the U.S. West Coast Commercial Fishing Fleet and Seafood Processors. The Research Group Report prepared for Pacific States Marine Fisheries Commission, February 2000. www.psmfc.org/efin/docs/fleetreport.pdf.

Rogers, J. 2000. *Sea-fish Marketing Channels in Northern Ireland*. Department of Agriculture and Rural Development, Economics and Statistics Division. A Government Statistical Publication, December 2000.

Tarnowski, J. and W. Heller. 2004. The Super 50. *Progressive Grocer*, May 2004.

United States Census Bureau (USCB). 2007. Census of Wholesale Trade, Wholesale Trade: Subject Series – Establishments and Firm Size: Summary Statistics by Concentration of Largest Firms for the United States. Available at www.census.gov/econ/isp/sampler.php?naicscode=4244&naicslevel=4

United States Census Bureau (USCB). 2011. Census of Wholesale Trade, Annual Report, 2011. Available at www.census.gov/econ/isp/sampler.php?naicscode=42&naicslevel=2

United States Department of Agriculture, Economic Research Service (USDA-ERS). 2007. Wholesaling. Available at www.ers.usda.gov/topics/food-markets-prices/retailing-wholesaling/wholesaling.aspx#.VAilfPldV8E. Accessed September 4, 2014.

United States Department of Agriculture, National Agricultural Statistics Service (USDA-NASS). 2014. *Trout Production*. ISSN: 1949-1948. Available at usda.mannlib.cornell.edu/usda/nass/TrouProd//2010s/2014/TrouProd-03-06-2014.txt. Accessed September 11, 2014.

United States International Trade Commission (USITC). 2004. Certain Frozen or Canned Warmwater Shrimp and Prawns from Brazil, China, Equador, India, Thailand, and Vietnam. Investigations Nos. 731-TA-1063-1068 (Preliminary). Washington D.C., Publication 3672, February 2004.

University of Alaska, Anchorage. 2001. *A Planning Handbook*. Institute of Social and Economic Research, University of Alaska, Anchorage, Alaska, May 2001.

Wright, J. 2014. Consolidation, high prices, branding bump sales for top North American seafood suppliers. *Seafood Business* 33:18–21.

CHAPTER 6

Seafood and aquaculture product processing

Much of the seafood to be consumed by the end consumer must be processed into a more customer-friendly and usable product prior to sale. There are many complexities involved in the processing sector but it is a major step in supply and value chains and a vital part of the marketing process. This chapter will first describe the various types of processing for different types of seafood and aquaculture products, the structure of the seafood and aquaculture processing sector, and the degree of concentration and integration in the sector. It will discuss plant location and capacity utilization as well as the important concept of the law of market areas. Innovation and branding in the processing sector will be described, along with challenges in this market sector. The chapter concludes with a synopsis of the U.S. catfish processing sector.

Processing

Fish processing takes several forms depending on the fish species being processed, type and scale of processing operations, and product outputs. Marine fish accounts for more than 90% of fish production in the United States, while freshwater fish and farmed fish account for the remaining 10%.

Processing in the seafood and aquaculture industry encompasses all the steps that food goes through, from the time of harvest to the point at which seafood and aquaculture products reach the consumer. As in other sectors of the food industry, processing of seafood and aquaculture products is meant to provide products that are safe and meet consumers' demand requirements for quality and convenience. Thus, processing of these products must aim at increasing shelf life, reducing microbial content, preserving the products, and providing convenience. The trend in processing is generally driven by consumer demands and technological advances. In 1999, several manufacturers of retail seafood products reported a significant shift by consumers from buying higher-priced

Seafood and Aquaculture Marketing Handbook, Second Edition. Carole R. Engle, Kwamena K. Quagrainie and Madan M. Dey.
© 2017 John Wiley & Sons, Ltd. Published 2017 by John Wiley & Sons, Ltd.

premium items, such as grilled fish and specialty items, to lower-priced items such as basic breaded fish products and minced fish products. Now the trend in food processing is towards the production of ready-to-eat and ready-to-serve products that only need heating in the oven or microwave.

Processing of seafood and aquaculture products is very diverse and depends on species and products. Processing ranges from simple cleaning, dressing, and icing to elaborate grading and processing schemes. A simple dressing process typically entails removing viscera and gills of fish but leaving the head on. Dressing produces semi-products in rudimentary condition that usually undergo further processing into ready-to-cook and ready-to-serve products. Secondary processing treatments include heading and gutting, cutting products into chunks, de-boning, filleting, buttering, breading, stuffing, canning, and packing.

Generally, when raw seafood/fish has arrived at the facility the processing operations involve washing, deheading, peeling/skinning, grading, blanching, cooking, cooling, freezing (IQF), glazing, glaze freezing, packaging, and placing in cold storage at about −20 °C. Specific examples will help to illuminate the processes involved.

The basic processing of catfish, the largest aquaculture product in the U.S., is a whole-dressed fish (headed, gutted/eviscerated, and skinned or, simply, HGS). Sometimes whole fish may be headed and gutted (H & G) with tail and fins intact. This rudimentary product then undergoes further processing whereby it is cut into a variety of forms that include fillets with belly flap, shank fillets with belly flap or nugget removed, fish strips/fingers or fillet strips (boneless finger-size pieces cut from shank fillets), nuggets (belly flap section removed from fillet), and steaks (Fig. 6.1). Secondary processing also includes the production of breaded fillets and nuggets, portions and nuggets, marinated fillets, heat-set, breaded fillets, and smoked fillets and dressed fish (Silva and Dean 2001).

Figure 6.2 is a flow chart of the production technology for processing trout and carp fillets in Poland and Fig. 6.3 is a flow chart for processing marine white fish in the U.S. From Fig. 6.3, the fish are first gutted and washed and may be deheaded on board the fishing vessel before landing. The fish are kept on ice until they are delivered to the processing plant. At the plant, pre-treatment of the fish involves the removal of ice, washing, grading, and deheading, if not done previously. Large fish may also be scaled before further processing. Filleting is the next process; this is done by mechanical filleting machines. A typical fillet-ing machine has pairs of mechanically operated knives for cutting the fillets from the backbone and for removing the collarbone. This stage of processing may also involve skinning of fish fillets. Trimming involves removal of pin bones. Fillets are then inspected and defects and portions that are deemed to be of lower qual-ity are removed. Offcuts are also collected separately and minced. The fillets are cut into portions and weighed depending on the final product, or the fillet may be separated into parts such as loin, tail, and belly flap. The final step is inspec-tion of the fillets to ensure they meet market specifications. Final products are

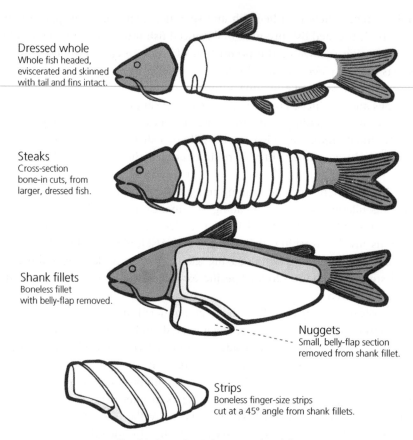

Dressed whole
Whole fish headed,
eviscerated and skinned
with tail and fins intact.

Steaks
Cross-section
bone-in cuts, from
larger, dressed fish.

Shank fillets
Boneless fillet
with belly-flap removed.

Nuggets
Small, belly-flap section
removed from shank fillet.

Strips
Boneless finger-size strips
cut at a 45° angle from shank fillets.

Fig. 6.1 Cuts of U.S. farm-raised catfish.

then packed for shipment. Fresh products are packaged with ice, which is separated from the fish products by a layer of plastic sheet. Frozen products are packed in different ways. For example, fillets may be individually frozen and wrapped in plastic. Processed fish can also be packed as 6 to 11 kg blocks in waxed cartons, frozen, and kept in cold storage.

Processed seafood products may be sold fresh, frozen, smoked, seasoned, canned, dried, or dehydrated. Inedible and substandard portions of processed products are usually used to produce fishmeal products used for animal feed.

Processing may also involve an extremely complex set of techniques and ingredients that transform raw products into products that are tasty, nutritious, and ready-to-eat, requiring minimal preparation and cooking, or formulated food products such as surimi. Surimi is an important fish product. Most marine fish catches for some species are used solely for surimi production.

Aquaculture and seafood products that have undergone complex processing are generally known as value-added products. Adding value in seafood processing generally implies a degree of processing that makes the seafood product

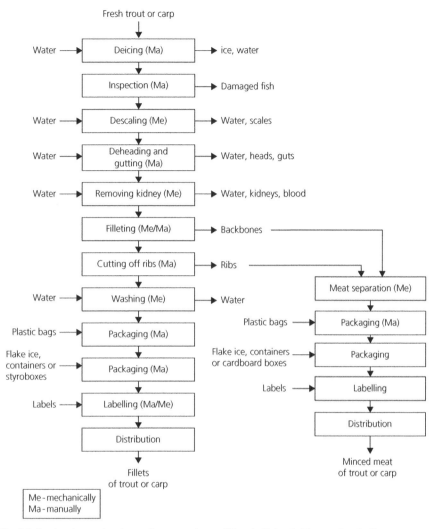

Fig. 6.2 Production technology of trout and carp fillets in Poland. Me, mechanically; Ma, manually. Source: Bykowski and Dutkiewicz (1996).

more desirable to consumers, which may relate to better appearance, taste, texture, flavor, or greater convenience. Value adding may also relate to processing products to improve shelf stability and functionality.

Value-added seafood could include glazed and coated portions, burgers, and fish tender products (coated fillets). For glazed and coated fish portions, the processing operation involves removal of fish collar and skin, filleting, and injection with a marinade. Fish products can also be cut into portions of loin, center cut, or tails. Products are weighed and sorted automatically into 4–6 ounce portions. They are then frozen and either packaged for shipment or further processed by the addition of glazes or transformed into other value-added products

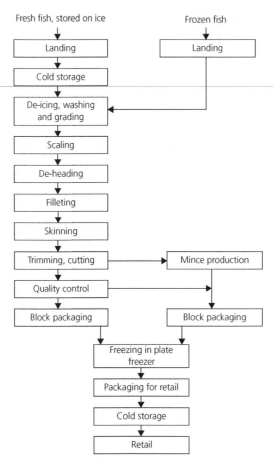

Fig. 6.3 Process flow diagram for the filleting of white fish. Me, mechanically; Ma, manually. Source: Bykowski and Dutkiewicz (1996).

(Mermelstein 2002). For further processing of fish portions, a press stamps out the portions that feed into the breading machine, or a steaking machine produces steaks. A glazing line applies flavored glazes to the fish portions. Some of the flavorings used are garlic, butter, and honey-sesame ginger to enhance the flavor. Flavored marinades are either injected directly into the fillets or incorporated by vacuum tumbling.

The process of making salmon burgers involves chopping frozen fillets in a bowl chopper with other ingredients and then forming into 1–2 ounce tender, battered and breaded burgers, which are then par-fried. The burgers require heating in an oven at 375–400 °F for 10–15 minutes before they are ready to be eaten. The product can be vacuum infused with marinade. Liquid carbon dioxide can also be added as a fog to lower the temperature to below freezing to facilitate forming (Mermelstein 2002). Packaging of the salmon products includes vacuum packaging for retail sales as well as bulk packaging. Natural portions can be packaged in expanded polystyrene trays with a clear overwrap or vacuum packed and frozen for retail sale.

Freezing and storage are important in maintaining the quality of processed products. High-quality storage provides the processor with a means of controlling its products to ensure consistency in supply, quality, and shipments to distributors or retailers. Supply of fish raw materials either from aquaculture facilities or natural catches from the oceans is seasonal, therefore freezing and storage provide a means of stabilizing product temperatures and accumulating complete lots or loads for direct shipment to buyers with minimal repacking, transfer, and temperature fluctuation (Kolbe and Kramer 1997). The recommended temperature, relative humidity, and approximate storage life for selected seafood products are presented in Table 6.1.

Table 6.1 Recommended temperature and relative humidity, and approximate transit and storage life for seafood.

Product	Temperature		Relative humidity (%)	Approximate storage life
	°C	°F		
Haddock, cod, perch	−1 to 1	31 to 34	95 to 100	12 days
Hake, whiting	0 to 1	32 to 34	95 to 100	10 days
Halibut	−1 to 4	31 to 34	95 to 100	18 days
Herring, kippered, smoked	0 to 2	32 to 36	80 to 90	10 days
Mackerel	0 to 1	32 to 34	95 to 100	6 to 8 days
Menhaden	1 to 5	34 to 41	95 to 100	4 to 5 days
Salmon	−1 to 1	31 to 34	95 to 100	18 days
Tuna	0 to 2	32 to 36	95 to 100	14 days
Frozen fish	−29 to −23	−20 to −10	90 to 95	6 to 12 months
Clams (shucked meats)	−1.7	29	85 to 90	5 days
Crabmeat, pasteurized	0 to 1.1	32 to 34		6 months
Crabs, king, snow, cooked, frozen	−18	0		12 months
Crabs, Dungeness, cooked, frozen	−18	0		3 to 6 months
Scallop meat	0 to 1	32 to 34	95 to 100	12 days
Shrimp	−1 to 1	31 to 34	95 to 100	12 to 14 days
Lobster, American, live	5 to 10	41 to 50	in water	indefinite
Lobster, American, fresh meat	−1.1 to 0	30 to 32	90 to 95	3 to 5 days
Lobster, American, frozen, shell	0	−18		3 to 6 months
Lobster, meat, cooked, frozen	0	−18		6 to 9 months
Lobster, spiny, frozen, shell	0	−18		10 to 12 months
Oysters, meat	0 to 2	32 to 36	100	5 to 8 days
Oysters, clams, in shell	5 to 10	41 to 50	95 to 100	5 days
Frozen shellfish	−29 to −20	−20 to −4	90 to 95	3 to 8 months

Source: The Refrigeration Research and Education Foundation (1996); American Society of Heating, Refrigeration, and Air Conditioning Engineers, Inc. (1994).

Structure of the seafood and aquaculture product processing industry

Processing of seafood and aquaculture products primarily takes place in processing establishments, but some large fishing vessels that operate in deep waters have facilities on board where seafood/fish are processed. Some fishing vessels both catch and process seafood/fish while other vessels are mainly processing ships.

The National Marine Fisheries Service (NMFS) periodically reports the results from annual surveys of all seafood processors that operate in the U.S. The primary operations of these establishments are in one or more of the following: (1) eviscerating fresh fish by removing heads, fins, scales, bones, and entrails; (2) shucking and packing fresh shellfish; (3) manufacturing frozen seafood; and (4) processing fresh and frozen marine fats and oils. Processed fresh and frozen products include fish fillets, steaks, fish sticks, and portions as well as breaded shrimp. In 2001 there were 994 processing plants engaged in processing fresh and frozen seafood and aquaculture products. They employed 48,900 workers and produced fishery products valued at about US$8.1 billion (NOAA-NMFS 2003). There were 824 plants reported in 2011 with 37,079 employees and production valued at about US$9.9 billion (NOAA-NMFS 2013). This suggests that between 2001 and 2011 the number of processing plants decreased by 21% and employment decreased by 32%, while production of fresh and frozen products increased by 22%. The trend is a reflection of the expansion through mergers and acquisitions in the food processing industry in the U.S. since 2000.

Economists and policy makers are often interested in the structure of these companies since it could have implications for market performance. The structure of the seafood and aquaculture products industry in the U.S. relates to the concentration of the industry, the degree of vertical integration, product characteristics, and freedom of entry and exit (Fig. 6.4). Each of these industry features is discussed below.

The number and quality of firms competing in an industry are sometimes thought to determine the nature of competition in the industry, depending on the industry concentration. When the number of firms operating in the industry is sufficiently large, and the product handled by the industry is homogeneous or standardized, the industry is said to be perfectly (purely) competitive. Pure monopoly applies to an industry in which there is only one firm with a unique product that has no close substitute. In between pure competition and pure monopoly are monopolistic competition and oligopolistic competition. Monopolistic competition is a blend between monopoly and perfect competition. The oligopolistic structure has few firms and products may be differentiated (Table 6.2). The structure of the seafood and aquaculture products industry is typical of a monopolistic competitive industry, with a relatively large number of firms operating competitively in

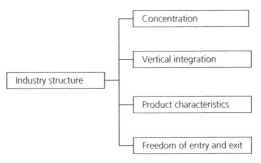

Fig. 6.4 Characteristics of industry structure.

Table 6.2 Types of product market structures.

Type of structure	Characteristics			
	Number of firms	**Type of product**	**Control over price**	**Freedom of entry and exit**
Perfect competition	Numerous	Homogenous	None	Very easy
Monopolistic competition	Many	Differentiated	Some	Relatively easy
Oligopoly	Few	Homogenous or differentiated	Some	Partially restricted
Monopoly	One	Unique	Considerable	Absolutely restricted

the production of differentiated products (USDC-NOAA 2004). However, Dillard (1995) and Kouka (1995) concluded that the catfish-processing sector is somewhat oligopolistic because the industry is dominated by a few relatively large companies with frequent price wars that have tended to keep prices received by processors below cost. Weise (2004) reported that 85 catfish processing plants had entered the industry from 1981 to 2003, while 72 plants exited the industry during that time period, with an average of 5 years in the catfish processing business. However, the U.S. catfish processing industry has stabilized since 2004 with few firms entering or exiting since then. Most plants have been in the business for more than 20 years. While some seafood processing industries may be relatively concentrated, easy entry makes the industry quite competitive and less concentrated because of relative prices, abundance of raw materials, and government policies.

The market structure of the processing industry includes both the market for processed fish products as well as raw fish material, in which processors serve as buyers. The logical counterpart of a monopolistic market is the monopsonistic market where there is only one buyer. Similarly, an oligopsonistic market implies few buyers of homogenous or differentiated products, and costly entry or exit of buyers. The market structure in the market for raw seafood products may be somewhat different. Hackett and Krachey (2002) suggest that various markets for landed fish ranged between being moderately concentrated and concentrated on the buyer side. In a study of U.S. West Coast processors, Radtke and Davis (2000) reported that in 1997, California processors could be characterized as oligopsonists in the market for fish because the 15 largest processing companies or parent groups processed 65% of the fish by volume and 46% of the total fish by value. The authors reported that the processing industry in California had experienced additional consolidation since 1997.

Concentration

Industry concentration is the percentage of business (share of total value of shipments) accounted for by a number of businesses in the industry. An industry with a large number of firms may not necessarily be competitive. The U.S. Census Bureau (USCB) reports the concentration ratios for various industries in terms of share of value added accounted for by companies. The 4, 8, 20, and 50 largest fresh and frozen seafood processing companies in 2002 had ratios of 19%, 28%, 45%, and 69%, respectively (USDC-CB 2002). In 2007 the corresponding ratios were, respectively, 32%, 44%, 61%, and 78% (USDC-CB 2007). The data indicate an increasing trend to concentration in the seafood processing sector. Similar trends can be seen in the animal (except poultry) packing industry where the concentration ratios in 2007 for the 4, 8, 20, and 50 largest companies were 43%, 64%, 80%, and 89%, respectively (USDC-CB 2007).

Concentration ratios have implications for competition and economic performance in the industry. While a concentrated industry may be viewed by some as less economically efficient, high operational and financial performance of firms has often led to expansions, mergers, and acquisitions that result in a concentrated industry. Food processors often specialize by product line, but the trend has been to diversify and add additional product lines. Brands allow processors to differentiate product and certify product quality.

Larger firms may have the advantage of economies of scale where more can be produced at a lower per unit cost, but mergers and acquisitions are usually made with the intention of increasing market share. In 2000, Trident Seafoods Corporation acquired the seafood division of Tyson Foods, while Bumblebee Seafood Incorporated acquired Tyson's surimi seafood business. In 2004, Trident Seafoods acquired Norquest Seafoods but Trident's effort to acquire Ocean Beauty did not materialize (Duchene 2005; Wright 2006). Bumblebee Seafood Inc. and Trident Seafoods Corp. are ranked among North America's top five seafood suppliers by *SeaFood Business* magazine. Similarly, the merger in 1999 between Stolt SeaFarms and International Aqua Foods made Stolt SeaFarms the largest farmed salmon producer in North America. Marine Harvest purchased Stolt SeaFarms in 2005, although they had shed their salmon farms at that point. Currently, Cook Aquaculture owns most of the salmon farms in Maine, and Icicle Seafoods owns the salmon farms on the West Coast.

Antitrust laws exist, however, to promote competition. The federal government usually challenges any cooperation and merger between firms that will result in a monopoly or near monopoly. Seafood industry observers suggested that the failure of the Trident Seafoods and Ocean Beauty merger to materialize in 2006 could have been the result of the threat of a federal antitrust review of the deal (Wright 2006). Exceptions occur when one of the merging firms is on the verge of bankruptcy, in which case the federal government may allow such a merger. In some cases, however, action has been taken towards existing concentrations if there is evidence that the firm that has more than 60% of the market used deliberate conduct to achieve dominance (e.g., the federal government's antitrust actions against Microsoft in 1998 alleging violations of the Sherman Act §§ 1 and 2[1]). The merger of Stolt SeaFarms and International Aqua Foods was the result of the Fisheries Act[2], which required a minimum of 75% ownership by Americans.

---

[1] *§1 Sherman Act, 15 U.S.C. §1 (Trusts, etc., in restraint of trade illegal; penalty)* Every contract, combination in the form of trust or otherwise, or conspiracy, in restraint of trade or commerce among the several States, or with foreign nations, is declared to be illegal. Every person who shall make any contract or engage in any combination or conspiracy hereby declared to be illegal shall be deemed guilty of a felony, and, on conviction thereof, shall be punished by fine not exceeding $10,000,000 if a corporation, or, if any other person, $350,000, or by imprisonment not exceeding three years, or by both said punishments, in the discretion of the court.

§2 Sherman Act, 15 U.S.C. §2 (Monopolizing trade a felony; penalty).

Every person who shall monopolize, or attempt to monopolize, or combine or conspire with any other person or persons, to monopolize any part of the trade or commerce among the several States, or with foreign nations, shall be deemed guilty of a felony, and, on conviction thereof, shall be punished by fine not exceeding $10,000,000 if a corporation, or, if any other person, $350,000, or by imprisonment not exceeding three years, or by both said punishments, in the discretion of the court.

[2] *SEC. 202. Standard for Fishery Endorsements* (a) STANDARD.—Section 12102(c) of title 46, United States Code, is amended to read as follows—

"(c)(1) A vessel owned by a corporation, partnership, association, trust, joint venture, limited liability company, limited liability partnership, or any other entity is not eligible for a fishery endorsement under section 12108 of this title unless at least 75 per centum of the interest in such entity, at each tier of ownership of such entity and in the aggregate, is owned and controlled by citizens of the United States.

"(2) The Secretary shall apply section 2(c) of the Shipping Act, 1916 (46 App. U.S.C. 802(c)) in determining under this subsection whether at least 75 per centum of the interest in a corporation, partnership, association, trust, joint venture, limited liability company, limited liability partnership, or any other entity is owned and controlled by citizens of the United States. For the purposes of this subsection and of applying the restrictions on controlling interest in section 2(c) of such Act, the terms 'control' or 'controlled'—

"(A) shall include—

"(i) the right to direct the business of the entity which owns the vessel;

"(ii) the right to limit the actions of or replace the chief executive officer, a majority of the board of directors, any general partner, or any person serving in a management capacity of the entity which owns the vessel; or

"(iii) the right to direct the transfer, operation or manning of a vessel with a fishery endorsement; and

"(B) shall not include the right to simply participate in the activities under subparagraph (A), or the use by a mortgagee under paragraph (4) of loan covenants approved by the Secretary.

Mergers often result in closure of some processing plants and cuts in labor to streamline the production base and improve overall operating efficiencies and competitiveness. The 21% decline in the number of seafood processing plants between 2001 and 2011 could largely be the result of mergers and acquisitions.

Vertical integration

Some seafood and aquaculture companies are vertically integrated because they operate fish farms as well as fish processing plants. When firms operate at more than one level of a series of levels in the food system from raw materials to the final consumer, they are considered to be vertically integrated. In some cases, firms become vertically integrated as a result of merging firms at different stages of the production process (vertical merger). Firms integrate vertically for several reasons including: (1) to lower their cost; (2) to achieve economies of scope through diversification; and (3) to strengthen the business. Firms may also integrate vertically to monitor and maintain quality along the production process.

With vertical integration, the entire production process of seafood and aquaculture products from harvest to the final consumer is divided and undertaken by a single firm. Trident Seafoods is a typical example of a vertically integrated seafood business that operates vessels in Alaska and in the Pacific Northwest. Seafood harvested is processed and canned or frozen for retail food and food service customers. Trident operates dozens of processing boats and trawlers, as well as onshore processing facilities in Alaska and the Pacific Northwest. The company also operates a retail store in Seattle. Carolina Classics Catfish, an aquaculture company, is a vertically integrated company that produces feed (operated under Carolina Fish Feeds), grows fish, and processes and delivers catfish products. Idaho Trout Processors Company operates the trout farms in Rim View Trout, Rainbow Trout Farms, and Clear Lakes Trout Farm as well as a processing company. Some of these vertically integrated companies involve a group of producers who collectively own the processing company.

Product characteristics

A monopolistic competitive industry often consists of firms producing a differentiated product, such that each firm's output is distinguishable from any other firm's output. Products may be differentiated through physical attributes, functional features, material make-up, packaging, advertising, and branding. In the seafood and aquaculture products industry, there are a variety of different types of products including shellfish, finfish, scaled fish, and other unclassified fish. The final product forms (buttered, breaded, stuffed, dried, marinated, or canned) and the different brands of these products make seafood and aquaculture products different from one another. The U.S. Department of Commerce lists as many as 120 different seafood and aquaculture products that are produced by fish processing establishments and over 1600 brands of products (Table 6.3).

Table 6.3 Inspected fishery products produced in USDC approved establishments.

Raw portions, sticks, nuggets, etc.	Marinated fillets
Raw steaks	Breaded raw portions, sticks, nuggets, etc.
Raw fillets	Breaded raw strips
Raw whole	Breaded raw fillets
Raw dressed head off/on	Breaded precooked portions, sticks, nuggets, etc.
Raw dressed and boned	Breaded precooked strips
Breaded raw portions, sticks, nuggets, etc.	Breaded fully cooked minced cakes, patties or burgers
Breaded raw fillets	Seafood frozen
Breaded precooked fillets	Breaded precooked minced cakes, patties or burgers
Fish frozen	Crab fresh/refrigerated
Raw portions, sticks, nuggets, etc.	Breaded raw cakes, patties or burgers
Raw steaks	Crab frozen
Raw stuffed	Raw cakes, patties or burgers
Raw cakes, patties or burgers	Raw soft shell
Raw fillets	Cooked soups
Raw blocks	Breaded precooked cakes, patties or burgers
Raw minced	Crab canned
Raw dressed and boned	Dips and spreads
Breaded raw portions, sticks, nuggets, etc.	Lobster fresh/refrigerated
Breaded raw steaks	Live
Breaded raw strips	Shrimp frozen
Breaded raw stuffed	Raw headless
Breaded raw fillets	Raw whole
Breaded raw meat	Marinated meats
Breaded precooked portions, sticks, nuggets, etc.	Breaded raw imitation
Breaded precooked croquettes	Breaded raw meats
Breaded precooked portions, sticks, nuggets, etc.	Breaded raw whole
Breaded precooked strips	Breaded precooked crisps
Breaded precooked cakes, patties or burgers	Breaded precooked dinners
Breaded precooked fillets	Breaded precooked meats
Breaded fully cooked portions, sticks, nuggets, etc.	Batter coated precooked meats
Breaded fully cooked fillets	Breaded precooked minced
Batter coated precooked portions, sticks, nuggets, etc.	Peeled raw meats
Batter coated precooked strips	Peeled raw deveined
Batter coated precooked fillets	Peeled raw whole
Breaded raw minced portions, sticks, nuggets, etc	Peeled cooked deveined
Breaded precooked minced portions, sticks, nuggets, etc.	Oyster fresh/refrigerated
Breaded precooked minced cakes, patties or burgers	Live
Breaded fully cooked minced portions, sticks, nuggets, etc.	Raw shucked
Batter coated precooked minced portions, sticks, nuggets, etc.	Oyster frozen
Farm-raised catfish fresh/refrigerated	Breaded raw meats
Raw bellies	Breaded raw whole

(Continued)

Table 6.3 (*Continued*)

Raw portions, sticks, nuggets, etc.	Scallop fresh/refrigerated
Raw steaks	Raw shucked
Raw strips	Breaded raw whole
Raw fillets	Scallop frozen
Raw dressed and skinned	Raw shucked
Marinated bellies	Breaded raw whole
Farm-raised catfish fresh/refrigerated (continued)	Breaded precooked whole
Marinated steaks	Batter coated precooked whole
Marinated fillets	Squid fresh/refrigerated
Marinated dressed and skinned	Raw whole
Farm-raised catfish frozen	Squid frozen
Raw bellies	Breaded raw tubes and/or rings
Raw portions, sticks, nuggets, etc.	Surimi fresh/refrigerated
Raw steaks	Cooked analog
Raw strips	Surimi frozen
Raw fillets	Cooked analog
Raw dressed and skinned	Breaded precooked minced analog

Source: United States Department of Commerce – National Oceanic and Atmospheric Administration (USDC-NOAA 2004).

Entry into the industry

New firms enter an industry if they expect to make profit. However, in a monopolistic competitive industry, firms make nominal profits. The level of profit is termed "normal" if the amount of profit gained is sufficient to induce the firm to stay in business but is neither excessive nor minimal. Natural barriers that could restrict free entry into the seafood and aquaculture products processing industry include economies of scale, large capital outlays, ownership of essential raw materials, advertising and product differentiation, sunk costs (incurred cost that cannot be recovered), and government policies. Knapp et al. (2001) suggested seven important "reality checks" for anyone planning a fish processing plant at the village or local level: (1) availability of fish; (2) current and future competition from other processors; (3) availability of good plant management; (4) availability of skilled production workers; (5) availability of water and power, and waste disposal; (6) marketing of products; and (7) availability of reliable transportation to take products to market. The authors concluded that it is only after considering these issues that one can move on to the planning phase. Planning involves addressing issues such as products to produce, markets for the products, kinds of building and equipment needed, financing and, most importantly, whether the plant can earn enough money to stay in business.

Economy of size relates to the efficiency of large firms; thus large-scale processing operations with associated large capital outlays could be a hindrance to the entry of new firms. Vertical integration allows firms to control the raw materials of captured or farmed seafood and fish needed for the processing establishment. Product branding is a major cue to consumer behavior, and processing companies have different lines of branding to differentiate their products from similar products produced by competing companies. Non-recognition of new brands by consumers could be a barrier to potential new entrants to the industry. All sorts of government policies including ownership requirements, licensing, trademark protection, and regulations can become barriers to entry. A key objective of the American Fisheries Act is the 75% minimum American ownership of fishing vessels operating in U.S. waters. Prior to the Act, some major fishing companies had majority ownership that was foreign. The seafood and aquaculture products industry has seen more mergers and acquisitions during the past decade than new entrants into the industry.

Plant location

Proximity to inputs, availability of services, and the type of marketing system needed by a company greatly determine the location and size of processing plants. Generally, firms would expect the costs associated with obtaining raw materials and essential services including technology, labor, communication, and transportation, and access to the markets for their outputs to be low. There is a web of linkages among industries because the output of some firms and industries constitutes the inputs of other firms and industries. This linkage allows firms to realize substantial cost advantages due to proximity. Thus, economies of location play a significant role in the choice of location for processing plants. Bykowski and Dutkiewicz (1996) suggest that the most important factor when considering the location of a processing plant is adequate size for both present needs and future development. The authors also suggest a location close to public transport such as rail or road, access to electricity, water and steam, and adequate waste disposal. The local authorities should be actively involved in the process in order to avoid problems in the future.

Regarding the design of a processing plant, Bykowski and Dutkiewicz (1996) recommended that the building should have sufficient work space: the space should be large enough to allow processing under hygienic conditions, and adequate space must be available for machinery, equipment, and storage. Separation of operations is necessary to avoid food contamination. There should also be adequate natural or artificial lighting, ventilation, and protection against pests.

In the U.S., seafood processing plants are commonly located along the seaboards, while processing plants for aquaculture products are located in major aquaculture production regions. The Quonset-Davisville Port and Commerce Park in Rhode Island is a location with extensive infrastructure and facilities including deep water access, an airport, and rail and highway connectors. Bridges on the rail lines accommodate double stack containers. Seafood companies such as Seafreeze Ltd., American Mussel Harvesters and others are located at the port.

The infrastructure and facilities at the port continue to attract new seafood processing companies. Major catfish processing plants are located within a 50-mile radius in the Mississippi delta region where over 80% of catfish production takes place. The region also has access to state and interstate highways, railways, and regional industrial parks.

Although fish processing operations are located close to commercial fishing areas, catches may be transported long distances or exported for processing in some cases. *EUROFISH Magazine* (January/February 2001 issue) reported that Danish exports of unprocessed fish had almost doubled since 1983 but there was a decline in exports of processed fish product within the same period. This is because more of the fish landed by the national fleet was exported unprocessed, resulting in the processing sector becoming increasingly dependent on imported raw materials. The Polish market is one of the fastest-developing European fish markets. Poland has a strong fish processing sector but a weak fisheries sector, therefore Poland imports fish for processing, including herring, mackerel, Alaskan pollock, hake, salmon, and cod (European Parliament 2005).

Law of market areas

The location for any business involves a consideration of the sales potential that exists in the area or region. The marketing areas of processors involve sales to wholesalers and retailers. Combinations of geographic, demographic, economic, and competitive factors will determine the market area that a processing company will service. In particular, human resources and costs of operation are important. Market areas have different demographic characteristics, competitive factors, and sales potentials. The size of the market area offers opportunities for or constraints to potential sales and expansion, and business development in general. Geographic factors such as mountain ranges, rivers, and road patterns influence the nature of transportation systems. These factors can influence market areas and are important to the distribution of processed products. Highway speed limits, nature of roads, highway access, bridges across rivers, and general topographic features determine trading patterns within and between market areas. Wholesale distribution of products would depend on the number and proximity of potential outlets to serve and the quality of transportation that would enable delivery to clients in near and distant market areas. Transportation characteristics affect transportation costs, delivery policies, and delivery structure of products.

Businesses usually target markets in urban areas and areas of larger population because of high demand. Apparently, these large market areas are where more direct competition among suppliers exists. Each major market area has its own unique characteristics in terms of age distribution, number and types of households, income levels, work patterns, shopping patterns, retail sales levels, and economic health. Therefore, demographic interest in assessing a market area would focus on personal income, education level, age, and lifestyles of potential workers as well as customers within the market area.

Capacity utilization

Economic theory suggests that low capacity utilization or excess capacity is one of the characteristics of a monopolistic competitive industry. Individual companies do not disclose their processing capacity due to confidentiality but the U.S. Census Bureau reports the rate of production capacity utilization for all industries (USDC-CB 2013). The rate of production capacity utilization is the ratio of total capacity utilized relative to the total processing capacity available. Industries with full utilization of processing capacity are characterized by two and three shifts of production workers, but this is usually not the case with the seafood and aquaculture processing products industry. In this industry, the rate of capacity utilization has averaged 70% over the last 10 years compared to 83% for the meat (beef, pork, and poultry) processing industry (USDC-CB 2013).

In Canada, optimistic projections for fish stocks in the 1970s led to a significant increase in fish processing capacity in the Atlantic provinces, particularly with the anticipated extension of Canada's economic zone to 200 miles. Excess processing capacity was also built to meet the peak landings from the seasonal inshore fishery (FIRB-Government of Canada 1997).

Product differentiation is another characteristic of a monopolistic competitive industry. The more products are differentiated, the less elastic the demand curve for the products. With inelastic demand, production does not take place at the minimum of the average cost curve, leading to low capacity utilization or excess capacity. The seafood and aquaculture industry produces differentiated products that provide a number of varieties of seafood products in terms of fish species, cuts and portions, preparations, and cooking forms. For example, American Pride Seafoods produces a variety of seafood products including fresh and frozen Atlantic salmon fillets; batter-dipped cod, Alaska pollock and whiting; baked, broiled, and breaded cod and pollock portions; fried natural shaped pollock; skinless boned cod loins and cod; pollock and haddock fillets; raw, breaded cod, haddock and flounder portions; minced cod; fresh and frozen catfish fillets, breaded and marinated catfish, whiting, sea scallops, and frozen whole sea scallops.

Innovation and branding

The art of marketing is recognizing the factors underlying the seafood preferences of buyers and producing to meet those preferences. Meeting the diverse preferences of buyers of seafood and aquaculture products requires innovation in the development of new products, modification of old products, and presentation of new products in better ways to buyers. Companies that usually want an edge over the competition find innovative ways of staying ahead through the development of new products, new processing operations, or formulation of new ideas for marketing products. Associated with this is branding of products for company identification and product differentiation.

Best new products at the 2016 Seafood Expo North America (formerly the International Boston Seafood Show) included:

- Kickin' Seafood Chili;
- Pacific Cod Bites;
- Gold Premium Pineapple-Teriyaki Sockeye;
- Mussels in a Creamy Stout Sauce;
- Seafood Toast;
- Honey Glazed Oak Roasted Salmon;
- Char Marked Barramundi.

Branding of products is achieved through the use of brand name, brand mark, logo, registered brand (®), or trademark (™). Registered brands and trademarks are protected by law and meant for the exclusive use of the registered owner. Many consumers use brand names as cues for purchasing products because they provide some sense of satisfaction and security. Thus, it requires that companies ensure consumer familiarity with their brands. In the seafood and aquaculture industry, brand names are associated with company reputation that may relate to specific products, quality, price, packaging, organic products, etc. Some companies and brand names are associated with one line of products. An example is StarKist® which offers different tuna products including Flavor Fresh Pouch™, Naturally Low Sodium-Low Fat Tuna, Chunk Light Tuna, Lunch To-Go, Solid White Albacore Tuna, Gourmet's Choice Tuna Fillets, Select Prime Light Fillets, and Tuna Creations™. Contessa® is associated with shrimp, which is available cooked or uncooked, tail-on or tail-off. Bumble Bee® offers several product lines including albacore tuna, salmon, shrimp, crab, oysters, clams, and ready-to-eat tuna salad. Bumble Bee prides itself on the quality of its products. Wild Oats Market, Inc., boasts of being a leader in the natural and organic food industry.

Traditionally, many manufacturing companies have adopted a new-product development cycle of 4–5 years. However, the dynamic nature of consumer preferences for food products has necessitated a rather accelerated process of new product development. Many new products introduced to the market have focused on attributes such as convenience and health. New products and ideas stimulate excitement and curiosity of consumers that can translate into purchase. Some of the new ideas and products by the seafood industry in the past decade include ready-to-eat products, resealable retail packs, reduced fat products, and fish products with no preservatives.

Challenges in aquaculture product processing

One of the objectives of the U.S. Department of Commerce is to increase the value of domestic aquaculture production from $900 million annually to $5 billion by year 2025 (USDC 2014). The major farmed foodfish in the U.S. include salmon, trout, catfish, striped bass, tilapia, clams, crawfish, mussels, oysters, and shrimps.

The production of aquaculture is expanding and intensifying in the U.S. and the total supply of fish available for consumption depends on future trends in the aquaculture industry. One of the challenges that confronts the aquaculture processing industry is realizing the potential growth in the market for aquaculture products and pursuing a stable, sustainable, and competitive processing sector. The U.S. imports about 90% of its seafood needs, and with free trade and other bilateral and multilateral trade agreements in place, the domestic industry faces an increasing level of import competition. The major competition to processed foodfish has come from imported salmon and trout from Chile and Canada, catfish from Vietnam, crawfish from China, and oysters from South Korea.

Aquaculture products have traditionally been fresh and frozen raw products. However, consumer demand requirements for health, quality, and convenience necessitate a production process that is consumer-oriented in order to take advantage of the market. Consumer demand for fish and seafood products continues to be strong due to the many nutritional and health benefits of consuming them. Scientific reports and government food guides continually cite fish and seafood products as low in fat, easily digestible, and a good source of protein and important minerals and vitamins. Besides the safety and quality, consumers with their busy schedules are looking for fish and seafood products and meals that can be cooked and served fresh in a matter of minutes. The challenge to the processing industry is to produce these types of products cost effectively to ensure that the domestic industry competes effectively.

The development of chain formation in the distribution and marketing system is another challenge that confronts the processing sector. While some processing firms have developed their own sales organization and marketing arms that are responsible for selling their products, others have established strategic alliances and partnerships with retail and restaurant outlets in order to give better guarantees in terms of continuity and quality. For example, Idaho Trout Processors Company supplies fresh and frozen dressed whole trout specially produced for the grocery giant Albertson's, and Alaska Seafood International produces frozen Cheese Salmon Tenders packaged in 2.5-lb standup polybags specifically for Sam's Club. Such partnerships help to achieve better coordination between links in the marketing chain while strengthening the distribution function of products. The challenge is to systematically utilize the information on the dynamics of the markets for the purpose of coordinating supply and demand, which could translate into the development of new products and product concepts.

The aquaculture industry is confronted with a series of environmental and health concerns relating to fish feed and pollution. Environmental groups and advocates of wild-caught fisheries have raised concerns relating to effluents from aquaculture production facilities and the quality of aquaculture products. The challenge for the processing industry is to address these issues and provide consumers with guarantees relating to the general quality, the functional quality, and the healthy image of aquaculture products as well as the eco-sustainability of aquaculture production practices. The main environmental issues associated with fish processing are high water usage, effluents, energy consumption, and generation of byproducts. For some plants, noise and odor may also be concerns.

A few seafood processing plants concentrate on a single species, such as tuna, salmon, or shrimp, but most plants process several different species to take advantage of the different fisheries in their region. This is not the case for aquaculture processing plants, which process mainly single species of fish. Diversification into multi-species processing would afford the aquaculture processing sector the opportunity to better utilize processing capacity, become cost effective, and reduce marketing risks.

Aquaculture market synopsis: U.S. channel catfish

In the U.S. production of channel catfish takes about 18 months from fry to foodfish size that weigh from 0.25 to 2.25 kg. However, hybrid catfish (♂Ictalurus punctatus × ♀Ictalurus furcatus) reach market size in 5–6 months. The use of hybrid catfish has increased rapidly across the industry. Ponds are partially harvested when there are 4500–18,000 kg of market size fish. Fish are placed in aerated tank trucks and shipped live to the processing plant. Samples of catfish are first checked for flavor through a taste test 2 weeks, 1 week, and the day before the fish are harvested, and checked again at the processing plant before the fish are unloaded. About 95–98% of all catfish production goes to processing plants with a small percentage sold through other channels that include livehaul, fee and recreational fishing facilities, government agencies, or direct to consumers, retailers, and restaurants. At the processing plant, catfish are kept alive, shocked, processed, and placed on ice or frozen to temperatures of −40 °F, using a quick-freeze method. Quick freezing retains the flavor, taste, and quality for longer periods of time.

In the U.S., Mississippi, Alabama, and Arkansas are the major catfish production states, with much of the catfish processing occurring in the state of Mississippi. The industry has contracted since its peak production level in 2003

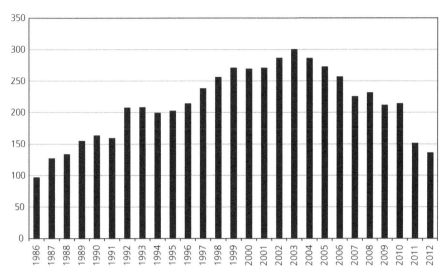

Fig. 6.5 Quantity of catfish processed (million kg). Source: USDA Economics and Statistics System.

when 300 million kg of round weight catfish were processed (Fig. 6.5). In 2012, 136 million kg were processed, an approximately 55% decrease since 2003. This is the result of reduction in the acreage of water devoted to catfish production. Total water surface acres declined from 187,200 in 2003 to 83,020 in 2013. Over the past decade, the industry has experienced rising costs of feed and fuel, volatile product prices, low demand, and reduced market share for final products due to competition from low-priced imports. As a consequence many producers in Alabama, Arkansas, Louisiana, and Mississippi have converted their pond acreage to corn and soybean production.

The size of catfish processors ranges from very small enterprises to relatively large businesses that produce various fresh, frozen, and value-added catfish products for wholesale and retail sales. Processed catfish products take the form of dressed whole fish, fillets, nuggets, steaks, or value-added products (Fig. 6.6). Processed products are also sold fresh, frozen, breaded, marinated, or in some other value-added form such as patties, smoked, and precooked frozen dinners or entrées prepared as heat-and-serve items to provide catfish buyers with a variety of products. Generic advertising and promotional activities for catfish are the responsibilities of The Catfish Institute, which is funded through a feed checkoff program. However, individual processing companies also try to differentiate their products through advertising, packaging, services associated with sales, and use of trademarks and brand names.

Theoretically, the catfish processing sector conforms to a monopolistic competitive industry because the industry comprises a fairly large number of processing companies that compete with each other to produce a differentiated product. There has not been a significant number of companies entering and

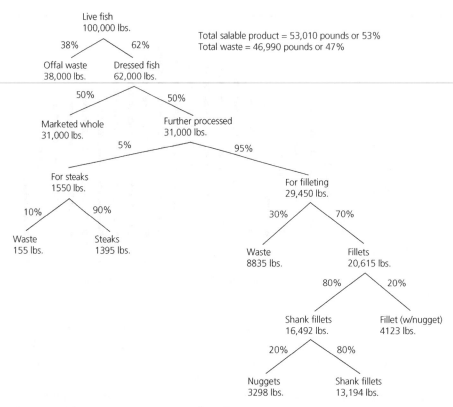

Fig. 6.6 Catfish processing input-output chart illustrating a hypothetical product mix. Source: Silva and Dean (2001).

exiting the catfish processing industry in recent years. In the 1980s and early 1990s, major food companies such as ConAgra Foods, Cargill, and Hormel Foods sold their catfish processing operations to other processing companies. Some of the major catfish processing operations were owned by groups of catfish farmers, farm families, and individuals. The fairly large number of firms ensures the independence of companies without the possibility of collusion to restrict quantity in order to boost price.

In 1981, a study of the catfish processing sector concluded that, structurally, the industry was characterized by a high degree of market concentration with 5 of the 14 processing firms reporting to USDA at the time handling 98% of the total pounds of live weight fish processed (Miller et al. 1981). Miller et al. (1981) also observed that there was a high degree of mutual interdependence among the processing companies in their pricing and other business policies as well as excess processing capacity in the sector.

Dillard (1995) suggested that the structure of the catfish processing sector fell somewhere between oligopoly and monopolistic competition, but perhaps more towards oligopoly because catfish processors were mutually interdependent, that is, each processor recognizes that its output and pricing decisions influences its rival's decisions, and vice versa. However, Dillard was quick to add that the catfish processing sector does not strictly conform to all the characteristics of an oligopoly. Dillard (1995) also suggested that there were no short-run economic profits accruing to the catfish processing industry and estimated the average processing cost for catfish to be $5.31 per kilo in 1994. However, Dean and Hanson (2003) estimated preliminary average cost of catfish processing in Mississippi to be $1.84 per kilo in 2002. The difference in cost was probably due to improved processing technology, but cost of production largely depended on the product mix in the production process. Figure 6.6 is an illustration of a breakdown of approximate yields and product mix of various catfish product forms based on the conversion of 10,000 pounds of live catfish to processed products (Silva and Dean 2001). Further processing of catfish from the whole fish product results in lower yields and more waste and increased cost per kg of marketable product. The product mix varies with processors and depends largely on the processor's marketing strategy and customer demands.

Catfish processors cannot be considered as mutually interdependent, because each processor's output and pricing decisions are independent of the other processors. Processors have limited control over the prices of their products given that products are not sold directly to consumers but to individual and chain retail grocery store outlets, food service distributors and brokers, as well as to individual and chain restaurants. Some processors have received higher prices by servicing niche markets and providing special customer services. In general, however, the price and output results of the catfish processing sector could be similar to those of pure competition because of the intense competition among processors to supply processed products, mainly fillets, to the food service sector. Furthermore, the highly elastic nature of the demand curve for individual catfish processing companies suggests that pricing and quantity results are near pure competition. Catfish processors commonly lose customers to one another.

The trends in processing capacity and sales of catfish fillets over the past decade have been affected by trade competition. Specifically, Vietnam became the largest exporter of frozen catfish fillets to the U.S. beginning in 1998. Consequently, the market share of U.S. farm-raised catfish fillets peaked in 1997, with the introduction of catfish fillets from Vietnam to the U.S. market. Thereafter, the market continued to decline in the share of domestic fillets along with processor and producer prices. The declining market share and the associated price decline of catfish prompted various actions from the industry that resulted in the imposition of tariffs on the catfish species imported from Vietnam.

On the consumption side, however, per capita demand for catfish increased from about 0.39 kg in 1995 to about 0.52 kg in 2003, an increase of 64% and at a rate of 5% per year (Fig. 6.7). The increase in catfish consumption during

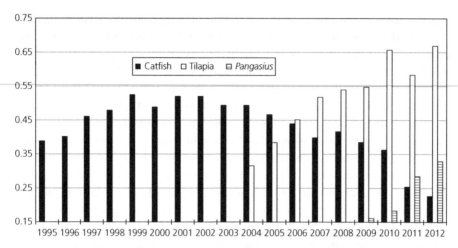

Fig. 6.7 U.S. annual per capita consumption of catfish, tilapia, and *Pangasius*. Source: National Fisheries Institute: www.aboutseafood.com/top-ten-seafood/.

that period could be attributed partly to the general increase in fish consumption, intensive marketing efforts within the industry, and changes in consumption patterns with respect to demand for new fish products. The southern region of the U.S., which includes Oklahoma, Texas, Arkansas, Kentucky, Tennessee, Mississippi, and Alabama, is the traditional market area for catfish. However, per capita consumption has consistently declined to about 0.23 kg in 2012, which industry observers have attributed to substitution by other white fish such as tilapia and *Pangasius* (Fig. 6.7). Catfish ranked among the top five seafoods consumed in the U.S. from the 1990s through the mid-2000s. In 2012, catfish ranked ninth as the most consumed seafood in the U.S. while tilapia and *Pangasius* ranked fourth and sixth, respectively (National Fisheries Institute 2014).

Summary

Processing in the seafood and aquaculture industry encompasses all the steps that food goes through from the time of harvest to the time seafood and aquaculture products reach the consumer. The structure of the seafood and aquaculture products industry relates to the concentration of the industry, the degree of vertical integration, product characteristics, and freedom of entry and exit. The industry is typical of a monopolistic competitive industry, with a relatively large number of firms operating competitively in the production of differentiated products. The U.S. Census Bureau reports the concentration ratios for the 4, 8, 20, and 50 largest fresh and frozen seafood and aquaculture products processing companies in 2007 as 32%, 44%, 61%, and 78%, respectively, compared to the 2002 ratios of 19%, 28%, 45%, and 69%. The trend indicates an increasing concentration in the seafood processing sector similar to trends that can be seen in the animal (except poultry) packing industry in which the concentration ratios in 2007 for the 4, 8, 20, and 50 largest companies were 43%, 64%, 80%, and 89%, respectively. The seafood and aquaculture product processing industry is much less concentrated than other meat processing industries. Also, the rate of capacity utilization has averaged 70% since 2003 compared to 83% for the beef, pork, and poultry processing industries. This indicates relative underutilization of processing capacity. Some of the challenges confronting the aquaculture processing industry include realizing the potential growth in the market, since the U.S. imports about 90% of its seafood needs, and producing healthy, high-quality and convenient seafood products in a cost-effective way.

Study and discussion questions

1 What is the importance of seafood processing in the food system?
2 List two factors that drive seafood and aquaculture product processing.
3 What is value-added processing?
4 What factors characterize the structure of an industry? What is the importance of each of the factors?
5 Explain what "concentration" of an industry means.
6 What is the most concentrated food industry in the U.S.?
7 Distinguish between a competitive market and a monopolistic market.
8 What are antitrust laws?
9 What is the difference between vertical integration and horizontal integration?
10 Give two advantages of vertical integration.
11 What is production capacity utilization rate?
12 What are some of the challenges facing the U.S. seafood and aquaculture processing industry? Suggest various ways by which the industry can address these challenges.

References

Bykowski, P. and D. Dutkiewicz. 1996. Freshwater Fish Processing and Equipment in Small Plants. Food and Agriculture Organization Fisheries Circular No. 905. FAO, Rome.
Dean, S. and T. Hanson. 2003. Economic Impact of the Mississippi Farm-Raised Catfish Industry at the Year 2003. Mississippi State University Extension Service, Mississippi Agricultural and Forestry Experimental Station, Publication 2317, Mississippi State University, MS.

Dillard, J.D. 1995. Organization of the Catfish Industry: A Comment on Market Structure-Conduct-Performance. Paper presented at the Catfish Farmers of America Annual Meeting, Memphis, Tennessee, February 16.

Duchene, L. 2005. Sales for top 10 reach $7.75 billion in 2004. *SeaFood Business* 24(5).

EUROFISH Magazine. 2001. The fish processing industry in Denmark – Europe's major fish exporter imports raw materials. January/February 2001. Accessed October 22, 2004. www.eurofish.dk/indexSub.php?id=570&easysitestatid=-1801202820.

European Parliament. 2005. Policy Department Structural and Cohesion Policies. Fisheries in Poland. Report IPOL/B/PECH/N/2005_03. PE 355.363. Available at www.europarl.europa.eu/RegData/etudes/note/join/2005/355363/IPOL-PECH_NT(2005)355363_EN.pdf.

Fishing Industry Renewal Board (FIRB), Government of Canada. 1997. A Policy Framework for Fish Processing. FIRB final report, January 3, 1997.

Hackett, S.C. and M.J. Krachey. 2002. California's Wetfish Industry Complex: An Economic Overview. Humboldt State University working paper, Humboldt State University, Arcata, California. users.humboldt.edu/hackett/wetfishpaper.htm. Accessed October 18, 2004.

Knapp, G., C. Wiese, J. Henzler, and P. Redmayne. 2001. A village fish plant: yes or no? – a planning handbook. Institute of Social and Economic Research, University of Alaska Anchorage.

Kolbe, E. and D. Kramer. 1997. Planning Seafood Cold Storage. Alaska Sea Grant College Program, University of Alaska, Fairbanks. Marine Advisory Bulletin, No. 46.

Kouka, P.J. 1995. An empirical model of pricing in the catfish industry. *Marine Resource Economics* 10:161–169.

Mermelstein, N.H. (2002). Processing in the largest U.S. seafood plant. *Food Technology* 56: 69.

Miller, J.S., J.R. Coner, and J.E. Waldrop. 1981. Survey of commercial catfish processors – structural and operational characteristics and procurement and marketing practices. Agricultural Economics Research Report No. 130. Mississippi State University, Mississippi.

National Fisheries Institute. 2014. Top 10 consumed seafoods. Available at www.aboutseafood.com/top-ten-seafood/. Accessed March 28, 2016.

National Oceanic and Atmospheric Administration, National Marine Fisheries Service (NOAA-NMFS). 2003. Fisheries of the United States – 2002. Office of Science and Technology, Fisheries Statistics Division, Silver Spring, Maryland.

National Oceanic and Atmospheric Administration, National Marine Fisheries Service (NOAA-NMFS). 2013. Fisheries of the United States – 2012. Office of Science and Technology, Fisheries Statistics Division, Silver Spring, Maryland.

Radtke, H. and S. Davis. 2000. Description of the U.S. west coast commercial fishing fleet and seafood processors. Pacific States Marine Fisheries Commission.

Silva, J.L. and S. Dean. 2001. Processed catfish – product forms, packaging, yields, and product mix. Southern Regional Aquaculture Center, Mississippi State University.

United States Department of Commerce (USDC). 2014. Aquaculture Policy. Available at www.nmfs.noaa.gov/aquaculture/docs/policy/doc_aq_policy_1999.pdf.

United States Department of Commerce, National Oceanic and Atmospheric Administration (USDC-NOAA). 2004. USDC Participants List for Firms, Facilities and Products, USDC-NOAA, Volume 19, No. 2, July 2004.

Unites States Department of Commerce, Census Bureau (USDC-CB). 2002. Concentration Ratios – 2002. 2002 Economic Census, Subject Series, EC0231SR1.

Unites States Department of Commerce, Census Bureau (USDC-CB). 2007. Concentration Ratios in Manufacturing: 2007. 2007 Economic Census, Subject Series, EC0731SR13.

Unites States Department of Commerce, Census Bureau (USDC-CB). 2013. Survey of Plant
    Capacity Utilization: 2014. Available at www.census.gov/manufacturing/capacity/historical_
    data/index.html. Accessed September 21, 2014.
Weise, N.J. 2004. Market characteristics of U.S. farm-raised catfish. Unpublished MS thesis
    dissertation, Aquaculture/Fisheries Center, University of Arkansas at Pine Bluff, Pine Bluff,
    Arkansas.
Wright, J. 2006. Trident-Ocean Beauty deal off the table. *SeaFood Business* 25(6).

CHAPTER 7

The international market for seafood and aquaculture products

The basis for trade

International trade is the exchange of goods and services between two countries. Trade can occur due to: (1) differences in technology between countries; (2) differences in resource endowments; (3) differences in consumer demand; (4) existence of economies of scale in production; or (5) existence of government policies (Suranovic 2010).

Trade is based on the benefits gained from specialized production and the relative advantage that the country has in the production of certain goods. For one nation to produce all the goods and services that its citizens desire would mean that some of the products would be produced less efficiently than if they were produced in other nations that have a particular advantage and have specialized in the production of those goods. Moreover, most countries do not have the climate or the resources needed to produce all the goods and services that their citizens might want.

The theory that has been developed as a basis for analyzing and understanding international trade shows that costs of production alone do not explain whether trade occurs or not. The lowest cost producers may not be competitive in international markets and some countries may still benefit from free trade if they are less efficient than other countries. In fact, less efficient companies can compete with foreign companies depending upon the relevant price ratios, the size of the countries involved, and the difference in domestic demand for the relevant products (Suranovic 2010).

If free trade conditions exist, countries import those goods for which the domestic supply in the country is less than its demand (Anderson 2003). A country will also look to export its excess supply. If price of the product on the world market is above the equilibrium price in that particular country (price that results in all the domestic quantity supplied clearing the market because

Seafood and Aquaculture Marketing Handbook, Second Edition. Carole R. Engle, Kwamena K. Quagrainie and Madan M. Dey.

it is being purchased by domestic buyers), then this "excess" supply would be exported. The importing country would import that quantity that corresponds to price levels on the world market that are below the importing country's equilibrium price. In other words, if the world market price is higher than the domestic country's market price, it will look to import those products for which the world market price is lower than the domestic market price. Thus, the excess supply from one country interacts with the excess demand curve of another and trade results if the price relationships are favorable. The volume of trade depends on the quantity–price relationships involved. (The reader may wish to review the demand, supply, and price determination sections in Chapter 2.) Changes in any of the factors that affect the excess supply of the exporting country and the excess demand of the importing country will result in changes in the price and quantity of the product traded.

Anderson (2003) demonstrated that seafood product supplies from open-access fisheries may or may not conform to traditional trade models, depending upon the shape of the fish supply curve from open-access fisheries. Seafood products supplied from aquaculture, however, will likely trade in a manner more similar to agricultural products due to property rights and well-understood production practices. Thus, products in seafood sectors that are becoming more dominated by aquaculture products are likelier to trade in a way that conforms to traditional trade models. On the other hand, aquaculture products in markets dominated by open-access fishery products may behave differently in international trade. In these cases, careful analysis and development of appropriate trade models will be necessary to understand the impacts on trade of various policy measures.

Dimensions of the international market

All countries engage in some sort of trade with other countries. Some countries, with more relaxed (laissez-faire) policies, engage in greater volumes of trade than do countries with more restrictive policies. However, as technologies of communication and travel have developed over time, there has been an overall increase in the volume of international trade (Fig. 7.1). The value of seafood exports has increased faster than the volume of exports.

International markets are more variable and difficult to predict than are domestic markets. The volume of seafood that is traded internationally is very large: $130 billion in 2011 (FAO 2014b). More than 200 countries traded over 800 species of fish in 2011. Moreover, the volume of trade in fish and seafood products has increased, particularly from 2004 to 2011 (Fig. 7.1; FAO 2014a).

Traded fish and seafood has played an important role throughout history. Kurlansky (1997) argued that the search for cod resulted in international trade that played a prominent role in both the exploration and development of the

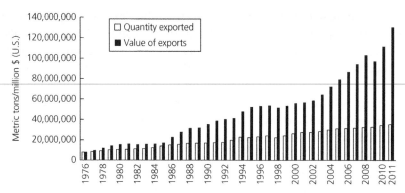

Fig. 7.1 Quantity and value of international exports of fish and seafood, 1976–2011. Source: FAO (2014b).

New World. Moreover, access to and control of cod fisheries around the world played an important role in the economies of a number of European nations. Conflicts over cod have resulted in wars and have influenced political strategy in several countries for several centuries.

Historically, international trade in seafood was enhanced by the establishment of exclusive economic zones (EEZs) from their inception in 1952. The EEZ is a 200-mile exclusive zone imposed by countries with coastal areas. Countries with important coastal fishery resources, such as Peru, Ecuador, and Chile, implemented EEZs in 1952 in reaction to foreign fishing vessels exploiting resources considered as belonging to their nation. The United States followed 37 other nations when it declared its EEZ in 1976. With the implementation of EEZs, countries such as Spain and Japan that had relied on distant-water fishing fleets for their domestic seafood supply were cut off from rich fishing areas. As a result Spain and Japan have become greater importers of seafood to meet their domestic demand.

Seafood products are among the most widely traded types of food categories (World Bank 2013). In countries such as Greenland, the Seychelles, and Vanuatu, fishery exports composed more than half the value of all traded commodities (FAO 2014a). Trade in fish products is particularly important in Asia, where fish products are the largest group of food products traded.

Overall, international trade in fisheries commodities increased by 8.3% per year from 1976 to 2012 (FAO 2014a). Approximately 37% (live weight equivalent) of world fish production was traded internationally in 2012.

Japan is the leading seafood-importing nation in the world, importing 14% of total world seafood production by value (Fig. 7.2). Its trade in seafood products is nearly twice as great as its trade in the all meats food group category, the category with the next highest value. The importance of Japan's volume of imports can be seen in statistics that show Asia, as a region, to be a net importing region. However, when Japanese statistics are excluded, Asia is a net exporter

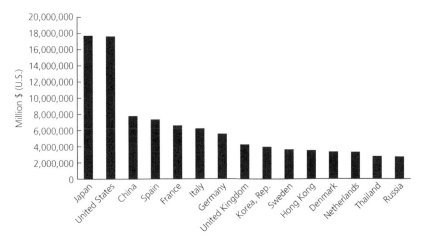

Fig. 7.2 Leading importers of fish and seafood products worldwide, 2012. Source: FAO (2014b).

with a large seafood trade surplus (World Bank 2013). Japan is followed by the U.S. in terms of the value of seafood imported. The value of trade in fish products is also important to the U.S. After the U.S., the following were the next most important countries: China, Spain, France, Italy, Germany, the United Kingdom, Republic of Korea, Sweden, Hong Kong, Denmark, The Netherlands, Thailand, and Russia. Japan, the U.S., and several European countries (Spain, France, Italy, Germany, and the United Kingdom) accounted for 50% of the world's imports of fishery products (FAO 2014b), but this percentage has declined from the 80% reported by Vannuccini (2003). Countries such as China, Republic of Korea, Sweden, Thailand, and Russia have proportionately increased their share of imports over time (Fig. 7.2).

The world's leading exporter of seafood is China, exporting 13% of the world's seafood in 2011 (Fig. 7.3). China is followed in descending order by Norway, Thailand, Vietnam, the U.S., Chile, Denmark, Spain, Canada, The Netherlands, India, Indonesia, and Russia. The leading exporting nations have changed over time. In 1976, Japan was the leading seafood exporter (in terms of value), followed by Norway, Canada, Denmark, and Taiwan. Only Norway and Canada have continued in the top 15 seafood-exporting countries over time.

Asia emerged as the most important international trading partner for North America (52%) and the European Union (63%) in 2012 (FAO 2014b). For the U.S., the second most important trading partner was Canada (25%), followed by South America (13%), while Australia was the second-leading trading partner for the European Union. The major trading partners for Asia were within the region (50%), 18% from the European Union, 12% from the U.S., and 11% from South America. South America mostly traded within the region (62%), with 15% from the European Union and 16% from Asia.

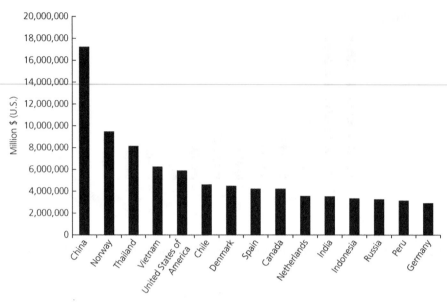

Fig. 7.3 Leading exporters of fish and seafood products worldwide. Source: FAO (2014b).

The seafood export trade is particularly important for developing countries, while the developed world is heavily dependent on imported seafood (World Bank 2013). Most seafood trade flows from lesser developed countries to more developed countries. Developing countries accounted for 53% of world fish exports in 2011 (FAO 2014a). Revenues from fisheries are crucial for many developing countries; fishery products are a major source of foreign exchange for a number of countries. Net export revenues (value of exports minus value of imports) of developing countries reached $35.3 billion in 2012, exceeding those of other major agricultural commodities such as coffee, rubber, cocoa, and sugar (FAO 2014a). Trade among developing countries has increased, reaching 33% of the value of seafood exports in 2012 (FAO 2014a). Most of the target markets for exports from developing countries are in Japan, the U.S., and the European Union, although China has emerged as the third largest seafood-importing country (Fig. 7.2). This has been particularly true in Asia, Central and South America, and, to a lesser extent, Eastern and Central Europe.

Shrimp is the major fishery commodity traded internationally and it accounted for about 7% of the total value of internationally exported fishery products in 2011 (Fig. 7.4). Other main types of species exported in 2011 were salmon, tunas, catfish, and mollusks. Fishmeal represented 3.1% of the value of exports, and fish oil was 1% of the total value of fishery exports. Export trade from developing countries is gradually changing from raw products to value-added products.

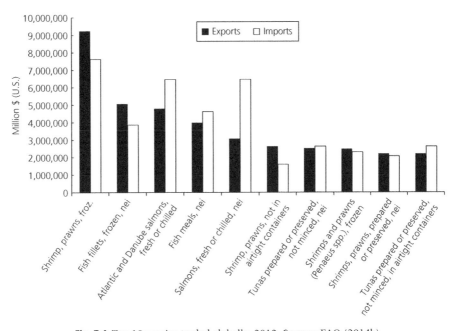

Fig. 7.4 Top 10 species traded globally, 2012. Source: FAO (2014b).

Much of the growth in the international trade in fish and seafood has come from aquaculture, primarily shrimp, salmon, trout, tilapia, oysters, and carp. The entry of aquaculture products into the international market has provided a mechanism for countries to become major suppliers internationally. Thailand (shrimp) and Norway and Chile (salmon) are good examples of countries that had only low levels of seafood exports until significant aquaculture industries developed there. Ecuador and Honduras have also become major traders in both shrimp and tilapia worldwide as these farm-raised industries developed.

Re-trade in seafood and aquaculture products is also growing. China, for example, imports large quantities of raw product to process, re-package, and re-export. Roe herring from Alaska is imported into China, extracted, processed, and exported to Japan (Anderson 2003). China also imports frozen cod, salmon, and other species, and then exports frozen fish fillets and value-added products. Re-export has also been used to circumvent trade restrictions.

Trade policy tools

Politically, most citizens expect their government to demonstrate responsibility toward producer groups that are hurt by competition from imported products. The Trade Reform Act of 1974 (in the U.S.) provides for adjustment assistance to companies and workers depending on whether the circumstances meet the specific program requirements. Trade policies can be considered as either "beggar-thy-neighbor" or as a "strategic trade policy." Beggar-thy-neighbor policies benefit one country by forcing losses on its trading partners. Strategic trade polices shift profit away from international competitors or consumers.

Countries may prioritize maintaining a self-sufficient food supply to avoid dependence on another country in case of war or other threats. Some countries use trade barriers to maintain employment at home. These barriers are implemented through control of foreign exchange, lending and borrowing, licensing, or use of state trading agencies. Barriers have been used to attempt to stabilize prices by developing a more consistent supply. U.S. import quotas on beef stimulate greater imports when U.S. slaughter quantities are lower and reduce imports when U.S. slaughter numbers are higher. New industries tend to have greater costs initially, and protection may enable the industry to grow until it is large enough to compete effectively.

Tariffs on imported products have been used to support farm prices and incomes, while exporting countries may use farm support prices, export subsidies, or tariffs. Tariffs can be levied as either a specific tariff or an ad valorem tariff (tax based on the value of the product). Specific tariffs are based on establishing a certain cost per unit of the good imported. For example, tariffs of $0.033/kg are charged in the U.S. for grapefruit imported from August to September from countries without most favored nation status. In the U.S., the U.S. Customs Service administers and collects the tariffs.

Tariffs often hamper development of the aquaculture sector in developing countries because they protect final consumer product industries in importing countries (Anrooy 2003). Moreover, tariffs discourage processing industries in exporting countries. Import tariffs still exist for aquatic products in many countries, including Taiwan, Vietnam, the European Union, the U.S., and China.

Quotas restrict the quantity or volume that can be imported. Quotas can be either absolute values or can be based on a quota tariff rate. Limitations on the quantity imported during a specified period of time can be established to affect all imports or only those from certain countries. However, quotas can also be set to import a certain quantity either at a reduced rate or a tariff rate quota. In the U.S., quotas are generally (with some exceptions) administered by the U.S. Customs Service.

Voluntary export restraints (VERs) regulate the volume of a good to be exported. VERs typically result from pressure from the importing country. With the emergence of the General Agreement on Tariffs and Trade (GATT), VERs emerged as a form of protectionism that did not violate GATT agreements. Suranovic (2010) provides examples of the U.S.-Japan automobile and textile VERs and the effects of implementing these restrictions.

There are also a number of non-tariff barriers that have been used to protect domestic industries. These include: (1) government participation in trade, including production subsidies; (2) customs that make it difficult to import certain products; (3) industrial, health, and safety standards that may include packaging and labeling regulations; (4) embargoes, bilateral agreements, and voluntary restraints; and (5) special duties and credit restrictions (Rhodes 1993).

Export taxes and subsidies often are used to collect duties to generate revenue for governments. For example, Malaysia, Kenya, Mali, Norway, Vietnam, and the Dominican Republic tax exports of aquaculture products (Anrooy 2003). Export taxes reduce the price received by producers as a consequence. However, if taxes are used for promotional campaigns, there may be some benefit returned to the aquaculture sector. Export subsidies benefit primarily the exporting companies. If vertically integrated, primary production sectors would also benefit from export subsidies.

Most major fish-exporting nations have established councils to promote their export products. These councils are typically supported by public funds to enhance sales and position products in world markets. The Vietnam Association of Seafood Exporters and Producers, the Norwegian Seafood Export Council, Dutchfish, SalmonChile, and the Scottish Salmon Board are examples of export promotion councils for seafood.

Export processing zones (EPZs) have been developed in a number of countries as industrial zones to encourage foreign investment in processing (Anrooy 2003). Incentives such as free trade zones, financial services zones, free ports, duty free imports, good infrastructure, easy market access, and others are used to attract investors to the EPZs. Companies that process aquaculture products have taken advantage of EPZs, and a number of countries have provided infrastructure to accommodate processing facilities for aquaculture products.

Exchange rate policies can affect international trade. Exchange rates determine the price received by exporters in local currency and the prices paid by importers, also in local currency. Devaluation of a country's currency results in reducing its export price and often increases overall production for export. Dey and Bimbao (1998) recommended currency devaluation for Bangladesh, the Philippines, and Thailand to improve export competitiveness. However, devaluation raises prices of imports and so can have negative effects on sectors that are heavily dependent upon them. High breakeven prices for Indonesia freshwater aquaculture products were reported to have resulted from devaluation of the Indonesian Rupee (Dey et al. 2001).

Trade policy in seafood and aquaculture

The General Agreement on Tariffs and Trade (GATT)
International trade occurred for many years within the framework of GATT. GATT grew out of the Bretton Woods Agreement and was organized in 1948 with more than 90 countries. GATT rules defined export subsidies and included commitments from countries to reduce export subsidies with the goal of reducing or eliminating tariff and non-tariff barriers to trade. An export was said to be

subsidized when the export price was lower than the comparable price charged for similar products in the domestic market.

GATT was traditionally concerned with trade measures and not domestic production policies unless trade was involved. The most recent set of tariffs was implemented after the Uruguay Round of GATT negotiations. The Uruguay Round of multilateral trade negotiations was begun in 1987 (FAO 2003). It was the first round of trade negotiations in which developing countries were directly involved. The Uruguay Round Agreement on Agriculture overruled the provisions of GATT. It continued to allow export subsidies on agricultural products, but constraints were imposed. Early data on export subsidy use under the Uruguay Round Agreement on Agriculture indicated that export subsidies for some products were small (especially as compared to a product such as wheat) and were allowed, although utilization rates for dairy products and various meats were quite high. Moreover, export subsidies on agricultural products were permitted, but were subject to the potential for antidumping and countervailing duties. Overall, the Uruguay Round resulted in lower import duties for many products, including fish and fishery products.

In addition, the U.S. has signed free trade agreements with Canada, Mexico, and Israel. GATT also provided for the Generalized System of Preferences (GSP) for many less developed countries. The GSP is a framework under which developed countries give preferential treatment to manufactured goods imported from certain developing countries. Following the Uruguay Round, the trade-weighted tariff on industrial products fell by 40% to 3.8% in 2000, while the average tariff decreased by 37% in developing countries (Ariff 2004). However, non-tariff barriers began to increase over time.

The World Trade Organization (WTO)

The WTO emerged from the Uruguay Round negotiations and the Marrakech Agreement in 1995 with its membership increasing to 160 in 2014. The WTO is a binding treaty that implements the GATT articles in support of free trade and has greater enforcement authority than did the GATT. The most recent round of trade negotiations is referred to as the Doha Development Round, or Doha Agenda, that began in 2001. The Doha Agenda includes issues such as improved access to markets for fish and fishery products, fisheries subsidies, environmental labeling, and the relationship between WTO trade rules and environmental agreements. One Doha proposal is to eliminate all import duties on fish. Russia and Vietnam were the only two major fisheries countries that did not belong to WTO in 2004, but both countries have since gained WTO membership.

Major provisions of the WTO agreement include the: (1) Agreement on Application of Sanitary and Phytosanitary (SPS) Measures; (2) Agreement on Technical Barriers to Trade (TBT); (3) Agreement on Subsidies and Countervailing Measures; (4) Anti-dumping Agreement; (5) Agreement on Safeguards; and (6) WTO Dispute Settlement Procedures. Membership is required for dispute settlement, but most cases are settled out of court. The WTO sets defined stages with set time limits during disputes. For example, there have been a number of international trade cases involving aquaculture that have been filed under one of the above provisions.

Sanitary and phytosanitary (SPS) measures focus on food safety protection for humans, animals, and plant life and from spread of diseases, contaminants, and toxins. The WTO recognizes that countries have a right to develop regulations that protect human, animal, and plant life. For example, import regulations that require hazard analysis of critical control point (HACCP) plans are considered SPS applications. SPS-related trade disputes have included: (1) a U.S. case against imported salmonids from Australia; and (2) rejection of Malaysian consignments of prawns and frozen seafood by the European Commission due to bacteria counts (Ariff 2004). The European Union has set a zero bacterial count instead of the minimum acceptable level defined in international standards. A study of compliance with EU food safety regulations (Palin et al. 2013) found weaknesses at the primary production level, with residues of veterinary medicinal products and contaminants. Notifications of food safety issues were greatest from Asia (1800), followed by Europe (1200). Fish and fish products were the third most reported commodity category. Border rejections from Vietnam were not significantly greater than those from other countries, but most concerned *Pangasius* fillets and included trifluralin, chlorpyrifos and the prohibited nitrofuran and nitrofurazone, *Listeria*, and *Salmonella*.

Technical barriers to trade deal primarily with labeling and testing disputes. For example, in 2003, The Netherlands discovered two shipments of salmon contaminated with malachite green. This substance has been banned in Chile since 1997. Shipments of Chilean salmon to Japan were also detained in 2003. An audit by the European Union found that Vietnam, Ecuador, and Chile had published dates for official sampling which allowed farmers to evade residue detection (Palin et al. 2013). The audit also found that the testing conducted did not include all required contaminants. In a study of U.S. seafood import refusals, Anders and Westra (2011) found the greatest percentages of refusals to be from Vietnam (17%), followed by Indonesia (13%) and China (12%). Detections and refusals appeared to be greater for consignments from economies with lower levels of development.

The subsidies and countervailing measures agreement sets procedures for determining whether countries subsidize their exports. Rules for fish products typically are more stringent than those for agriculture products.

The WTO antidumping provisions have been used several times with regard to aquaculture and seafood products and have been widely publicized. Recent antidumping cases are discussed later in this chapter, and Appendix 7A presents details of the process for antidumping lawsuits in the U.S.

Transshipment (unloading fishery products to another vessel) is a growing concern. The EU audit found "wide disparities and inconsistencies" with products in transit, with transshipments, and with indirect imports. No global system exists to prevent this (Palin et al. 2013).

U.S. Antidumping

In the U.S., countervailing and antidumping duties were authorized by the Tariff Act of 1930. Countervailing duties are used when imports receive an unfair subsidy from the foreign government (King and Anderson 2003). Antidumping duties are used when imports are sold or likely to be sold for less than fair value. The Trade Act of 1974 authorized an additional measure known as safeguard remedies. Safeguard remedies are used when increasing volumes of imports threaten to injure an industry in the U.S. Special measures can be taken if the products imported are being "dumped," which refers to selling product at prices lower than those in the home, or domestic, market. The Uruguay Round of GATT allowed countries to impose antidumping tariffs on products if it was determined that dumping had occurred.

The U.S. antidumping law (U.S. Department of Commerce 2004) is designed to provide relief to U.S. industries that are injured as a result of foreign goods being sold at unfairly low prices in the U.S. Antidumping investigations are conducted in two phases: one by the United States Department of Commerce (USDOC) and the other by the United States International Trade Commission (US-ITC). The USDOC determines whether the imports in question have been sold at less than fair value in the U.S. The US-ITC determines whether the imports in question are causing or threatening to cause material injury to a U.S. industry. An antidumping order is issued if both agencies reach positive determinations. Appendix 7A provides additional detail on the process and procedures of antidumping lawsuits in the U.S.

The greatest number of antidumping and countervailing duties listed in 2014 by the U.S. International Trade Commission were from China (121), followed by India (22), Taiwan (18), Korea (15), Japan (13), Brazil (10), Indonesia (10), Mexico (9), Vietnam (8), Thailand (7), and Italy (7) (US-ITC 2014).

Byrd Amendment, Continued Dumping and Subsidy Offset Act of 2000

The Byrd Amendment to U.S. antidumping law, known as the Continued Dumping and Subsidy Offset Act of 2000, provided for the distribution of revenue from antidumping tariffs imposed on foreign firms to the domestic firms that filed the dumping complaint (Collie and Vandenbussche 2004). It was justified by the expectation that it would lead to lower duties and greater welfare when compared to tax revenues if the weight on the profits of the domestic industry was sufficiently large. However, in response to complaints from the European Union and other countries in 2003, the WTO found that the Byrd amendment was inconsistent with the GATT antidumping agreement. In 2004, the European Union was approved to impose sanctions on goods from the U.S. in response to the Byrd Amendment. The Byrd Amendment was repealed by the U.S. Congress in 2005.

Salmon trade conflicts

United States and Norway

Pen-raised salmon aquaculture technologies were developed on a large commercial scale, primarily in Norway, in the 1980s. During this same time period, the U.S. wild-caught salmon industry began to divert a larger portion of its production from canned products to the fresh/frozen salmon market in Japan (Anderson 1994). As pen-raised salmon aquaculture production grew in Norway, the volume of Norwegian exports to the U.S. increased rapidly.

Salmon prices declined in 1990, largely due to the increased supplies from aquaculture, and led to an antidumping petition from the Coalition for Fair Atlantic Salmon Trade (U.S.). The petition alleged that Norwegian producers were dumping salmon in the U.S., materially damaging the domestic industry. The U.S. International Trade Commission ruled on February 25, 1991, that the Norwegians were selling below fair market value. A countervailing duty of 2.27% and antidumping duties ranging from 15.65% to 31.81% (depending upon the company) were imposed. The magnitude of these duties caused Norway to be uncompetitive in the U.S. market. By March, 1991, Norway's share of imports had sunk to less than 5% (Fig. 7.4). As Norwegian imports into the U.S. market fell in 1991, Canadian and Chilean imports began to increase.

Norwegian imports into the U.S. increased in the 2000s and briefly exceeded those of Chile in 2010. The primary reason was a major disease outbreak in the Chilean salmon industry. The Chilean industry subsequently recovered and Chilean imports have again become the most important salmon product in the U.S. market from 2012 to 2014.

United States and Chile

Another antidumping petition was filed in 1997 by the Coalition for Fair Atlantic Salmon Trade in the U.S. against Chilean salmon exporters. The USDOC-ITC investigation found insufficient evidence of government subsidies to support imposing countervailing duties on fresh Atlantic salmon from Chile (Asche and Bjorndal 2011). On June 2, 1998, the USDOC-ITC investigation determined that two of Chile's largest Atlantic salmon producers had traded salmon fairly

(Asche and Bjorndal 2011). Three other companies were found to have sold salmon at less than fair market value, and antidumping duties were set at 8.27%, 10.91%, and 2.24% with all others charged at 5.19%. The petition was later suspended by producers from the state of Maine (Sloop 2003).

USTR (2014) reported that Sernapesca (Chile's Ministry of Fisheries) had suspended imports of salmonid eggs from the U.S. several times beginning in 2010. The suspensions were contingent upon additional risk analyses, audits of U.S. oversight (by the Animal and Plant Health Inspection Service – APHIS), and the addition of totivirus that was not listed by the World Organisation for Animal Health (OIE) even though no specific health concern had been identified for U.S. products. The same report showed that the EU, through the Commission's Directorate General for Health and Consumers, began to ban imports of shellfish (other than scallops) from the U.S. in 2010, following the expiration of the United States-European Community Veterinary Equivalence Agreement in 2009. By 2014, imports of shellfish from the U.S. were still not allowed.

European Union and Norway
There have been several veterinary control issues at EU borders with Norwegian salmon. Exports of salmon from Norway to Russia were suspended due to concerns over food safety. The EU Commission has set minimum import prices for Atlantic salmon that particularly affected Norwegian salmon.

In 1989, 1991, and 1996, Scottish and Irish farmers filed dumping complaints against Norway (Asche and Bjorndal 2011). The Norwegian salmon industry voluntarily restrained production with feed quotas (Asche and Bjorndal 2011). The "Salmon Agreement" of 1997–2003 between the EU and Norway formalized production control, but the agreement was terminated in 2003. A 2004 complaint, again by Scottish and Irish farmers, led to antidumping findings in 2005 for a short period of time. A followup complaint by Norway in 2007 to the WTO led to the termination of measures against Norwegian salmon.

Blue crab conflict

Imports of swimming blue crab (*Callinectes sapidus; Charybdis hellerii; Portunus pelagicus*) meat in the U.S. increased in the 1990s, with the greatest concern on the part of the U.S. blue crab industry being the increase in volumes imported from Venezuela, Indonesia, Thailand, and Mexico. A number of domestic blue crab processors went out of business over this same time period. However, unlike salmon growers, the U.S. blue crab industry could not raise the funds to file an antidumping petition. In 2000, they filed a Section 201 petition for safeguard remedies instead that required only a surge in imports thought to cause material injury and did not require proof of unfair trade practices. The US-ITC found no threat to U.S. suppliers, citing evidence that increased imports did not result in idle plant capacity or greater underemployment.

U.S. crawfish and China

For many years, the U.S. freshwater crawfish industry had experienced limited competition from imported supplies. However, in 1994, China captured 58% of the market share of crawfish tail meat in that one year. Within three years, China's market share had increased to 87% (U.S. International Trade Commission 1997). The impacts to the Louisiana industry were substantial for a number of reasons. First, the market was a small geographic market centered in southern Louisiana. Second, the Cajun French culture of southern Louisiana was linked to crawfish as a food (Roberts 2000). Thus, the U.S. crawfish market was highly localized and consisted of numerous small businesses that produced undifferentiated tail meat. It was also seasonal and operated only in the first half of the year. Domestic supply fluctuated during this time period and was affected substantially by fluctuations in the wild catch from the Atchafalaya Basin. The combination of these factors created marketing opportunities during times of low supply, and its market characteristics made it a relatively easy target.

An antidumping petition was filed with the US-ITC in 1996 (Roberts 2000). The US-ITC found that the U.S. crawfish industry was being materially injured from crawfish tail meat imports from China being sold at less than fair value. Company specific antidumping duties ranging from 92% to 123% were published in 1997. Other companies that initiated shipments following the period of investigation were assessed a tariff of 201%.

Since China was deemed to be a non-market economy, a normal price could not be calculated and surrogate countries were selected (see Appendix 7.A for more details on procedures of antidumping lawsuits). Spain's imported price of live crawfish from Portugal was used as the surrogate country farm price. India was selected as the processing surrogate country because it had a market economy with a large seafood processing industry that utilized hand labor. U.S. importers and representatives from China challenged this decision (unsuccessfully) because the ungraded Spanish imports commanded a higher price than graded U.S. crawfish.

The ruling that resulted applied only to China. Since U.S. imports from other countries were free of the 123% import duty, China attempted to avoid the ruling by repackaging Chinese tail meat in Singapore. However, the Singapore company did not meet the substantial transformation test used by the U.S. Customs Service to determine country of origin, and duties were levied subsequently on the shipments from Singapore. A similar increase of imports from Spain during 1999–2000 was identified. As of 2014, an antidumping order was still in place for crawfish tail meat from China.

U.S. catfish and Vietnamese basa

The introduction of basa and tra (*Pangasius* spp.) from Vietnam as lower-priced alternatives to U.S. farm-raised production of channel catfish (*Ictalurus punctatus*) contributed to a severe and protracted downturn in U.S. catfish prices. The quantities of imports from Vietnam increased rapidly from 2000 to 2001 and reached 15% of total frozen fillets in just two years (Fig. 7.5). The Vietnamese imports very quickly captured a noticeable portion of the most profitable and fastest-growing segment (smaller frozen fillets) of the U.S. catfish market.

Within the same time period, catfish prices declined by over 30%. Since fillets account for about 60% of the total volume of processed catfish sold, the impact of imports was considered an important factor contributing to lower price levels and price instability.

Quagrainie and Engle (2002) found that the market for domestic frozen fillets played a significant role in the price determination of imported catfish. Thus, once the potential for competition had been established in the U.S. farm-raised catfish market, periods of higher prices were countered by increased supplies of imported product. This effect is particularly important during times when the U.S. dollar is strong. Ligeon et al. (1996) also concluded that the quantity of catfish imported into the U.S. will decline if the domestic price of catfish falls relative to the import price. These studies implied that, if the industry expects to see higher catfish prices, production and supply control strategies may be needed. Efforts to require labeling of Vietnamese fish fillets and strict inspections of imported fillets may help to reduce the quantity of imported product into the U.S.

The USDOC placed an antidumping order against imports of Vietnamese frozen basa and tra fillets in August 2004. Tariffs ranged from 36.84% to 63.88%. The average pond bank price also increased from $1.21/kg ($0.55/lb) to $1.52/kg ($0.69/lb) in July 2004. Under antidumping law, the exporters were entitled

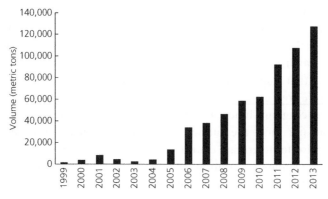

Fig. 7.5 Quantities imported of frozen siluriformes, 1990–2013. Source: ERS (2014).

to "administrative reviews" of the tariffs that focused on the duty rate applied. Moreover, one importer requested that the USDOC rule that live basa and tra from Vietnam processed into frozen fillets in Cambodia not be covered by the antidumping order. U.S. industry requested that this be ruled a "circumvention" of the antidumping duty order and that the Cambodian-processed fillets be covered by the antidumping order (Warren 2004). As of 2014, the antidumping order was still in place for frozen fish fillets from Vietnam.

Mussel conflicts

Great Eastern Mussel Farms of Tenants Harbor, Maine, filed an antidumping petition in January 2001 against mussel producers in Prince Edward Island (PEI), Canada. In October 2001, the USDOC assessed preliminary tariffs on two of the four PEI producers named in the antidumping petition at 4.7% dumping margin on one producer and 3.48% on the other. Shortly thereafter, PEI mussel producers increased prices twice, resulting in Great Eastern Mussel Farms withdrawing its antidumping petition against PEI mussel producers, and the US-ITC and USDOC terminated the suit.

Shrimp conflicts

In 2003, a shrimp antidumping petition was filed from a different perspective from those filed previously in the U.S. U.S. shrimp fishermen and processors specializing in wild-caught shrimp filed a dumping petition against importers who were purchasing farm-raised shrimp from China, Vietnam, Ecuador, Brazil, Thailand, and India. In July 2004, the USDOC imposed preliminary duties ranging from 7.67% to 112.8% on shrimp from China and Vietnam. The preliminary ruling required importers to post cash deposits or bonds equal to the preliminary dumping margins. The countries named in the suit lost market share in the U.S. that was captured by Mexico, Indonesia, Venezuela, Honduras, Guyana, and Bangladesh. As of 2014, antidumping duties were in place for frozen warmwater shrimp and prawns from Brazil, China, India, Thailand, and Vietnam.

The Convention on International Trade in Endangered Species (CITES)

The Convention on International Trade in Endangered Species (CITES) was adopted in 1973 with the goal of protecting species threatened by international trade. CITES listed over 30,000 species in 2001 and was supported by 154 member countries. CITES plays a more important role in capture fisheries than in aquaculture.

Aquaculture market synopsis: ornamental fish

Sri Lanka is credited with starting the collection and export of tropical marine fish in the 1930s (Wabnitz et al. 2004). The trade grew during the 1950s as other countries began to collect fish for export. Lewbart et al. (1999) estimated that, in 1998, 1.5 to 2 million people worldwide kept marine aquaria and the trade in marine ornamentals was valued at $200 to 300 million/yr (Chapman et al. 1997; Larkin and Degner 2001). Overall estimates of the international trade in ornamental fish in 2007 were that there were more than 4000 freshwater and 1400 marine species traded per year from 100 countries (Whittington and Chong 2007), while Rhyne et al. (2012) reported that more than 1802 species of marine fish from 125 families were imported into the U.S. as aquarium fish in 2004–2005.

Ornamental fish are traded in high dollar amounts worldwide, but the U.S. and Europe are the largest markets for aquarium fish (Conroy 1975; Hemley 1984; Andrews 1990; Basleer 1994; Chapman et al. 1997). The United States Coral Reef Task Force (2000) estimated that 50% of the marine fish traded as ornamentals are exported to the U.S. Japan, Australia, and South Africa also import measurable quantities of ornamental fish (Whittington and Chong 2007; Wood 2007). In Europe, Germany is the primary importer (22.5%), followed by the United Kingdom (18%), France (15%), The Netherlands (10%), Italy (8%), Spain (6%), and Belgium (5%) (OATA 1998–2004). Taiwan, Hong Kong, and Japan are also important markets for ornamental fish. The Philippines, Indonesia, the Solomon Islands, Sri Lanka, Australia, Fiji, the Maldives, and Palau provided 98% of the marine fish exported from 1997 to 2002 (Shuman et al. 2004; Wabnitz et al. 2004).

In terms of the total volume of imports into the U.S., however, 96% were freshwater (80% in terms of value) (Chapman et al. 1997). Of all the species, the guppy (*Poecilia reticulata*) and the neon tetra (*Paracheirodon innesi*) were the most popular. Guppies and neon tetra, along with the platy (*Xiphophorus maculatus*), the betta (*Betta splendens*), Chinese algae-eater (*Gyrinocheilus aymonieri*), and goldfish (*Carassius auratus*), composed half of the total number of ornamental fish imported. The blue-green damselfish (*Chromis viridis*) with 8.8% of imports into the U.S. and the sapphire devil (*Chrysiptera cyanea*) with 6.9% of imports were the top two marine species imported into the U.S. (Rhyne et al. 2012). Other popular species included the clown anemonefish (*Amphiprion ocellaris*), the whitetail dascyllus (*Dascyllus aruanus*), and the threespot dascyllus (*Dascyllus trimaculatus*) (Wabnitz et al. 2004).

The U.S. both imports and exports ornamental fish. Of the ornamental fish exported from the U.S., most went to Canada (29%), Southeast Asia (25%), Europe (20%), and Japan (18%). The U.S. produces, through aquaculture, more than 800 varieties of freshwater ornamental fish (Hill and Yanong 2013) with 95% of ornamental fish production in Florida (Fig. 7.6). Ornamental aquaculture

Fig. 7.6 Freshwater ornamental fish produced in Florida. Courtesy of Craig Watson.

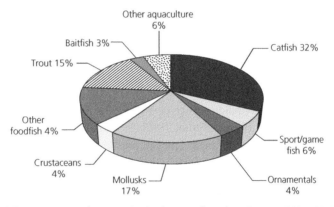

Fig. 7.7 Major aquaculture species in the U.S., by value. Source: USDA (2014).

production in the U.S. ranks fourth in importance in U.S. aquaculture in terms of value (Fig. 7.7).

Bruckner (2005) reported more than 45 countries supplying ornamental fish to the worldwide market. Globally, Indonesia and the Philippines were the largest suppliers followed by Brazil, Maldives, Vietnam, Sri Lanka, and Hawaii. The largest volume of imports of ornamental fish into the U.S. come from Southeast Asia and Japan, with the major ports of entry into the U.S. being Los Angeles (39%), followed by Miami (22%) and New York City (16%) (Chapman et al. 1997). The top five countries supplying the U.S. market were Singapore, Thailand, the Philippines, Hong Kong, and Indonesia. The next largest region supplying the U.S. was South America, with Colombia, Brazil, and Peru being the leading country suppliers.

Most of the freshwater fish imported from Southeast Asia are cultured, whereas those from South America are caught from the wild. Brazil is the leading exporter of freshwater ornamental fish (Monteiro-Neto et al. 2003). Most marine fish imported as ornamental fish are wild caught. Philippines and Indonesia supply the greatest volume of marine ornamentals into the U.S. Of Caribbean exports, the most popular species include royal gramma (*Gramma brasiliensis*), jawfish (*Opistognathus aurifrons*), queen triggerfish (*Balistes vetula*), blenny (*Ophioblennius atlanticus*), puddingwife (*Halichoeres radiatus*), bluehead wrasse (*Thalassoma bifasciatum*), and blue chromis (*Chromis cyanea*) (Bruckner 2005). Advances have also been made in culture of the popular seahorses for aquarium fish (Koldewey and Martin-Smith 2010). Seahorse (*Hippocampus* spp.) culture expanded in the late 1990s and 2000s, and by 2010 there were 11 different countries raising seahorses. The first records of cultured seahorses in international trade date from 2002.

Distribution channels for ornamental fish are complex. Hong Kong and Singapore are world purchase and transshipment centers for ornamental fish. Ornamental fish arriving in the U.S. through the ports of Los Angeles, Miami, New York, and Tampa arrive at broker-wholesale warehouses for subsequent delivery to single retail pet stores and warehouses (Chapman et al. 1997). The smaller ports for ornamental fish, such as Chicago, New Orleans, and San Francisco, are thought to service primarily pet dealers and wholesalers in the immediate localities.

Challenges for the ornamental fish industry include concerns over the ecological sustainability of wild capture of species for export. Those who oppose the international trade in ornamental fish cite damages caused by collecting techniques, over-harvesting of target species, and high levels of mortality along the supply chain (Wabnitz et al. 2004). Of increased concern is the harvest of coral reef fish (Wood 2001). Supporters maintain that proper conservation and management of coral reefs and other aquatic resources along with well-managed shipping and husbandry practices can alleviate these problems. Moreover, the capture of ornamental fish for export creates employment in rural areas (Bruckner 2005).

There is increased concern over the possible role of introduction of invasive species through the aquarium trade. Part of the concern relates to use of genetically modified organisms in the aquatic trade (Fossa 2004). Genetically modified fluorescent zebra fish and medakas were introduced into the market in 2003 and more will likely come. The concern is the consequences to wild populations if genetically modified ornamental fish escape into the wild. From the breeders' point of view, these techniques allow them to create new products to supply to the aquarium trade.

Interest in the culture of marine ornamental fish has grown. Aquaculture might be the solution to supplying the demand for these fish without undue pressure on natural populations and resources.

Summary

International trade in seafood and other types of fish products has become increasingly more important over time with improvements in transportation logistics and packaging as well as increased economic globalization. International trade in seafood has increased rapidly. The seafood trade is important to both exporting and importing countries. There are several countries whose economies depend heavily on seafood exports and others that depend heavily on imported seafood products to satisfy consumer demand.

Japan is the leading seafood-importing country, and China is the leading seafood exporter. Other important importers include the U.S., China, and the European Union. Other important exporting countries include Norway, Thailand, and Vietnam. The overall pattern of trade flows demonstrates that many of the leading exporting countries are developing countries, while most of the major importing countries are developed nations. China has emerged in recent years as a substantial importing country in addition to being the leading seafood-exporting country.

The increased trade in seafood has been accompanied by increasing trade conflicts. The World Trade Organization is the multinational treaty that serves as the major source of rules and agreements that govern international trade. Many of the conflicts related to the international trade in seafood have been related to the antidumping and countervailing duties provisions of the WTO and, more recently, to issues related to testing and regulation of the safety of seafood products in trade.

Study and discussion questions

1 Explain why countries engage in trade. Is the cost of production the only factor that determines whether trade will occur?
2 How important is international trade in aquaculture products? Describe at least five examples that justify your answer.
3 How has international trade in seafood changed over the past decade? Provide specific examples that illustrate your main points.
4 What are the top three exporting and importing countries worldwide? What are the main products exported and imported for each?
5 How large is the international trade in seafood? How does it compare with other types of agricultural commodities?
6 List and define the major types of trade policy tools.
7 Develop a timeline of the major international agreements on trade with bulleted lists of the major provisions of each.
8 Draw a diagram illustrating the sequence of the major trade disputes related to salmon, indicating the countries involved and the outcomes in terms of the effect on trade flows.

9 What surrogate countries were selected for the crawfish antidumping lawsuit in the U.S.?
10 What was different about the U.S. shrimp antidumping lawsuit as compared
  to other antidumping lawsuits?

Appendix 7A: The U.S. Antidumping Law

Both the Department of Commerce (USDOC) and the U.S. International Trade
Commission (US-ITC) are involved in antidumping petitions in the U.S. The
USDOC examines whether dumping has occurred while the US-ITC determines
whether the U.S. industry has been harmed. For duties to be levied, both dumping and harm to the domestic industry must be proved.

The petition must include information to support the allegation of sales at
less than fair value and to support the allegations of material injury, the threat
of material injury, and causation. This involves obtaining information on the
prices at which subject merchandise is being sold in the U.S., estimated costs
of production, and the estimated margin of dumping. The costs of production
(or factors of production) include estimates of the labor required to produce the
goods, the raw materials used for production, the energy and other utilities
consumed, and capital costs. The margin is calculated by first determining the
U.S. price and the normal, benchmark value.

In cases brought against non-market economies, a "surrogate" analysis is
used for normal value. The surrogate is a market economy country that is at a
level of economic development comparable to that of the non-market economy,
and is a significant producer of the goods that are the subject of the investigation
or other comparable merchandise. All factors of production, except labor, are
valued using costs in the market economy country. These are taken from publicly
available data.

The U.S. Department of Commerce
When an antidumping petition is filed, the USDOC has 20 days to initiate an
investigation. The USDOC normally bases its determination on data obtained in
response to detailed questionnaires sent to foreign producers and exporters. The
USDOC final affirmation is determined on day 235. The USDOC normally will
compare the price at which the good is sold to the U.S. (U.S. price) with the price
at which similar goods (the foreign like product) are sold in the foreign market
(normal value). The US-ITC makes its final determination on day 280. On day
287, the USDOC issues its order.

An antidumping order requires an importer to post a cash deposit equal to
the dumping margin. Actual duties are not paid until after there has been an
administrative review. If the USDOC issues an affirmative preliminary determination, all imports that enter after that date must be accompanied by bonds or cash

deposits equal to the assigned margin. The USDOC may suspend an investigation involving a non-market economy country if an agreement to restrict the volume of subject imports into the U.S. is reached. Different types of suspension agreements may be available in cases involving exports from market economy countries. A non-market economy country is a foreign country that does not operate on market principles of costs or pricing structures. The USDOC assumes that sales prices and costs in a non-market economy country cannot be used to determine the normal value of the goods.

In non-market cases, the USDOC will "construct" a normal value for the less than fair market value comparison. This constructed normal value is derived by: (1) identifying the cost elements involved in producing the foreign like product (factors of production); and (2) valuing those cost elements in a market economy country that is at a comparable stage of economic development as the non-market economy (surrogate country).

The USDOC identifies the first sale made for export to the U.S. to an unaffiliated purchaser in the calculation of the U.S. price. The U.S. price is compared to the normal value and the dumping margin calculated from that. The USDOC bases its determination on responses to questionnaires issued to foreign producers and exporters of subject merchandise. The questionnaires request detailed sales and cost information. Domestic producers who support the petition must represent at least 25% of total U.S. production and must account for more than 50% of the production of those domestic producers who take a position for or against the petition.

Between the preliminary and final determinations, the USDOC will conduct an on-site verification of each foreign producer. Based on the questionnaire responses, verification results, briefs filed by interested parties to the proceeding, and a formal hearing, the USDOC will make its final determination.

The U.S. International Trade Commission (ITC)

The International Trade Commission (US-ITC) makes its preliminary determination as to whether there is a "reasonable indication" of material injury or threat of material injury to a domestic industry "by reason of" the imports in question within 45 days of filing the petition. The US-ITC's staff will prepare a report based on questionnaire responses received from domestic producers, foreign producers, and importers. The focus of the US-ITC's analysis is on the material injury or threat of material injury and the causation of injury by imports. Material injury is defined as "harm which is not inconsequential, immaterial, or unimportant." The US-ITC considers whether the volume or any increase in imports is significant in absolute or relative terms compared with domestic production or consumption, whether imports are underselling the domestic product, or whether they have depressed or suppressed domestic prices, and the impact of imports on the domestic industry.

The impact of imports on the domestic industry is measured by: (1) the actual or potential decline in output, sales, market share, profits, productivity, return on investment, and capacity utilization; and (2) actual and potential negative effects on cash flow, inventories, employment, wages, growth, ability to raise capital, and investment. Even if the US-ITC determines that the domestic industry is not currently being injured by imports, it may nevertheless make an affirmative determination if it finds that the industry is being threatened with material injury.

If the US-ITC finds that the domestic industry is suffering from or threatened with material injury, it must then determine whether this is "by reason of the less than fair value of imports." This requires "adequate evidence to show that the harm occurred by reason of the less than fair value of imports, not by reason of a minimal or tangential contribution to material harm caused by less than fair value of goods."

The US-ITC's questionnaires request financial, production, shipment, and pricing information. Based on the staff report, a hearing before the staff, and briefs filed by those in favor or opposing the petition, the US-ITC will issue a preliminary decision. The US-ITC's final determination is made on day 280.

References

Anders, S.M. and S. Westra. 2011. A review of FDA imports refusals – US seafood trade 2000–2010. Selected paper prepared for presentation at the Agricultural & Applied Economics Association's 2011 AAEA & NAREA Joint Annual Meeting, Pittsburgh, Pennsylvania, July 24–26, 2011.

Anderson, J.L. 1994. *The growth of salmon aquaculture and the emerging New World Order of the salmon industry.* Presented at: Fisheries Management – Global Trends, University of Washington, Seattle, Washington.

Anderson, J.L. 2003. *The International Seafood Trade.* Woodhead Publishing Limited, Cambridge, UK.

Andrews, C. 1990. The ornamental fish trade and conservation. *Journal of Fish Biology* 37:53–59.

Anrooy, R. 2003. *Policy considerations linked to better marketing and trade of aquaculture products destined for poverty alleviation and food security.* Food and Agriculture Organization of the United Nations, Rome, Italy. Can be found at Library.enaca.org/AquaMarkets/presentations/OtherPapers/PolicyMatters.pdf.

Ariff, M. 2004. Dismantling trade barriers. Malaysian Institute of Economic Research.

Asche, F. and T. Bjorndal. 2011. *The Economics of Salmon Aquaculture.* John Wiley & Sons Ltd., Chichester, UK.

Basleer, G. 1994. The international trade in aquarium/ornamental fish. *Infofish International* 5:15–17.

Bruckner, A.A. 2005. The importance of the marine ornamental reef fish trade in the wider Caribbean. *Revista de Biologia Tropical* 53:127–138.

Chapman, F.A., S.A. Fitz-Coy, E.M. Thunberg, and C.M. Adams. 1997. United States of America trade in ornamental fish. *Journal of the World Aquaculture Society* 28:1–10.

Collie, D.R. and H. Vandenbussche. 2004. Anti-dumping duties and the Byrd amendment. Can be found at feb.kuleuven.be/public/ndbad40/publications/unlinked/byrdjune20.pdf.

Conroy, D.A. 1975. An evaluation of the present state of world trade in ornamental fish. FAO Fisheries Technical Paper No. 146, Food and Agricultural Organization of the United Nations, Rome.

Dey, M.M. and G. Bimbao. 1998. Policy imperatives for sustainable aquaculture development in Asia: lessons from Bangladesh, the Philippines and Thailand. In: *Network of Aquaculture Centers in Asia and the Pacific*. Proceedings of the Regional Workshop on Aquaculture and the Environment, Beijing, October 6–12, 1995. Bangkok Asian Development Bank and NACA.

Dey, M.M., M.F. Alam, and F.J. Paraguas. 2001. Production, accessibility, marketing and consumption patterns of freshwater aquaculture products in Asia: a cross-country comparison. FAO Fisheries Circular No. 973, FAO, Rome. 275 pp.

Economic Research Service (ERS). 2014. Aquaculture data. Available at www.ers.usda.gov/data-products/aquaculture-data.aspx.

Food and Agricultural Organization of the United Nations (FAO). 2003. *The Uruguay Round Agreement on Agriculture*. FAO, Rome. www.fao.org/docrep/003/x7353e/x7353e03.htm.

Food and Agricultural Organization of the United Nations (FAO). 2014a. *The State of World Fisheries and Aquaculture*. FAO, Rome.

Food and Agricultural Organization of the United Nations (FAO). 2014b. FishStatJ. FAO, Rome.

Fossa, S. 2004. Genetically modified organisms in the aquatic trade? *Ornamental Fish Journal* 44:1–16.

Hemley, G. 1984. U.S. imports millions of ornamental fish annually. *Traffic USA* 5:1.

Hill, J.E. and R.P.E. Yanong. 2013. Freshwater ornamental fish commonly cultured in Florida. University of Florida Extension Circular 54, Gainesville, Florida.

King, J. and J. Anderson. 2003. Institutions and measures of importance to international trade in seafood. In: J. Anderson, ed. *The International Seafood Trade*. Woodhead Publishing Limited, Cambridge, UK.

Koldewey, H.J. and K.M. Martin-Smith. 2010. A global review of seahorse aquaculture. *Aquaculture* 302:131–152.

Kurlansky, M. 1997. *Cod: A Biography of the Fish that Changed the World*. Walker and Co., New York.

Larkin, S. and R. Degner. 2001. The U.S. wholesale market for marine ornamentals. *Aquarium Sciences and Conservation* 3:13–24.

Lewbart, G., M. Stoskopf, T. Losordo, J. Geyer, J. Owen, D. White Smith, M. Law, and C. Altier. 1999. Safety and efficacy of the Environmental products Group Masterlow Aquarium Management System with Aegis Microbe Shield. *Aquacultural Engineering* 19:93–98.

Ligeon, C., C.M. Jolly, and J.D. Jackson. 1996. Evaluation of the possible threat of NAFTA on US catfish industry using a traditional import demand function. *Journal of Food Distribution Research* 27:33–41.

Monteiro-Neto, C., F.E. de Andrade Cunha, M.C. Nottingham, M.E. Araújo, I.L. Rosa, and G.M.L. Barros. 2003. Analysis of the marine ornamental fish trade of Ceará State, northeast Brazil. *Biodiversity and Conservation* 12:1287–1295.

Ornamental Aquatic Trade Association (OATA), 1998–2004. Ornamental Aquatic Trade Association Worldwide. Can be found at www.ornamentalfish.org.

Palin, C., C. Gaudin, Espejo-Hermes, and L. Nicolaides. 2013. Compliance of imports of fishery and aquaculture products with EU legislation. European Parliament Committee on Fisheries. Available at www.europarl.europa.eu/RegData/etudes/etudes/join/2013/513968/IPOL-PECH_ET%282013%29513968_EN.pdf. Accessed February 25, 2015.

Quagrainie, K. and C.R. Engle. 2002. Analysis of catfish pricing and market dynamics: the role of imported catfish. *Journal of the World Aquaculture Society* 33:389–397.

Rhodes, V.J. 1993. *The Agricultural Marketing System*. 4th edition. Gorsuch Scarisbrick, Publishers, Scottsdale, Arizona.

Rhyne, A.L., M.F. Tlusty, P.J. Schofield, L. Kaufman, J.A. Morris, Jr., A.W. Bruckner. 2012. Revealing the appetite of the marine aquarium fish trade: the volume and biodiversity of fish imported into the United States. *PLOS ONE* 7:e35808. Doi:10.137/journal.pone.0035808.

Roberts, K.J. 2000. Import and consumer impacts of U.S. antidumping tariffs: freshwater crawfish from China. In: *Microbehavior and Macroresults: Proceedings of the Tenth Biennial Conference of the International Institute of Fisheries Economics and Trade*, July 10–14, 2000, Corvallis, Oregon. IIFET, Corvallis, Oregon.

Shuman, C.S., G. Hodgson, and R.F. Ambrose. 2004. Managing the marine aquarium trade: is eco-certification the answer? *Environmental Conservation* 31:339–348.

Sloop, C.M. 2003. Chile Fishery Products Annual 2003. Foreign Agricultural Service GAIN (Global Agriculture Information Network) Report #C13022.

Suranovic, S. 2010. International trade theory and policy. The International Economics Study Center. internationalecon.com/v1.0.

United States Coral Reef Task Force. 2000. *International trade in coral and coral reef species: the role of the United States*. Report of the Trade Subgroup of the International Working Group to the USCRTF, Washington, D.C.

United States Department of Commerce. 2004. Federal Register Notice. http://ia.ita.doc.gov.

United States International Trade Commission. 1997. *Crawfish Tail Meat from China*, Investigation no. 731-TA-752 (final), publication 3057, Washington, D.C., 1997. www.usitc.gov.

United States International Trade Commission (US-ITC). 2014. AntiDumping Duties (AD/ CVD): Finding Non-Tariff Duties. Available at www.usitc.gov/tariff_affairs/nontariff_duty_ finder.htm. Accessed March 16, 2015.

United States Trade Representative. 2014. 2014 Report on sanitary and phytosanitary measures. Office of the United States Trade Representative, Washington, D.C.

Vannuccini, S. 2003. Overview of fish production, utilization, consumption and trade. *Fishery Information, Data and Statistics Unit, Food and Agriculture Organization of the United Nations*, FAO, Rome. Available at www.fao.org.

Wabnitz, C., M. Taylor, W. Green, and T. Razak. 2004. *From ocean to aquaria: the trade in marine ornamentals worldwide*. UNEP/WCMC, Cambridge, UK.

Warren, H. 2004. Update on Vietnam antidumping case. *The Catfish Journal* 19:6.

Whittington, R.J. and R. Chong. 2007. Global trade in ornamental fish from an Australian perspective: The case for revised import risk analysis and management strategies. *Preventive Veterinary Medicine* 81:92–116.

Wood, E. 2001. Global advances in conservation and management of marine ornamental resources. *Aquarium Sciences and Conservation* 3:65–77.

Wood, E. 2007. Collection of coral reef fishes for aquaria: global trade, conservation issues and management strategies. *Marine Conservation Society*, Ross-on-Wye, UK. 80 pp.

World Bank. 2013. *Fish to 2030: Prospects for fisheries and aquaculture*. World Bank Report Number 83177-GLB, Washington, D.C.

CHAPTER 8

Marketing by aquaculture growers

Aquaculture growers must make choices with regard to what type of markets to pursue for their products. One of the leading causes of failure of aquaculture businesses is the failure to spend time planning for the marketing component of their business. Successful businesses are market-driven, and successful aquaculture businesses are those that have spent time analyzing their marketing options.

Fish species and markets

Aquaculture includes an astonishing diversity and complexity of cultured organisms. There are more than 210 species of aquatic finfish, crustaceans, mollusks, and aquatic plants raised (Engle and Stone 2005). Of these, the vast majority (99%) are raised for human consumption (FAO 2002). Other animal protein sources of food are far more limited and are composed of products from just a few species of animals (cows, swine, chickens) and a few other specialty livestock crops. Fish and seafood consumption has tended to be driven by the species that are available locally from the wild. Consumer preferences and the seafood markets that have developed over time are as diverse and complex as the number and types of species raised.

The diversity of species raised and the resulting diversity of specific markets developed for these products present a different type of challenge to aquaculture growers. Business growth and development requires market growth, but seafood consumer preferences have tended to be provincial and regional in nature. Thus, market expansion of aquaculture products often requires aquaculture companies to change attitudes and preferences of consumers located outside the region where their species has traditionally been sold.

Market development is complicated by the biological differences among the multitude of aquatic species cultured. For example, aquatic organisms are

Seafood and Aquaculture Marketing Handbook, Second Edition. Carole R. Engle,
Kwamena K. Quagrainie and Madan M. Dey.
© 2017 John Wiley & Sons, Ltd. Published 2017 by John Wiley & Sons, Ltd.

poikilothermic, meaning that they cannot control their body temperature. All other animals that supply animal protein (cattle, swine, and poultry) for human consumption can control their body temperature. While temperature levels play a role in production of terrestrial livestock, they are not a key production parameter. In aquaculture, cultured organisms are divided into warmwater and coldwater organisms. Trout, for example, are coldwater fish that cannot tolerate the average water temperatures under which channel catfish or most shrimp species are cultured. While chickens can be raised throughout the world in a variety of climates and temperature zones, trout can only be raised where there is a source of cold water. The majority of cultured shrimp are warmwater species and can only be raised where there is a source of warm water. There are also coldwater shrimp species, but coldwater shrimp cannot be raised in warm waters. Aquaculture supply will, thus, be more partitioned based on temperature than will terrestrial livestock production. Supply of particular aquaculture species will be concentrated in regions of the world that present optimal temperature conditions for that specific species.

Catfish farming in the United States, for example, is generally profitable in those regions in which ambient temperatures are suitable. Pond water temperatures (which closely follow air temperature) should be above 18 °C (65 °F) for at least 180 days of the year and above 24 °C (75 °F) for at least 125 days. Fluctuations in water temperature trigger key phases of the reproductive cycle, making it difficult to raise channel catfish outside regions with temperate climates.

The majority of aquaculture species are also restricted to either freshwater or saltwater. Thus, carps, trout, and channel catfish must be grown in freshwater, while salmon, sea bass, yellowtail, and flounder require saltwater for culture. Some species, such as tilapia and shrimp (*Litopenaeus vannamei*), have been cultured successfully in both saltwater and freshwater although the majority of tilapia continue to be raised in freshwater and the majority of shrimp in brackishwater. However, it remains to be seen whether culturing species under salinities different from those of their natural environments can be the basis for large industrial sectors.

Production systems and intensification

The earliest recorded aquaculture production was practiced in earthen ponds in China (Avault 1996) and in Egypt (Bardach et al. 1972). Much of the early expansion of aquaculture occurred in freshwater earthen ponds (common, grass, bighead, and silver carps, and tilapia) throughout the world. Today, the majority of aquaculture production worldwide continues to come from earthen ponds.

From a marketing perspective, ponds are a convenient way to store and hold fish for long periods of time. However, production is heavily dependent upon the climate and ambient water temperatures. In temperate climates, for example, warmwater species such as tilapia will grow only over the warm months and must be brought indoors or sold before temperatures fall too low. Thus, supplies of warmwater aquaculture product will be seasonal and highest prior to the onset of the cold season in temperate regions. Tropical areas also experience varying temperatures, but not to the same extent as temperate climates. Even so, there are important differences in salinity, sunlight intensity, and cloudiness imposed by the rainy and dry seasons of the tropics.

Tilapia are raised in ponds throughout the world at a wide variety of densities and levels of intensification (Pullin et al. 1987; Pillay 1990). Subsistence farmers in many countries stock tilapia at low densities with composted vegetative matter as the primary input to produce fish for family consumption (Boyd and Egna 1997). Near-subsistence farmers culture tilapia using manure and supplemental feeds to maintain a "savings account" for future cash needs or to generate some cash through local sales (Smith and Peterson 1982; Little 1995; Engle 1997; Setboonsarng and Edwards 1998) (Fig. 8.1). Commercial small-scale tilapia production requires regular feeding but produces higher yields that allow growers to supply domestic markets in larger urban areas (Fitzsimmons 2000; Green

Fig. 8.1 Roadside sales of tilapia in Honduras, Central America. Source: Carole Engle.

and Engle 2000; Hanley 2000; Popma and Rodriquez 2000). Tilapia are also produced intensively in several countries to meet the high quality standards required to export to markets in the U.S. and the European Union (Engle 2006).

The majority of shrimp produced in the world are raised in earthen ponds along brackishwater estuaries. In Honduras, the leading producer of *Litopenaeus vannamei* in the Central American region (Rosenberry 1999), for example, pond shrimp production is classified as semi-intensive in which farmers raise two crops a year with stocking densities that vary from 5 to 20 post-larvae (PLs) per m² (Dunning 1989; Stanley 1993; Valderrama and Engle 2001). Typical yields of shrimp from semi-intensive production systems in Honduras range from 400 to 2000 kg/ha/yr of shrimp tails, while feed quantities range from 1 to 8 metric tons (MT)/ha/yr (Valderrama and Engle 2001). However, shrimp are also produced extensively on artisanal farms with low stocking densities and feeding rates. About one third of the world's pond-raised shrimp continues to be produced in extensive systems (Rosenberry 1999).

Asia has developed more intensive shrimp production methods, probably due to the higher costs of land in areas with greater population pressure. For example, in Thailand, shrimp are stocked at rates up to 30–35 PLs per m² with yields of 7000–8000 kg/ha/yr (Lin 1995). The higher yields are necessary to spread the higher fixed costs of land over greater amounts of production to be price competitive.

Over 98% of the catfish produced in the U.S. are grown in earthen ponds. Production costs are generally lower for catfish grown in ponds than in other culture systems. Over 98% of levee-style ponds use groundwater pumped from shallow wells (less than 391 m). These wells yield abundant water at low cost. The most common stocking strategy in the catfish industry is to stock fish in multiple batches to supply processing plants year-round. In this system, fingerlings are stocked each year in the spring, but multiple harvests are made throughout the year to sell market-sized fish. Since the production cycle is approximately 18 months for channel catfish production, varying sizes of fish are present in the pond at the same time. Single-batch stocking strategies are more profitable, but farmers stock in multiple batches due to the necessity of spreading the market risk of delayed sales associated with the presence of off-flavor (Engle and Pounds 1993). Multiple-batch stocking strategies provide the cash flow pattern necessary to meet financial obligations on catfish farms. The rapid adoption of hybrid catfish (♂*Ictalurus punctatus* × ♀*Ictalurus furcatus*) allows farmers to harvest them in 5–6 months.

Cage production has grown in importance in certain areas of the world. A number of countries with high population densities, small land areas, and ample access to marine resources have developed successful cage-based aquaculture industries. Cage and net pen technologies developed rapidly in the latter decades of the 1900s. Large-scale net pen operations were developed to culture marine species such as salmon (Norway, Chile, Canada), sea bream and bass (Greece), yellowtail (Japan), cobia (Taiwan), cod (Norway, Canada), and others.

Marketing fish from cage operations presents unique transportation challenges. Net pen companies have developed technologies that include use of helicopters, barges with fish pumps, and others, to transfer fish from net pens to hauling tanks to markets. Significant market coordination is required to minimize losses when transporting large quantities of fish from net pen operations.

Atlantic salmon (*Salmar salmo*) are the fastest-growing of the salmon species and typically reach market size in 2 years. However, the high costs of the marine cage systems used for production create economies of scale that have resulted in large, vertically integrated companies. These companies compete globally, largely on price. As salmon prices have decreased over time, largely due to the increased supply from aquaculture, the industry has moved into new product development. Currently there are a wide variety of salmon products available on the market that include gourmet, smoked and canned products, salmon burgers, salmon jerky, salmon bits (as a substitute for bacon bits), and many others.

Abundant flowing surface waters have been used in some areas to develop flow-through raceway production systems. Raceways are the predominant production system for coldwater trout and, in more recent years, have become the main production system for large-scale warmwater production of tilapia in tropical regions. Raceways provide for ease of harvest that provides flexibility to accommodate processing schedules and sales programs. Size classes of trout can be maintained in separate tanks within a raceway. This size separation facilitates marketing specific sizes to specific markets.

Indoor recirculating aquaculture systems have been developed with advances in engineering technologies and have become a reliable, but high-cost, production system. Recirculating systems, if constructed indoors, provide a way of controlling water temperatures that allows for year-round growth of fish in temperate climates. Moreover, indoor recirculating systems can be constructed closer to major seafood market areas than pond-based production systems. One major disadvantage of recirculating systems, however, is the continuing high cost of production in spite of the extent of improvement in the reliability of the systems.

The primary species raised successfully in indoor recirculating systems has been tilapia, mostly due to its ability to withstand adverse water quality fluctuations. Other species, such as turbot and sole (in Spain), yellow perch (in U.S.) and even more recently, shrimp (*Litopenaeus vannamei* in U.S. and Israel), have been raised successfully in indoor recirculating systems.

Sizes of producers

The diversity and complexity of aquaculture species, production systems, and markets is accompanied by an equal diversity in the size of aquaculture businesses. A tilapia "farm" in Rwanda may consist of one pond with a surface area of only 0.01 ha, while a company that exports tilapia fillets may have 60–70 ha of land.

Most of the shrimp produced on a commercial scale come from large-scale, private farms (>50 ha). Valderrama and Engle (2001) identified (Honduran survey data) groupings of commercial shrimp farms clustered around farm sizes of 73 ha, 293 ha, and 966 ha. However, a total of 68 artisanal producers were operating 239 ha of ponds in Honduras in 1997 (ANDAH 1997). A typical artisanal farm is operated by a family group and is composed of from 1 to 30 ha of ponds. While most of the semi-intensive farms in Honduras were vertically integrated and marketed their product in international markets, artisanal producers sold to local processing plants and/or shrimp markets. Saborío (2001) reported a total of 90 artisanal shrimp farming cooperatives operating in the Estero Real area of Nicaragua by the year 2000. These loosely defined community-based groups consisted of dozens of families holding a site concession (Jensen et al. 1997). Their extensive methods utilized tidal inflows to stock ponds, exchange water, and supply nutrients. The cooperative organization allowed artisanal shrimp farmers to supply markets more consistently.

Unlike pond production industries that have tended to demonstrate a wide variety of sizes of individual farms across the major species raised, net pen salmon farming is highly concentrated (Bjorndal et al. 2003). In Norway, the four largest net pen salmon farms controlled 28% of the country's production capacity while the 10 largest controlled 46%. The Chilean net pen salmon industry is even more highly concentrated with the four largest firms accounting for 35% and the 10 largest 60% of exports in 2001. Moreover, there are growing Norwegian interests in the Chilean salmon industry.

Indoor recirculating systems may range from one 2-m diameter tank in a greenhouse that produces 100 kg of fish a year to large, industrial facilities with a production capacity of several million kg a year. The smaller systems typically are targeted toward home consumption while the larger indoor systems frequently target the higher-priced live fish markets.

Supply response and biological lags

Given the variety of aquaculture species cultivated and their varying biological characteristics, the supply response will vary a great deal from species to species and even with different production systems. Tilapia and shrimp raised in the tropics will reach market size in 6 months whereas salmon and catfish may require 18–24 months for individual fish to reach market size.

Lengthy biological lags for animals to reach market size can cause supply to be inelastic (unresponsive to price changes). Inelastic supply makes it difficult for growers to respond to changing market conditions. Channel catfish, for example, when stocked at 5 inches or less, require 18 months to reach market size. Kouka and Engle (1998) showed that response to price changes occurs at the hatchery/fry stage of production and that overall catfish farm supply is

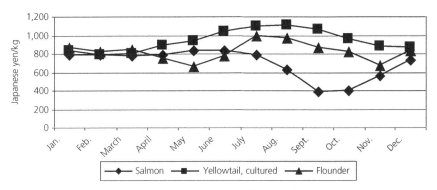

Fig. 8.2 Price seasonality for three finfish species, wholesale markets in Japan. Source: NOAA-NMFS (2003).

inelastic. Farmers must stock production units and begin feeding before knowing what the price will be when fish reach market size. Thus, it is difficult for farmers to adjust quickly to changing market conditions. However, the recent adoption of a hybrid catfish allows fish to reach market size within 6 months.

Prices of many aquaculture crops demonstrate seasonality effects across the year. For example, Figure 8.2 illustrates wholesale price seasonality by month of salmon, cultured yellowtail, and flounder in Japan (NOAA-NMFS 2003). Salmon prices were lowest in September and October in Japanese wholesale markets, while prices of cultured yellowtail were highest in July through September, and flounder prices were highest in July and August. Price seasonality may vary by year. It is best to look at monthly prices over several years. Higher revenues may be generated if the production cycle or production plus holding can be managed to target a percentage of the crop toward marketing in months when price is high.

Commodities, markets, and niche markets for differentiated products

The species of aquatic organism to be raised, its biological requirements, the resulting supply characteristics, and the production system selected must all be appropriate for the specific markets to be targeted in the marketing plan. Commodity markets will be feasible only for larger businesses targeting high-volume markets in industries with processing capabilities and the ability to compete on price. A commodity is an economic good that can be legally produced and sold by almost anyone (Rhodes 1993). Niche markets for specialty, differentiated products may be the only feasible market outlets for small-scale growers. A differentiated product is an economic good that belongs to a single seller and that often may be patented, copyrighted, or trademarked to the exclusive use of that seller.

Large firms may also enter markets with differentiated products. Development of product brands can assist businesses to differentiate farmed and captured supply chains. Branding can occur at the species level, country-of-origin generic level, or as private brands within individual retailer chains (Burt 2000). Consumer perceptions related to origins of supply can be used as a basis for the development of brands for aquaculture product lines.

Farmers' marketing alternatives

Sales to processors

The majority of aquaculture products that are sold commercially are sold by farmers to processing plants. The larger seafood markets are those with higher-income consumers who have little interest in cleaning or dressing fish for consumption at home. Fillet products are the primary product sold and the processing plant plays the role of changing the form of the product into that preferred by the majority of consumers. Chapter 6 goes into more detail about seafood and aquaculture product processing.

Processing plants will schedule delivery of loads of fish depending upon their current and anticipated orders. Those farms that are able to regularly supply the volume and size of fish or shrimp desired by the processor or packing plant will be those scheduled for regular deliveries to the plant. Thus, farmers need to have a clear idea of what the plant's specifications are for size tolerances, delivery volumes, timing of deliveries, quality control checks, and flavor checks. Even firms that are vertically integrated may have separate cost centers such that the grow-out business will have to meet the processing center's delivery specifications.

Processing plants typically purchase the greatest overall volume of product, when compared to other potential market outlets, but also tend to pay the lowest price. This is because processing plants are frequently price takers in the market and there are many good substitutes for most fish and seafood products. Thus, many seafood products sold to processing plants are commodities that compete on price with other similar seafood products.

Some seafood processing plants are cooperatives to which the grower must belong to sell fish. Sales to cooperatives typically are in proportion to shares held by the member. New generation cooperatives often use delivery rights as a means to raise capital. If the cooperative is a successful business, the delivery rights acquire value through the ability to trade them with new members or with members who are seeking to expand their businesses.

Sales to livehaulers

In the U.S., EU, and other countries, fish farmers can sell fish to livehaulers, individuals who own large transport vehicles to haul fish to other distributors, or to retail outlets. Livehaulers typically purchase fish from farmers and re-sell them either to wholesalers or directly to retailers. Common carp in Europe are hauled extensively from fish farms in Hungary, Slovakia, and other Central European countries to markets in Germany and elsewhere. In the U.S., hybrid striped and largemouth bass are hauled from the southern part of the country to Asian ethnic grocery markets in New York City, Chicago, and San Francisco. Livehaulers frequently pay cash, but prices can be volatile.

Livehaulers may haul fish to paylakes, government-owned fishing lakes, and community and urban fishing programs. The scope of these programs frequently requires substantial quantities of fish to be purchased for stocking. Some of these programs prefer larger fish than do processing plants. Fish that are off-flavor may also be sold through these outlets.

Selling directly to end consumers

Small-scale growers frequently have higher production costs due to the economies of scale common in many forms of aquaculture production. The higher production costs mean that there will be fewer years in which it is profitable to sell to a processing plant. Even when the price paid by processing plants is sufficiently high to allow for profit to be made by small-scale growers, the profit margin will be much lower than for larger-scale farms that can produce fish at a lower cost. Thus, it can be more feasible for the small-scale grower to develop markets based on direct sales to end consumers.

Sales to end consumers require the grower to do all the marketing by him- or herself. Thus, market-sized fish will need to either: (1) be transported to customers in other locations (transportation marketing function); or (2) be held on the farm in cages or tanks for farm-bank sales, or in ponds for fee fishing (storage marketing function). Which of these is more feasible would be determined in the market analysis and should be detailed in the marketing plan. Chapter 9 includes more details on how to develop and use the business's marketing plan.

Holding fish for sale allows small-scale producers to take advantage of higher prices obtainable through direct sales to the public. Adequate holding facilities and proper handling can make a big difference in a producer's profits. However, it is critical that fish be readily available and in good condition. Dead fish will turn customers away as will lengthy waiting periods to catch fish.

Cages can be used for fish to be sold at irregular intervals, but tanks are best for supplying customers on a regular basis. Cages can be used to hold from 121 to 240 kg/m³ (7.5–15 lb/ft³) of cage, depending on the temperature (Rode and Stone 1994). For long-term holding, fish will need to be fed a maintenance diet (1% of fish weight), and supplemental aeration may be necessary. A pier often will be needed to retrieve fish quickly from cages to attend to customers. Cages anchored off shore without a pier will require a boat for both feeding and to retrieve fish, often with lengthy waits. Off-flavor problems in the pond where the fish cages are located will affect all fish and force sales to be curtailed until fish come back on flavor. Disease treatment can be difficult in cages.

Tanks can be used to hold fish and are preferable for shorter time periods for regular sales (Rode and Stone 1994). Fish can be held for several days at a rate of about 68 g/l (0.6 lb of fish/gallon) of water. A typical tank would be 3–13 m long, 1– 2 m wide, and about 1 m deep (10–40 feet long, 3–6 feet wide, and 2–3 feet deep). Tanks require a concrete slab with side walls of concrete block or poured concrete. Round or rectangular fiberglass tanks can be purchased. The tank facility is best covered with a roof so that sales can continue during inclement weather. An aeration or air blower system is needed to maintain adequate oxygen levels in the tank.

Fish held in tanks are susceptible to theft. Since fish are typically not fed while held in tanks, they can lose weight. More weight will be lost at higher temperatures. Channel catfish can lose as much as 4.5% of body weight in 2 days at a temperature of 22 °C (71 °F). Water in the tank will need to be exchanged at a rate of about 10% of the tank volume an hour to avoid buildup of ammonia. Well water is preferred because public waters have chlorine that is toxic to fish.

If the location of the farm is such that there is not a sufficient customer base to attract enough people to the farm for direct sales, fish will have to be transported to where the customers are. Fish can be hauled in hauling tanks to farmers' markets, street corners, parking lots, or directly to restaurants or grocery stores that purchase and sell live fish. The simplest type of hauling tank is a box made of marine plywood. Alternatively, aluminum or fiberglass tanks can be purchased in a range of sizes and dimensions to haul in the bed of a pickup truck, on the back of a bob truck, or for use with a flatbed or 18-wheeled tractor trailer rig (Rode and Stone 1994). Whichever type of tank is selected, internal dividers or baffles are necessary to reduce sloshing. Most fish species can be safely transported at a rate of 0.6 kg/l (5 lb of fish/gallon) of water (after fish added) for 16 hours. Thus, a tank with capacity of 800 l (200 gallons), loaded at 0.6 kg/l (5 lb/gal), can haul only 273 kg (600 lb) fish, with 480 l (120 gallons) of water remaining. Fish should be purged (held overnight to empty stomachs to avoid water quality problems during hauling). Some ice added to reduce the temperature of the water will reduce stress on the fish during transport. The transport tank will need a source of air: either an oxygen tank with diffuser bars, a blower, or small, 12-volt agitators.

The marketing plan for direct sales requiring transportation should indicate the dates and times of sales. Fish cannot be held indefinitely in a transport tank; customers will need to be reliable and in condition to accept regularly scheduled deliveries of fish so that fish can arrive in good condition. The scale must be a certified scale.

Fee fishing

Fee-fishing or paylakes are businesses in which water bodies are stocked with fish and customers pay to fish in them. Some fee-fishing operations are run by growout farmers, while others are strictly in the business of buying fish to stock for re-sale by angling, an aquatic pick-your-own operation. Fish farmers can sell fish to a livehauler to transport to paylakes elsewhere or can develop their own fee-fishing business. This section will discuss the basics of what is needed to develop, own, and operate a fee-fishing business. Sales to livehaulers are discussed in another section of this chapter.

A fee-fishing business sells recreation. As with any other type of business, the location is essential. Successful operations are those located within 50–83 km (30–50 miles) of a population center with at least 50,000 or more people. Fee fishing will not work well in areas where good fishing in natural water bodies is readily available. However, a location near a major urban area with few opportunities to fish might be a prime spot for a fee-fishing business.

People go to fee-fishing operations because they are looking for a family outing with good fishing, but also with amenities that ensure a fun, but safe activity. Typical clientele for fee-fishing operations often are families or grandparents with small children, elderly people, or physically handicapped people who find it difficult to get out on natural water bodies to fish. Adequate amenities for family activities are important and the site must be aesthetically pleasing and comfortable. A natural setting that is screened from urban distractions, with easy access, good parking, and adequate security, is best. Clean restrooms are essential. Concessions can generate important revenue to the business and contribute overall to the sense of a quality family activity. Sales or rental of bait and tackle are necessary to attract first-time anglers who may not own fishing gear. Chairs and umbrellas can be rented for the comfort of the customers. Snacks and drinks will help keep people there longer. Sales of coolers with ice to take fish home can be supplemented with cookbooks, fish batter, and seasonings. Many customers will prefer to have their fish cleaned and will pay a fee to do so. Sunscreen and first aid supplies can be sold along with hats and shirts with attractive designs that also serve as advertisement.

Word-of-mouth advertisement from patrons who have had a pleasant experience can be some of the best advertising. Roadside signs as well as ads in newspapers, on the television, radio, or in local shopper guides can be effective. Local fliers can be distributed at youth and community events.

Ponds should have banks with good grass cover or sodded if necessary. Small ponds are better than larger ones to ensure good fishing. With multiple ponds, patrons can be moved to those ponds where fish are biting better that day. Given that 30–50% of the fish in a pond may have learned to avoid hooks, ponds need to be seined regularly to remove fish that are not biting. An alternative market will need to be developed for those fish that will not bite a hook. Regular supplies of other fish must be added regularly to maintain good fishing.

Clearly-placed signs should direct customers to parking and provide all necessary information. Prices, fishing regulations, rules related to various activities, and the times of operation need to be clearly visible on attractive signs in multiple locations. Activity rules must be posted clearly. Swimming, alcohol use, and abusive language should be prohibited. These activities are not conducive to the family atmosphere that is important to the majority of patrons of fee-fishing businesses. Any fishing gear restrictions need to be posted clearly.

The marketing plan for a fee-fishing business may include offering group rates to youth groups. Boy and Girl Scout, church youth, 4-H, or school groups may be potential target market segments who may afterwards convince parents, grandparents, or other family members to return. Night fishing activities may offer an off-peak, exciting adventure to some organized youth groups. Stocking trophy-tagged fish provides an opportunity to generate excitement by advertising special prizes. Occasional additions of new species of fish will keep the experience new. Posting and selling instant pictures of customers with their catch can serve as both advertisement and to generate revenue.

As in any other type of business, certain permits may be required. These vary with the state and country, but may include permits related to starting a business, building ponds, effluents from ponds, cleaning fish, and sales from the concessions.

There are a wide variety of pricing mechanisms used in fee-fishing businesses. Some charge an entrance fee. The advantage to the entrance fee is that the revenue is generated immediately. However, costs may be high if the patrons catch many fish. A limit on the number or weight of fish caught can alleviate potential problems. Daily, or seasonal, entrance fees can be charged. An entrance fee will also discourage loitering and help to maintain the business's attractiveness to family-oriented groups. Other businesses charge by the unit weight of fish harvested.

Marketing by fisher/farmer groups

Consolidations into larger processing plants and companies in the seafood industry are the result of new economies of scale and competition in the marketplace. Economists have suggested that, when market demand for a product is growing slowly, increased consolidation can lead to increased concentration. Therefore, the structural changes occurring in the seafood industry could harm small-scale and medium-scale farmers. On the other hand, increased consolidation sometimes may be beneficial to consumers and society at large. This is because the economies of scale reduce costs, which can translate into reduced prices to consumers.

An option that is often suggested for small-scale aquaculture growers to market their products is to form a cooperative. Pooling production from different farms creates the larger volumes of product required to fulfill larger, longer-term contracts. More stable prices may be obtained with longer-term contracts. Successful farmer groups are those that have strong, well-respected management that is viewed as being fair to all and insists that all members follow the policies established by the group.

The structural changes occurring in the seafood industry can affect small-scale and medium-scale farmers in a variety of ways. These can include: (1) disparity of bargaining power; (2) use of production contracts; and (3) competition from imports (Torgerson 2000). Each of these will be discussed in more detail below.

1 *Disparity in bargaining power*: Increased mergers and concentration of the companies that purchase farm products results in farmers facing fewer, but larger buyers of their products. This often results in an apparent disparity in market and bargaining power between farmers and buyers. The large companies often have the greater bargaining power, and farmers can lose market access as a result. By organizing into cooperatives or farmer groups, farmers can control a larger portion of their products, have greater bargaining power than an individual farmer would have, and can approach several different potential buyers in different regions.

   The greater product resources available through the organizations provide opportunities to negotiate and develop larger supply contracts. Such farmer cooperatives or groups are likely to be able to negotiate higher prices for their products than an individual farmer with more limited available product supplies. With a cooperative, farmers or other marketing groups can collectively exercise some influence in the market place and begin to correct for market failure. The U.S. Department of Agriculture (2004) lists 12 associations that collectively bargain for processing fruit and vegetable commodities including apricot, cling peach, Bartlett pear, processing tomato, olive, prune, raisin, potato, peas, barley, flax, processing apples, plums, red tart cherries, asparagus, feeder pigs, raw milk, hazelnut, sugarbeet, and perennial ryegrass. The number has declined from previous years, suggesting a possible decline in bargaining activity, probably due to increased use of contract production.

2 *Use of production contracts*: Large companies and chains increasingly tend to use production contracts. For example, over 90% of broilers and processed vegetables have been produced under contract for several decades. The concept of a production contract is becoming popular and has been adopted for the production of hogs and cattle as well as for other commodities. Contract production involves relationships and activities between an owner of the plant or animal (often the major buyer or large company) and the services of the farmer. For example, in a typical poultry contract, Tyson Foods, Inc., provides the chicks and inputs while the farmer provides the labor, management, and facilities required to raise the Tyson-owned chicks to the appropriate processing weight. The farmer receives an agreed price per bird in addition to some performance incentives. In most coastal fisheries, some fish processing companies contract with fishermen by providing them with inputs such as nets, boats, motors, and gear while fishermen supply their catch to processors at some negotiated price less the cost of the inputs (University of Alaska 2001).

Production contracts can be beneficial but can also be disastrous for the farmer. Farmers are concerned when the market is dominated by a few, large, integrated buyers. Some farmers find that contracting limits their opportunity for growth, restricts entrepreneurship, and pressures them to keep up with technological changes. Moreover, farmers may lack the leverage to negotiate for better contract terms. Farmer groups can effectively participate in the development of agricultural contracts and provide the opportunity to work with buyers and processors to eliminate unfair or unreasonable terms. The group or cooperative can negotiate terms of sale, prices, and payment arrangements to share any financial risks between the members and the buyers.

3 *Competition from imports*: There has been growing competition in the U.S. between domestically produced food products and low-cost imported products. Many of these imported food products are produced under relatively fewer or no environmental guidelines, food safety, and labor controls. Production is subsidized in many other countries and labor costs are very low. Therefore, the price of many imported food commodities is low compared to the price of domestically produced products. The domestic seafood industry in the U.S. especially has faced very stiff competition from low-priced imported seafood products in recent years.

Agricultural growers have found it necessary to assume more control of their industry by working together, developing effective cooperatives, and coordinating cooperative systems for collective actions in marketing. With pooled resources, a cooperative can provide better market information and data for members to utilize in their management decisions. The cooperative can serve as a clearinghouse for trade information, promote the product on both the domestic and foreign markets, develop partnerships with other groups in foreign countries, and serve as a voice for producers. With some market power, cooperatives can influence terms of trade on the domestic and/or international markets. Terms of trade relate to price, timing, form, and other quality or quantity specifications. Cooperatives can provide mechanisms for resolution of trade disputes and enforcement of trade regulations and standards to ensure a fair playing field in the marketplace.

As discussed in Chapter 6, antitrust laws exist to promote competition, but organization of farmers into groups is not against antitrust laws. In fact the Capper-Volstead Act of 1922 (Appendix 8A) provides immunity to farmers who organize into groups for purposes of developing some bargaining power to: (1) better deal with other competitors; and (2) address supply chain issues from a cooperative and coordinated position of strength. The Act essentially grants farmers the legal right to pool their bargaining and marketing resources to place them on an equal footing with the large buyers of their raw agricultural products.

Marketing cooperatives

A marketing cooperative is a farmer organization with the purpose of collectively selling their farm products. The cooperative provides farmers the opportunity to perform some joint marketing responsibilities including assembling of products, negotiating with large buyers of farm products, exercising some power in the marketplace, spreading risks and costs, and in some cases processing farm commodities. Thus, a marketing cooperative may function as a contract and price bargaining cooperative, or it may be involved in processing or manufacturing of specific agricultural commodities. By joining the cooperative, farmers are provided a guaranteed outlet for their farm products. Another advantage to farmers in joining a cooperative is the benefit of economies of size and scale. There is also sharing of marketing risk and costs among farmer-members. This risk sharing plays an important role in development of individual farm enterprises, and in developing markets.

There are four classes of marketing cooperatives based on how they are organized, membership affiliation, control, and geographical area. These classes are: (1) local cooperatives; (2) centralized cooperatives; (3) federated cooperatives; and (4) mixed cooperatives.

Local cooperatives

Local cooperatives are usually farmer groups at the local or community level. They perform a limited number of marketing activities for the group such as assembling and grading of farm products. Most of the cooperatives for fruits, vegetables, specialty crops, and fisheries are local in nature because of the localized nature of the production of the commodities involved. Consequently, membership is almost exclusively restricted to farmers engaged in producing the commodity.

Centralized cooperatives

Centralized cooperatives, unlike local cooperatives or associations, operate over larger geographic areas and have members in several states. In addition to assembling farm products, they often provide more vertically integrated services such as processing. This is the common form of cooperatives in agriculture. In the livestock sector, for example, several small producers in Nebraska, Kansas, and Oklahoma have organized into marketing associations that ship livestock to central markets. Such cooperatives enabled small livestock producers to pool their small sale lots for more efficient shipment to terminal markets.

Another example is the Staple Cotton Cooperative Association (Staplcotn) which is America's largest and oldest cotton marketing cooperative, based in Greenwood, Mississippi. It is owned by 2500 cotton growers in Mississippi,

Arkansas, Louisiana, Tennessee, Missouri, Alabama, Florida, and Georgia, and handles about 15% of the U.S. cotton crop. Cotton growers have the option of storing with Staplcotn but not necessarily marketing through the cooperative, or vice versa. Under the cooperative's Mill Sales Program, members have two marketing options for their cotton. Seasonal Option and Call Option. Many of the members prefer the Seasonal Option, in which the cooperative makes the pricing decisions. The Call Option allows grower-members to make their own pricing decisions; the grower makes decisions relating to the futures market while Staplcotn markets the basis decision. The two decisions are major components of pricing decisions made for cotton.

Federated cooperatives

The federated cooperative consists of local associations or cooperatives. Leaders from member local cooperatives or associations elect directors and provide general operating guidelines for the federation. Federated cooperatives perform more complex and expensive marketing activities that the member local associations or cooperatives cannot perform, such as manufacturing, involvement in financial markets, or international marketing. CherrCo is an example of a federated marketing cooperative that has 28 member cooperatives in the U.S. and Canada. It represents significant portions of cherry production in New York, Michigan, Washington, Utah, Wisconsin, and Ontario, Canada. Members of the cooperative have production that ranges from about 600,000 pounds to more than 10 million pounds annually. Ocean Spray is also a federated cooperative; it is owned by more than 800 cranberry growers and 126 grapefruit growers located throughout the U.S. and Canada. It is the largest cranberry marketing organization in the U.S. and North America's leading producer of canned and bottled juices and juice drinks. It has been the best-selling brand name in the canned and bottled juice category since 1981.

Mixed cooperatives

Mixed cooperatives serve both the local cooperatives and the individual farmer members. The structure combines the features of local, centralized, and federated cooperatives as well as individual memberships. Mixed cooperatives are not common and are usually formed to fit particular industry situations. Dairy Farmers of America represents this type of cooperative. It is the largest milk cooperative in the U.S., representing more than 22,924 producer-members who market their milk through the cooperative. Dairy Farmers of America was formed in 1998 as a result of a merger of four leading dairy cooperatives: Associated Milk Producers, Inc., Mid-America Dairymen, Inc., Milk Marketing, Inc., and Western Dairymen Cooperative, Inc. Other cooperative organizations joined after 1998, including Independent Cooperative Milk Producers Association, Valley of Virginia Milk Producers Association, and California Cooperative Creamery. The cooperative represents 13,445 dairy farms in

49 states, and markets over 25.7 billion pounds of milk giving it a market share of 33% of the total U.S. milk supply. The cooperative also has nine bottling joint ventures, three manufacturing joint ventures and 25 cooperative-owned processing plants. Brands of products produced by the cooperative are Borden Cheese, Golden Cheese, Mid-America Farms, Jacobo, Enricco, CalPro, Sport Shake, and VitalCal.

Marketing cooperatives as marketing agents

Most marketing cooperatives operate as a marketing agent by collecting products of members for sale, grading, and packaging, and performing other marketing functions. Livestock cooperatives, milk cooperatives, and grain elevator cooperatives are examples of marketing agents. For example, CHS Cooperatives, formed in 1998, was a merger between two regional cooperatives, Cenex, Inc., and Harvest States Cooperatives. CHS markets substantial amounts of member-produced grain. However, in recent years, the trend has been toward affiliation with global grain marketing companies such as Archer Daniels Midland – ADM (Dunn et al. 2002). Some milk marketing cooperatives in Wisconsin, for example, do not process or physically market their members' milk but instead represent members only in pricing or establishing other terms of trade with processors on their members' behalf. The Alaska inshore pollock bargaining association historically has utilized exclusive delivery contracts between a surimi plant and the fleet delivering to that plant (Matulich and Sever 1999).

Marketing cooperatives as processing groups

Some marketing cooperatives are organized to perform processing functions. This typically includes packaging products of members as well as wholesaling final products. Examples of such marketing cooperatives can be found in vegetable canning, fruit packing, and cheese and butter manufacturing. These functions are part of the overall marketing activities performed by these cooperatives in an attempt to control their products as they move to the marketplace.

Farmers' bargaining groups

Agricultural bargaining groups are a special type of marketing cooperative. These bargaining groups do not own, process, or market the farm commodities of farmers. Instead, they negotiate with processors or buyers on behalf of the members.

The cooperative negotiates for price (including premiums and discounts), quality standards, and time and method of payment. In some cases, the bargaining group coordinates the distribution of product and timing of delivery. It may also negotiate other terms of transaction that may include grading, duration of contract, production rights and responsibilities, and transportation. A bargaining cooperative represents the occupational interests of farmers in the policy arena and in the marketplace. The association is mainly financed through checkoff programs. A checkoff could be a flat fee per unit of sale or some specified percentage of sale value of the products sold by members that is retained by the association. Other methods of financing include service charges to processors, annual dues, and membership fees.

A bargaining cooperative usually does not physically handle the farm produce. Members sell farm products directly to processors at the price negotiated by the cooperative. With control over large volumes and supplies of farm products, bargaining associations have more market power than do individual growers and are able to negotiate price more effectively. Bargaining associations are common in processing sectors of fruit, vegetable, specialty crop, dairy, and sugarbeet industries. Iskow and Sexton (1992) conducted a comprehensive survey of all active bargaining associations in markets for processing fruits and vegetables. The authors reported that bargaining associations bargain for raw product price, the terms of trade, including time and method of payment, and quality standards. Only 25% of associations surveyed reported negotiating for the quantity of raw product to be purchased by processor/handlers. In most cases, the total volume of raw product to be purchased was determined prior to price negotiations.

In the fisheries sector, perhaps one of the most successful bargaining groups is the Alaska pollock At-sea Processors Association. The pollock and the West Coast pacific whiting processing sector is highly concentrated with catcher vessels delivering over 80% of inshore allocation to onshore surimi processors, and the remainder delivered to motherships. In the 1990s, members formed harvesting cooperatives under the umbrella of the Association, which negotiated formal contracts, involving price, with each of the processors prior to each season and represents a countervailing monopolistic bargaining association (Matulich and Sever 1999). The association also coordinates harvesting efforts among the fishing vessels to reduce incidental catches.

A bargaining association could be an effective bargaining agent for farmers engaged in production contracts. A cooperative bargaining association can work legislatively toward establishing institutional rules that augment the bargaining process. This could include provisions for good faith negotiations, dispute resolution mechanisms, and enforcement procedures. The cooperative can effectively represent farmers negotiating marketing contracts and those negotiating production contracts.

Marketing orders

A marketing order is a legal instrument authorized by the U.S. Congress through the Agricultural Marketing Agreement Act of 1937. The Act authorizes the Secretary of Agriculture to establish "marketing orders" for milk, fresh fruits, vegetables, tobacco, peanuts, turkey, and specialty crops (such as almonds, walnuts, and filberts). The primary objective of the order is to stabilize market conditions and provide benefits to producers and consumers by establishing and maintaining orderly marketing conditions.

Many states also have parallel legislation modeled after the Federal Act (Agricultural Marketing Agreement Act of 1937) to provide for state marketing orders. Federal marketing orders may apply to an industry within a state boundary, a sub-region within a state, or encompass a production region covering more than one state. With state marketing orders, the jurisdiction is limited to individual states or sub-regions within the states.

The legal provisions of federal marketing orders fall into three broad classifications: (1) quality control provisions which involve specifying standardized packages or containers, and establishing uniform, mandatory quality standards, such as size, color, or minimum maturity; (2) quantity control methods which include smoothing the flow of the product to market and volume management provisions, such as permitting only a certain portion of the crop to move into specified outlets (e.g., reserve pools or market allocation), and producer allotments; and (3) market-facilitating provisions which include production research, market research and development, market information, and market promotion and advertising.

The enabling legislation of the Agricultural Marketing Agreement Act allows producers to form marketing orders that comprise elements from all three of the above types of provisions. However, in practice, commodity groups generally prefer to include only some of the provisions when designing a marketing order for their product. Most commodity groups forming federal marketing orders have tended to focus on quality regulations (such as grade, size, and packing or container regulations), research, and promotion.

Most of the state marketing programs have been utilized for research and/or promotion and advertising because of the support from farmers and policy makers. State marketing orders have been used more for quality regulations and not for quantity controls that have been controversial.

Marketing orders are established for commodities by a vote of the producers in the geographic area for which the order is proposed. Once the marketing order is established, committees of producers develop the details of enforcement. The details cover items outlined in the three broad areas above. The detailed regulations are forwarded to the U.S. Secretary of Agriculture and, upon approval, the order is published in the Federal Register, whereupon it becomes law and legally binding on all producers and handlers. Once approved, the provisions of the marketing order become mandatory across the industry.

In Texas, the Federal Marketing Order for oranges and grapefruit established specifications for the grades and sizes of fruit that could be shipped, container size, and packaging. There are provisions for inspections to ensure compliance and funding of market research and development, including paid advertising. The regulations focused on quality with its implementation which allows citrus to be shipped in regulated trade channels. These regulations are subject to change from season to season or even within a given marketing season as market conditions change. One of the main objectives of the regulation was to increase satisfaction and confidence of buyers and consumers of the product to motivate demand for citrus.

Federal Market Orders also regulate the importation of some commodities. In Texas, the Federal Marketing Order regulated the importation of all fresh fruit into the U.S.; in Florida, the Federal Marketing Order for grapefruit regulated all fresh grapefruit imports.

Futures markets for aquaculture products?

Futures markets have been used to hedge against price fluctuations by farmers for many years and, hence, can be used to reduce market risk for both buyers and sellers of, for example, shrimp. Futures contracts are standardized, legally binding agreements to either deliver or receive a certain quantity and grade of a specific commodity during a designated delivery period (the contract month). The contract includes information on where it would be delivered and any adjustments on price from substituting a different species or size. Commodities need to be standardized so that they can be exchanged. This makes it easier for anyone to enter into a futures contract and know exactly what they are buying.

Futures markets augment cash markets. No one actually has to deliver or receive the product. Feeder and fed (live) cattle, hogs, pork bellies, cotton, canola, and wheat are traded at exchanges such as the Chicago Board of Trade, New York Cotton Exchange, Winnipeg Commodity exchange, Minneapolis Kansas City, and Winnipeg Exchanges.

The Minneapolis Grain Exchange began trading futures contracts for farm-raised and wild white shrimp in 1993 and added a contract for farm-raised giant tiger shrimp in 1994. The two main shrimp contracts offered were: (1) 5000 lb of raw, frozen, headless, shell-on 41–50 count white shrimp (*Litopenaeus vannamei, P. occidentalis, P. schmitti, P. merguiensis,* and *P. setiferus*); and (2) 5000 lb of raw, frozen, shell-on 21–25 count farm-raised giant tiger shrimp (*P. monodon*). These shrimp futures contracts in the U.S. were discontinued after 2000.

The Fish Pool exchange, a futures market for salmon, opened in Norway in 2007. In 2013, it executed financial salmon contracts for 3.64 billion

Norwegian krone. Fish Pool works closely with the NASDAQ Clearing House and its cleared contracts approached 100% in 2015. Fish Pool provides the market with information on spot prices, forward prices, and historical information on salmon prices.

Generic advertising of seafood and aquaculture products

One of the major programs of coordinated cooperative action in marketing is generic advertising. A generic marketing campaign is typically conducted to benefit a generic product or grouping of similar products without identifying brand names or product origins. Generic advertising campaigns for individual commodities have often been supported and funded by producer groups, food companies, food organizations, and/or state governments. State governments have engaged in generic promotion programs to enhance the state's agricultural product sales. For example, generic state promotion programs include those conducted for Washington apples, Florida citrus, and Idaho potatoes. Generic marketing campaigns are run by organizations such as the Alaska Seafood Marketing Association, The Catfish Institute, Virginia Marine Products Board, and the National Fisheries Institute. Successful generic campaigns have been run by non-seafood producer organizations such as the American Dairy Association, American Egg Board, Beef Industry Council, California Raisin Advisory Board, International Apple Institute, National Honey Board, National Pea and Lentil Association, National Yogurt Association, Peanut Advisory Board, Popcorn Institute, the Wine Institute, and others.

These advertising programs are designed to stimulate consumers' demand for the related commodity. Consequently, in 1996, the U.S. Congress mandated that all commodity promotion programs utilizing price checkoff programs be evaluated at least once every five years under Section 501-(c) of the 1996 Farm Bill. Ward and Lambert (1993) found that generic advertising increased beef demand, and their results have been used to support additional funding on generic advertising. In contrast, Brester and Schroeder (1995) and Kinnucan et al. (1997) found that generic beef and pork advertising had little effect on demand. Lenz et al. (1998) reported that the effect of advertising by New York dairy farmers was minimal on fluid milk demand with an advertising elasticity of 0.06 for New York City. Chung and Kaiser (1999) confirmed this with an advertising elasticity estimate of 0.07. For catfish, Zidack et al. (1992) reported a benefit–cost ratio of about 13:1. The ratio suggests an enormous benefit, which the authors attributed largely to the inelastic supply of catfish. However, the authors reported an advertising elasticity of 0.007. Kinnucan et al. (1995) concluded that generic advertising was always beneficial to catfish farmers, that is, incremental benefits exceed incremental costs if producers and feed mills share the levy equally.

Evaluation of the effectiveness of other promotion programs includes yogurt and chicken (Mugera et al. 2016), produce (Burnett et al. 2011), vegetables (Govindasamy et al. 2003), cotton (Capps et al. 1996), soybeans (Williams et al. 1998), avocados (Carman and Green 1993), eggs (Reberte et al. 1996), and apples (Richards et al. 1997).

All generic promotional campaigns promote the generic product and do not promote one brand of product over another. Examples of generic promotional campaigns by fisheries and aquaculture related agencies are identified below.

Advertising of seafood – the National Fisheries Institute (NFI)

The NFI primarily promotes the interests of the general seafood industry in Congress and before regulatory agencies. It also promotes and defends the industry and its products to the media and consumers through generic advertising of fish and seafood in general. The NFI sponsors advertising programs that cover various species including catfish, sea bass, cod, crab, halibut, lobster, menhaden, oyster, pollock, quahog, salmon, scallops, shrimp, skate, tilapia, and tuna, among others. It frequently provides advertising materials relating to seafood recipes, seafood safety, and the health benefits of eating fish and seafood. One of the major advertising campaigns of the NFI is the "Eat Fish and Seafood Twice a Week" campaign. The campaign focuses on the message that fish and seafood are economical, delicious, and quick and easy to prepare and that eating seafood at least twice each week can go a long way toward helping to achieve healthy dietary goals. Fish oil also provides significant health benefits, especially in combating heart disease.

NFI is a non-profit trade association representing more than 1000 companies involved in all aspects of the fish and seafood industry. Membership includes U.S. firms that operate fishing vessels and aquaculture facilities; buyers and sellers, processors, packers, importers, exporters, and distributors of fish and seafood; and operators of retail stores and restaurants that sell fish and seafood.

Salmon advertising – the Salmon Marketing Institute (SMI)

The salmon industry had explored the possibility of developing a generic advertising promotion program but the difficulty had been the issue of funding for the program and whether the program should include both farmed and wild salmon. Efforts were made in 1997 to form the SMI, which was funded by salmon farmers in Chile, Canada, and Norway. SMI developed radio advertisement programs that promoted the consumption of fresh salmon and aired in some major U.S. cities at various times of the year. However, SMI fell apart when the U.S. filed an antidumping and countervailing duty case against Chilean salmon farmers in mid-1997. The Chilean farmers terminated their funding to the Institute, which eventually led to the collapse of SMI.

Catfish advertising – the Catfish Institute (TCI)

The Catfish Institute is mainly responsible for the generic promotion of catfish. The Institute is a non-profit organization established in 1986 by a group of catfish farmers and feed manufacturers to raise consumer awareness of the positive qualities of U.S. farm-raised catfish. It is a producer-controlled organization, and it receives its funding from catfish feed mills located in Alabama, Arkansas, Louisiana, and Mississippi in the form of a voluntary $5.00 checkoff per ton of catfish feed sold.

TCI's activities have mainly involved public relations, providing services to food service operators, and advertising. The food service and marketing program is designed to educate chefs and food service operators about the use of U.S. farm-raised catfish. Activities include workshops at culinary schools and sponsoring booths at chefs' and caterers' conferences. Regarding advertising, the focus of the Institute has been on enhancing the image of U.S. farm-raised catfish as a versatile, high-quality, convenient, and mild-flavored fish.

Early advertising themes by TCI focused on the quality of U.S. farm-raised catfish, its availability, versatility, low cost, taste, and relevance as part of the new American cuisine. Later programs highlighted varieties of preparation methods. A "Made in America" theme emphasized the stringent food quality regulations in the U.S. A "Spice it Up" campaign was developed in collaboration with a spice company with an emphasis on demonstrating grilling recipes and promoting summer sales of catfish. More recent campaigns have focused on the growing trends toward buying food locally, with programs targeting specific types of consumers. Other recent themes have included the tagline of "100% American," "Catch of the Every Day," and "Delicious Any Way You Cook It."

The Catfish Farmers of America (CFA) is a national association that is also engaged in some promotional activities. Much of CFA's advertising activities are offered through TCI programs. The association also provides some promotional materials on catfish on their website as well as in their monthly publication *The Catfish Journal*. Besides these established agencies involved in generic advertising and promotion of catfish, individual catfish processors advertise and promote their company's brands and product lines of catfish products.

Tilapia advertising – the Tilapia Marketing Institute (TMI)

The Tilapia Marketing Institute (TMI) was formed to develop generic advertising programs in the U.S. similar to those of TCI. However, the TMI did not become well established and has not functioned as well as TCI. It was started in 1997 as a consortium of producers and suppliers of goods and services to the tilapia industry. The founding members provided the initial funding for advertising programs to increase U.S. consumer awareness of tilapia.

The early emphasis was on a marketing communications program. Some of the earlier activities of TMI included working with journalists and a variety of print media to create familiarity and awareness of tilapia to U.S. consumers. TMI worked actively to obtain coverage of food stories that included recipes and food reviews, business stories that covered the growth of the tilapia industry, technology stories that discussed production practices, and travel stories that enlightened consumers about the international status of tilapia. Much of TMI's generic campaign focused on tilapia's mild flavor, recipe versatility, and widespread availability. The campaign used media such as food magazines and newspapers, well-known television personalities, and respected chefs. They also sponsored events at conferences of chefs.

Initial funding for TMI activities was for 2 years. Lack of funding since 2000 prevented the Institute from continuing any meaningful advertising campaigns. However, individual tilapia companies have developed brand advertising programs, promoting their brands and products in the seafood marketing trade literature.

There also is the American Tilapia Association (ATA), which engages in a minimal amount of promotion of tilapia. Advertising of tilapia by ATA is mainly in the form of providing information including production, supply, prices, trade, markets, and recipes on the Internet.

Trout advertising – the United States Trout Farmers Association (USTFA)

The USTFA is the main mouthpiece of the trout industry. The major objective of the Association is to promote all aspects of the trout industry and especially to establish a high-quality image of trout products in the marketplace. Membership is offered to all individual farmers and companies engaged in or associated with the trout industry, including major suppliers of products or services. The Association promotes trout in the form of providing information through its website as well as through a 40-page book of recipes. The book contains over 80 complete recipes plus an additional 10 recipes for sauce and stuffing for trout. General information about trout – its nutritional qualities, tips on handling, best basic preparation methods, and step-by-step instructions on how to bone a trout, whether cooked or uncooked – is included in the recipe book.

Aquaculture market synopsis: oysters

Oysters are produced under different aquatic systems: natural, managed, and cultured. The natural production cycle involves oysters that grow in the wild and are harvested for the market. It does not involve any human or artificial interventions in the growth process. With a managed production system, management entails periodic scraping of oyster beds to reduce clustering. The cultured system involves cultivation of wild-collected stock; the wild stock is

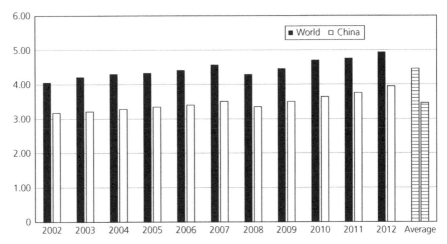

Fig. 8.3 World production of oysters, 2002–2012 (million tons). Source: FAO (2013).

used either as broodstock for spawning and subsequent hatchery and growout, or as early lifestage stock that is consequently used for growout into marketable sizes.

Total world oyster production in 2012 was about 4.94 tons with the dominant countries being China, Korea, Japan, the U.S., and France. China is the largest producer, accounting for about 77% of total world production over the past decade (Fig. 8.3). China is also the largest market for oysters, with domestic supplies accounting for much of the demand. Other markets include Korea, Japan, the U.S., and Canada. International trade in oysters is not as well developed as that for other seafood products because of public health and food safety concerns. Consequently, countries have strict regulations on the importation of live, fresh, and frozen oysters.

The U.S. has regulatory guidelines on oysters. A National Shellfish Sanitation Program (NSSP) certification is required to market oyster products. The NSSP is a federal/state cooperative program that ensures sanitary control of shellfish production and sale for human consumption. The U.S. Food and Drug Administration (FDA) also has international agreements with foreign governments to participate in the program. Under the program, processing plants and dealers are inspected and certified by individual states for both intrastate and interstate shipments of oysters as well as for import and export. The EU also has strict regulations on bivalve mollusks, which include oysters. Any oysters placed on the market should come from an EU-approved fishery product establishment or premises or approved bivalve mollusk production areas. This applies to both oysters from EU and non-EU countries. For non-EU countries exporting to the EU, each consignment should have an appropriate signed health certification.

In the U.S., capture-based oyster aquaculture is practiced in the Northwest and Northeast, while oyster production in the Gulf of Mexico is largely natural.

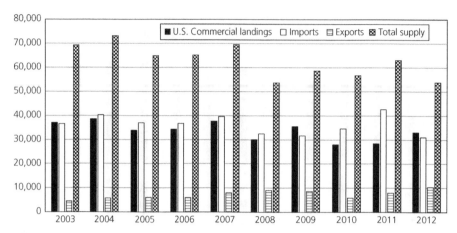

Fig. 8.4 U.S. supply of oysters, 2003–2012 (1000 lb meat weight). Source: NOAA-NMFS (2013).

The main oyster species produced from the Northwest is the Pacific oyster (*Crassostrea gigas*), while the oyster species found in the Gulf of Mexico region and the Chesapeake Bay region in the Northeast is the Eastern oyster (*Crassostrea virginica*). The Eastern oyster accounts for a majority of total U.S. oyster harvests. In 2012, the total U.S. oyster commercial landings yielded 33 million pounds of meat, out of which the Gulf of Mexico region accounted for 20.4 million pounds, representing 62% of the national total. The Northeast region produced 1.9 million pounds, which represented 5% (NOAA-NMFS 2013). The Northwest region produced 9.4 million pounds, representing 28%.

The U.S. is also an importer of oysters, averaging 36.3 million pounds from 2003 through 2012 compared to 33.7 million of domestic commercial landings over the same period (Fig. 8.4). Imported oyster products come in a variety of forms including canned, smoked, and fresh/frozen. China, South Korea, and Canada are the major exporters of oysters to the U.S. China and South Korea exports are canned and smoked oysters, while Canada exports fresh/frozen oysters to the U.S. Canned oysters constitute a significant portion of total U.S. imports. The main competitive product form imported to the U.S. is fresh/frozen oysters from Canada. Muth et al. (2000) reported that Canadian oyster products target the half-shell market and are distributed widely throughout the U.S.

The main players in the U.S. oyster value chain are harvesters, wholesalers, processors, and retailers. Mature oysters from natural, managed, and cultured systems are harvested and generally sold to wholesalers and processors, and sometimes directly to independent restaurants and food retail outlets. Harvesting of oysters occurs throughout the year although the meat yield is affected by season of the year (Lutz 2012). Higher meat yields are obtained from oysters harvested in colder months compared to oysters harvested in warmer months.

Oysters may be sold by the dozen or in packages of sacks, boxes, and/or bushel bags. The primary processing involves manual shucking of shellstock oysters and grading the meat into different sizes. Oysters are sold as live in the shell, raw shucked, or frozen. Retail packages come in the form of 8-ounce and 12-ounce cups, pint, quart, and gallon (Lutz 2012). Other processed products available on the market are half-shell and value-added products such as smoked, cooked, canned, and breaded oysters.

Regarding demand, size, flavor, and meat content are important attributes to consumers. The demand for oysters in Asia is growing, particularly from the hotel and restaurant sectors. In China, half-shell oysters are commonly used to make oyster sauce. French Gillardeau varieties are especially popular in upscale Chinese restaurants (Godfrey 2013). In the U.S., oysters are consumed both at home and in restaurants and may be in the raw form or cooked (steamed).

Summary

Aquaculture growers produce widely diverse types of aquatic plants and animals. Demand for seafood and aquaculture products tends to vary with species that traditionally have been captured in local waters. Markets and marketing systems for aquaculture products reflect this diversity.

The biology of the species raised and the production system used play major roles in the volume and seasonality of supply with implications for prices received by farmers. The chapter summarizes a number of widely different examples of aquaculture production and how these relate to marketing.

Marketing alternatives for aquaculture growers may include sales to processors, to livehaulers, or directly to end consumers. Some of the market requirements unique to each of these market outlets are discussed. Forming cooperatives to compete for larger contracts may be a viable option, but marketing cooperatives can be difficult to manage and the failure rate is high.

Trends in the U.S. food and agricultural industries point toward concentration, which means fewer companies competing with each other. The trend applies to fish processing plant sizes, as well as the food retailing and wholesaling industries. These changes can affect farmers in a variety of ways including disparity of bargaining power, use of production contracts, and competition from imports. The large companies often have the greater bargaining power, and farmers can lose market access as a result. Therefore, by organizing into cooperatives or farmer groups, farmers can control a larger portion of their products, and have greater bargaining power than an individual farmer would have.

The cooperative provides farmers the opportunity to perform some joint marketing responsibilities including assembling of products, negotiating with large buyers of farm products, exercising some power in the marketplace, spreading risks and costs, and in some cases processing farm commodities. There are four classes of marketing cooperatives based on how they are organized, membership affiliation, control, and geographical area: local cooperatives; centralized cooperatives; federated cooperatives; and mixed cooperatives. Most marketing cooperatives operate as a marketing agent by collecting products of members for sale, grading, packaging, and performing other marketing functions. Other cooperatives are organized to perform processing functions, or negotiate with processors or buyers on behalf of the members. One of the major programs of coordinated cooperative action in marketing is generic advertising. Marketing orders help to stabilize market conditions, and provide benefits to producers and consumers by establishing and maintaining orderly marketing conditions.

Study and discussion questions

1 Explain, using examples, how the choice of species and production system can affect the marketing alternatives available to an aquaculture grower.
2 What are the important questions to ask when considering selling to a processing plant?
3 What are the major difficulties associated with forming a marketing cooperative?
4 What advantage is it to a small-scale producer to hold fish that are market size?
5 What are the advantages and disadvantages of holding fish in cages?
6 What are the advantages and disadvantages of holding fish in tanks?
7 What are the keys to success for fee-fishing operations?
8 What is a marketing order and what key provisions are allowed under one?
9 What is the Capper-Volstead Act and why was it enacted?
10 What are the key differences among the various forms of cooperatives?

Appendix 8A: The Capper-Volstead Act

(Public-No. 146-67th Congress)
An Act to Authorize Association of Producers of Agricultural Products

> Be it enacted by the Senate and House of Representatives of the United States of America in Congress assembled, That persons engaged in the production of agricultural products as farmers, planters, ranchmen, dairymen, nut or fruit growers may act together in associations, corporate or otherwise, with or without capital stock, in collectively processing, preparing for market, handling, and marketing in interstate and foreign commerce, such products of persons so engaged. Such associations may have marketing agencies in common; and such associations and their members may make the necessary contracts and agreements to effect such purposes; Provided, however, That such associations are operated for the mutual benefit of the members thereof, as such producers, and conform to one or both of the following requirements:

> First. That no member of the association is allowed more than one vote because of the amount of stock or membership capital he may own therein, or,

Second. That the association does not pay dividends on stock or membership capital in excess of 8 per centum per annum.

And in any case to the following:

Third. That the association shall not deal in the products of nonmembers to an amount greater in value than such as are handled by it for members.

Sec. 2. That if the Secretary of Agriculture shall have reason to believe that any such association monopolizes or restrains trade in interstate or foreign commerce to such an extent that the price of any agricultural product is unduly enhanced by reason thereof, he shall serve upon such association a complaint stating his charge in that respect, to which complaint shall be attached or contained therein, a notice of hearing, specifying a day and place not less than thirty days after the service thereof, requiring the association to show cause why an order should not be made directing it to cease and desist from monopolization or restraint of trade. An association so complained of may at the time and place so fixed show cause why such order should not be entered. The evidence given on such a hearing shall be taken under such rules and regulations as the Secretary of Agriculture may prescribe, reduced to writing and made a part of the record therein. If upon such hearing the Secretary of Agriculture shall be of the opinion that such association monopolizes or restrains trade in interstate or foreign commerce to such an extent that the price of any agricultural produce is unduly enhanced thereby, he shall issue and cause to be served upon the association an order reciting the facts found by him, directing such association to cease and desist from monopolization or restraint of trade. On the request of such association or if such association fails or neglects for thirty days to obey such order, the Secretary of Agriculture shall file in the district court in the judicial district in which such association has its principal place of business a certified copy of the order and of all the records in the proceeding, together with a petition asking that the order be enforced, and shall give notice to the Attorney General and to said association of such filing. Such district court shall thereupon have jurisdiction to enter a decree affirming, modifying, or setting aside said order, or enter such other decree as the court may deem equitable, and may make rules as to pleadings and proceedings to be had in considering such order. The place of trial may, for cause or by consent of parties, be changed as in other causes.

The facts found by the Secretary of Agriculture and recited or set forth in said order shall be prima facie evidence of such facts, but either party may adduce additional evidence. The Department of Justice shall have charge of the enforcement of such order. After the order is so filed in such district court and while pending for review therein the court may issue a temporary writ of injunction forbidding such association from violating such order of any part thereof. The court may, upon conclusion of its hearing, enforce its decree by a permanent injunction forbidding such association from violating such order or any part thereof. The court may, upon conclusion of its hearing, enforce its decree by a permanent injunction or other appropriate remedy. Service of such complaint and of all notices may be made upon such association by service upon any officer or agent thereof engaged in carrying on its business, or any attorney authorized to appear in such proceeding for such association, and such service shall be binding upon such association, the officers, and members, thereof.

Approved, February 18, 1922 (42 Stat. 388) 7 U.S.C.A., 291–192

References

ANDAH. 1997. Boletín InformativoTécnico. Asociación Nacional de Acuicultores de Honduras. ANDAH, Choluteca, Honduras.

Avault, J.W. 1996. *Fundamentals of Aquaculture*. AVA Publishing Company, Inc., Baton Rouge, Louisiana.

Bardach, J.E., J.H. Ryther, and W.O. McLarney. 1972. *Aquaculture: The Farming and Husbandry of Freshwater and Marine Organisms*. Wiley-Interscience, New York.

Bjorndal, T., G.A. Knapp, and A. Lem. 2003. Salmon – A Study of Global Supply and Demand. FAO/Globefish Volume 73, Food and Agricultural Organization, United Nations, Rome.

Boyd, C.E. and H. Egna (eds.). 1997. *Dynamics of Pond Aquaculture*. CRC Press, New York.

Brester, G.W. and T.C. Schroeder. 1995. The impacts of brand and generic advertising on meat demand. *American Journal of Agricultural Economics* 77:969–979.

Burnett, P., T.H. Kuethe, and C. Price. 2011. Consumer preference for locally grown produce: An analysis of willingness-to-pay and geographic scale. *Journal of Agriculture, Food Systems, and Community Development* 2:269–278.

Burt, S. 2000. The strategic role of retail brands in British grocery retailing. *European Journal of Marketing* 34:875–890.

Capps, O., Jr., D.A. Bessler, G.C. Davis, and J.P. Nichols. 1996. Economic evaluation of the cotton checkoff program. In: Ferrero, J.L. and C. Clary, eds. Economic evaluation of commodity promotion programs in the current legal and political environment. NEC-63 and NICPRE (pp. 119–115). Cornell University, Ithaca, New York.

Carman, H.F. and R.D. Green. 1993. Commodity supply response to a producer-financed advertising program: the California avocado industry. *Agribusiness* 9:605–621.

Chung, C. and H.M. Kaiser. 1999. Distribution of gains from research and promotion in multistage production systems: comment. *American Journal of Agricultural Economics* 81:593–597.

Dunn, J.R., A.C. Crooks, D.A. Frederick, T.L. Kennedy, and J.J. Wadsworth. 2002. Agricultural Cooperatives in the 21st Century. United States Department of Agriculture, Rural Business–Cooperative Service, Cooperative Information Report 60, November 2002.

Dunning, R.D. 1989. Economic optimization of shrimp culture in Ecuador. Master's thesis. Auburn University, Auburn, Alabama.

Engle, C.R. 1997. Economics of tilapia (*Oreochromis sp.*) aquaculture. In: Costa-Pierce, B.A. and J.E. Rakocy, eds. *Tilapia Aquaculture in the Americas*, vol. 1. World Aquaculture Society, Baton Rouge, Louisiana.

Engle, C.R. 2006. Marketing and economics. In: Lim, C. and C.D. Webster, eds. *Tilapia: Biology, Culture, and Nutrition*, pp. 619–644. Food Products Press, New York.

Engle, C.R. and G.L. Pounds. 1993. Trade-offs between single- and multiple-batch production of channel catfish: an economics perspective. *Journal of Applied Aquaculture* 3:311–332.

Engle, C.R. and N.M. Stone. 2005. *Aquaculture: Production, Processing, and Marketing. Encyclopedia of Animal Science*. Marcel Dekker, Inc., New York.

Food and Agriculture Organization of the United Nations (FAO). 2002. The State of World Fisheries and Aquaculture 2002. FAO, Rome.

Fitzsimmons, K. 2000. Tilapia aquaculture in Mexico. In: Costa-Pierce, B.A. and J.E. Rakocy, eds. *Tilapia Aquaculture in the Americas*, vol. 2, pp. 171–183. World Aquaculture Society, Baton Rouge, Louisiana.

Godfrey, M. 2013. Oyster farmer struggles to fill China demand. Available at www.seafoodsource.com/news/aquaculture/oyster-farmer-struggles-to-fill-china-demand.

Govindasamy, R., B. Schilling, K. Sullivan, C. Turvey, L. Brown, and V. Puduri. 2003. Returns to the Jersey Fresh Promotional Program: The Impacts of Promotional Expenditures on Farm Cash Receipts in New Jersey. Department of Agricultural, Food and Resource Economics, Food Policy Institute, Rutgers University, New Brunswick, New Jersey.

Green, B.W. and C.R. Engle. 2000. Commercial tilapia aquaculture in Honduras. In: Costa-Pierce, B.A. and J.E. Rakocy, eds. *Tilapia Aquaculture in the Americas*, vol. 2, pp. 151–170. World Aquaculture Society, Baton Rouge, Louisiana.

Hanley, F. 2000. Tilapia aquaculture in Jamaica. In: Costa-Pierce, B.A. and J.E. Rakocy, eds. *Tilapia Aquaculture in the Americas*, vol. 2, pp. 204–214. World Aquaculture Society, Baton Rouge, Louisiana.

Iskow, J. and R. Sexton 1992. Bargaining Associations in Grower-Processor Markets for Fruits and Vegetables. Research Report 104. USDA Rural Development, 19 pp.

Jensen, G.L., G. Treece, and J. Wyban. 1997. Shrimp farming in Nicaragua: market opportunities and financing options for new investments. *World Aquaculture* 28:45 –52.

Kinnucan, H.W., R.G. Nelson, and H. Xiao. 1995. Cooperative advertising rent dissipation. *Marine Resource Economics* 10:373 –384.

Kinnucan, H.W., H. Xiao, C.J. Hisa, and J.D. Jackson. 1997. Effects of health information and generic advertising on U.S. meat demand. *American Journal of Agricultural Economics* 79:13 –23.

Kouka, P.J. and C.R. Engle. 1998. An estimation of supply in the catfish industry. *Journal of Applied Aquaculture* 8:1 –15.

Lenz, J., H.M. Kaiser, and C. Chung. 1998. Economic analysis of generic milk advertising impacts on markets in New York state. *Agribusiness* 14:73 –83.

Lin, K.W. 1995. Progression of intensive marine shrimp culture in Thailand. In: Browdy, C.L. and J.S. Hopkins, eds. *Swimming through Troubled Water. Proceedings of the Special Session on Shrimp Farming. Aquaculture '95*, pp. 13–23. World Aquaculture Society, Baton Rouge, Louisiana.

Little, D. 1995. The development of small-scale poultry-fish integration in Northeast Thailand: potential and constraints. In: Symoens, J.J. and J.-C. Micha, eds. *The Management of Integrated Freshwater Agro-piscicultural Ecosystems in Tropical Areas*, pp. 265–276. CAT/RAOS, Brussels, Belgium.

Lutz, C.G. 2012. Oyster Profile. Agricultural Marketing Resource Center, A National Information Resource for Value-added Agriculture. Available at www.agmrc.org/commodities-products/aquaculture/aquaculture-non-fish-species/#Oysters. Accessed March 28, 2016.

Matulich, S.C., and M. Sever. 1999. Reconsidering the initial allocation of ITQs: the search for a pareto-safe allocation between fishing and processing sectors. *Land Economics* 75:203–219.

Mugera, A., M. Burton, and E. Downsborough. 2016. Consumer preference and willingness to pay for a local label attribute in Western Australian fresh and processed food products. *Journal of Food Products Marketing* 1–21. DOI: 10.1080/10454446.2015.1048019.

Muth, K.M., D.W. Anderson, S.A. Karns, B.C. Murray, and J.L. Domanico. 2000. Economic Impacts of Requiring Post-Harvest Treatment of Oysters. Research Triangle Institute, Research Triangle Park, North Carolina.

National Oceanic and Atmospheric Administration – National Marine Fisheries Service (NOAA-NMFS). 2003. Average wholesale prices and sales volume of selected fishery products at 10 major central wholesale markets in Japan. Southwest Regional Office, National Marine Fisheries Service, Long Beach, California.

National Oceanic and Atmospheric Administration – National Marine Fisheries Service (NOAA-NMFS). 2013. Fisheries of the United States – 2012. Office of Science and Technology, Fisheries Statistics Division, Silver Spring, Maryland.

Pillay, T.V.R. 1990. *Aquaculture: Principles and Practices*. Fishing New Books, Oxford, UK. 575 pp.

Popma, T.J. and F. Rodriquez, B. 2000. Tilapia aquaculture in Colombia. In: Costa-Pierce, B.A. and J.E. Rakocy, eds. *Tilapia Aquaculture in the Americas*, vol. 2, pp. 141–150.

Pullin, R.S.V., T. Bhukaswan, K. Tonguthai and J.L. Maclean (eds). 1987. The second International Symposium on Tilapia in Aquaculture. Manila, Philippines. ICLARM Contribution No. 530. 623 pp.

Reberte, J.C., T.M. Schmit, and H.M. Kaiser. 1996. An ex-post evaluation of generic egg advertising in the U.S. In: Ferrero, J.L. and C. Clary, eds. *Economic Evaluation of Commodity Promotion Programs in the Current Legal and Political Environment*, pp. 83–95. NEC-93 and NICPRE. Cornell University, Ithaca, New York.

Rhodes, V.J. 1993. *The Agricultural Marketing System*. 4th edition, Gorsuch Scarisbrick Publishers, Scottsdale, Arizona, 484 pp.

Richards, T.J., P. van Ispelan, and A. Kagan. 1997. A two-stage analysis of the effectiveness of promotion programs for U.S. apples. *American Journal of Agricultural Economics* 79:825–827.

Rode, R. and N. Stone. 1994. Small scale catfish production: holding fish for sale. FSA 9075. University of Arkansas Cooperative Extension Service, Little Rock, Arkansas.

Rosenberry, R. 1999. *World Shrimp Farming, 1999*. Aquaculture Digest, San Diego, California.

Saborío, A. 2001. La Camaronicultura en Nicaragua – Año 2000. Centro de Investigación del Camarón. Universidad Centroamericana, Managua, Nicaragua.

Setboonsarng, S. and P. Edwards. 1998. An assessment of alternative strategies for the integration of pond aquaculture into the small-scale farming system of North-east Thailand. *Aquaculture Economics & Management* 2:151–162.

Smith, L.J. and S. Peterson. 1982. *Aquaculture Development in Less Developed Countries: Social, Economic, and Political Problems*. Westview Press, Boulder, Colorado.

Stanley, D.L. 1993. Optimización economica y social de la maricultura hondureña. In: Memorias del Segundo Simposio Centroamericano Sobre Camarón Cultivado, pp. 94–118. Federación de Productores y Agroexportadores de Honduras y Asociación Nacional de Acuicultores de Honduras, Tegucigalpa, Honduras.

Torgerson, R. 2000. Cooperative marketing in the new millennium. *Rural Cooperatives* January/February 2000.

United States Department of Agriculture (USDA). 2004. Directory of Farmer Cooperatives. Service Report 22. Rural Business-Cooperative Service. Washington, D.C. Revised January 2004.

University of Alaska, Anchorage. 2001. *A Planning Handbook*. Institute of Social and Economic Research, University of Alaska, Anchorage, Alaska.

Valderrama, D. and C.R. Engle. 2001. Risk analysis of shrimp farming in Honduras. *Aquaculture Economics & Management* 5:49–68.

Ward, R.W. and C. Lambert. 1993. Generic promotion of beef. Measuring the impact of the beef checkoff. *Journal of Agricultural Economics* 44:456–465.

Williams, G.W., C.R. Shumway, H.A. Love, and J.B. Ward. 1998. Effectiveness of the soybean checkoff program. Texas Agricultural Market Research Center Commodity Market Research Report No. CM 2-98. Texas A&M University, College Station, Texas.

Zidack, W., H. Kinnucan, and U. Hatch. 1992. Wholesale- and farm-level impacts of generic advertising: the case of catfish. *Applied Economics* 24:959–968.

CHAPTER 9

Marketing strategies and planning for successful aquaculture businesses

The most successful aquaculture businesses are those that are market-oriented, have diverse markets, and are committed to their customers. Many farmers who wish to develop or expand an aquaculture business have little interest in spending time on a market analysis. Those who are successful in aquaculture are those who have spent time talking to potential customers before beginning to design their production operation.

A carefully developed marketing strategy is important even for growers whose primary market is a processing plant. If the plant is already operating at full capacity, it will not be in a position to purchase additional fish supplies, and the farm will need to identify alternative market outlets. Alternatives may include sales to a different processing plant that targets different markets, live sales to pay lakes, or perhaps changing the production plan to grow a different size or even a different species of fish. The market analysis, plan, and strategy should be the basis from which to make decisions on species, harvest size, and volume. This chapter will present background information for each component necessary for an effective marketing strategy and plan. A sample market plan is presented at the end of the chapter.

Current market situation analysis

Market research

The risk associated with any business decision can be reduced by obtaining comprehensive information on the primary factors involved. However, research can be complex and expensive and should not be done if the cost of the study exceeds the value expected from any resulting business action. For example, a small catfish farm that would generate an annual net profit of $50,000 should not accept a consultant's proposal for a $250,000 study to research the size and structure of the catfish market. This chapter will include a short summary of the role

Seafood and Aquaculture Marketing Handbook, Second Edition. Carole R. Engle, Kwamena K. Quagrainie and Madan M. Dey.
© 2017 John Wiley & Sons, Ltd. Published 2017 by John Wiley & Sons, Ltd.

of market research in planning and implementing market strategies. Chapter 10 provides a more detailed description of marketing research methodologies.

Research will provide the most useful information when the research objectives are defined clearly. Questions for research can be developed more specifically when the company is well into the planning process and has compiled detailed information on overall market conditions and trends.

Gathering secondary (already published) information is much less expensive than generating new information from primary research. Much can be gleaned from the Internet, government reports, U.S. Cooperative Extension Service, and university resources. While it takes time to pull the information together, thorough compilation of secondary data is an essential first step for any size of company. Information on total supply of aquaculture products worldwide is available from the Food and Agricultural Organization (FAO) of the United Nations on their web site (www.fao.org/fishery/statistics/software/fishstatj/en). The total quantity produced and its value can be obtained for individual or groups of species by country, region, and ecosystem by year to determine the overall size of the global market. These data can be used to identify long-term trends in supply of competing species or countries. Information on trade in seafood species and products can also be obtained from the FAO to identify trends for specific types of export markets or to identify potential sources of competition from increased imports. Similar information can be found from published sources within individual countries. In the United States, for example, the National Agricultural Statistics Service (NASS) of the U.S. Department of Agriculture (USDA) publishes statistics on acreage, number of farms, quantities produced, prices paid to producers, and value of the major aquaculture species produced in the U.S. by species and by state (NASS 2004). Some limited information on imports and exports of aquaculture products is also included. Information on the overall seafood market in the U.S. is available from the National Marine Fisheries Service (NMFS 2004) and through its hard copy publications. Similarly, the Department of Agriculture in Australia and the Directorate-General of Agriculture and Rural Development in the European Union post statistics on aquaculture online. The Annotated Bibliography and Webliography at the end of this book include a variety of sources of this type of information.

There are a number of other useful ways to gather information on specific fish markets that may shed light on potential competitors and their marketing strategies. Some buyers and sellers post their requirements, offers, and advertisements on web sites. The advertisements shed light on how competitors are positioning their products, what markets they are targeting, and what their overall marketing strategy might be. Trade magazines such as *Seafood International*, *Fish Farming News*, *Fishing News*, *FiskeribladetFiskaren*, *Seafood Processor*, *Fishing News International*, and *SeaFood Business Magazine* provide similar information through paid advertisements by competing businesses. Seafood shows provide an excellent opportunity to see the array of products, product forms, pricing, and marketing strategies of competitors in the overall seafood market as well as within specific species or product type categories. The Seafood Expo North America, Seafood Processing North America (formerly the International Boston Seafood Show and Seafood Processing North America) is the largest, oldest, and best attended seafood show in the U.S. More specialized shows, such as the Fancy Foods Shows that are held several times a year in various cities in the U.S., provide insight into the higher-priced, value-added, gourmet food category. In Europe, the Anuga (Cologne, Germany) Show, the Bremen Seafood Show (Bremen, Germany), and the European Seafood Show (Brussels, Belgium) target European markets for seafood. Shows that target the major Asian seafood markets include the Japan International Seafood Show, China Fisheries and Seafood Expo, Singapore Seafood Exposition, and Seafood Asia (Hong Kong), among others.

While secondary information sources should be thoroughly mined before expending funds on direct research, secondary data and information should be scrutinized carefully. Much information on the Internet does not undergo peer scientific review or any other type of quality control. Individual companies promote their specific products and trade associations represent the interests of their membership. Neither is obligated to provide a balanced view. Adequate efforts need to be made to ensure that information obtained represents an accurate total view of the market and its trends.

Once a company has investigated secondary sources thoroughly, a decision may be made to initiate formal market research. Research can be done on a variety of levels. The first and necessary step is to spend time to observe potential target markets directly. Direct observations will provide many potential insights into market opportunities and can be used to develop hypotheses for subsequent formal testing. Internet directories, telephone listings, and word-of-mouth suggestions can be used to identify individuals who are knowledgeable of or engaged in the specific markets under consideration. Retail markets and suppliers are excellent sources for current information on their specific customers. Conversations with these individuals can provide an overall view of pricing structures, competing products, and a sense of what is most important in that market.

Direct observations provide clues as to market conditions, but their usefulness is limited to that specific situation. Identification of relationships, trends, and quantification of relationships requires more formal scientific testing and research that becomes more expensive.

An intermediate step can be to hold focus groups. Focus groups can be a cost-effective means of identifying product concepts, un-met customer needs, and market opportunities. However, focus groups should be conducted by an experienced facilitator who is also skilled in selecting participants who represent the target groups.

Once decisions have been made on larger questions related to products and target markets, more formal research may be required. Market experiments and surveys may be useful once very specific research questions have been developed for which secondary data are not available. Chapter 10 in this book provides more detail on methodologies related to formal market research, and Chapter 11 discusses seafood demand analysis.

Market survey research can provide guidance on trends and preferences to guide fish farmers and processing plants as to which types of products will have the greatest chance of success in different types of supermarkets and restaurants. For example, Olowolayemo et al. (1993) found that stores that were members of a chain, had a specialized fish market section, and had sales over $100,000 were those that had a higher likelihood of selling catfish. The study indicated that substantial potential existed for catfish market expansion if obstacles such as a negative consumer image, supply problems, freshness, off-flavor, and competition from other seafood products could be overcome. Hanson et al. (1996) found that stores with floor space greater than 40,000 square feet, a high-income customer base, and belonging to regional chains were likelier to have seafood counters. Stores with weekly sales of $40,000 to $99,000 were more likely to have a seafood counter than grocery stores with sales of $39,000 or less.

Perhaps even more importantly, market research can identify differences in quantity demanded and demand elasticities by season of the year and by region of a country (Singh et al. 2014). Other efforts have used market research to identify the potential to sell locally-caught shrimp at a premium price to restaurants if it were peeled and deveined (Nash and Sharpless 2011).

Competition

Open-market economies prevail throughout the world. The main defining characteristic of open markets is that there is competition among companies and products that results in the availability of choices for consumers. Successful products are those most often selected by consumers, and successful companies are those that do the best job of satisfying needs and wants of consumers by producing products with the most desired characteristics at prices that consumers are willing and able to pay. Thus, understanding the competition is a critical first step in developing an analysis of the current market situation. It is not enough to have identified market opportunities; these opportunities must be assessed in terms of the strength of the competition (Shaw 1986). The fundamental question that the business owner or manager must answer is what their business can provide to customers that is better than anything currently offered by their competitors.

The analysis of the competitive situation should include definition of the size, goals, market share, product quality, and marketing strategies of potentially competing products and companies. The company must identify those areas in which it has a particular strength and can compete successfully within the current competitive situation.

Consumer attitudes/preferences

It is essential to understand the attitudes and preferences of consumers in design-ing market strategies. Development of either new markets for existing products or finding a market for a new product often follows a pattern of: (1) developing awareness by consumers; (2) increasing availability of a new product; (3) changing attitudes toward the product; (4) changing preferences for products; and (5) developing new consumption patterns. Thorough study of market charac-teristics and trends during the planning process should reveal to what extent the product is known, how available it and similar or competing products are, and what the prevailing attitudes, preferences, and purchasing patterns are within the market segments under consideration.

Chapter 11 in this book covers seafood demand analysis in greater detail. This section is included in the context of applying knowledge about consumer attitudes and preferences to develop plans and strategies for more effective marketing.

Consumer surveys conducted over time are helpful to identify regional and national differences in consumer attitudes and preferences and can assist in identification of new, emerging markets and potential strategies for entry of new products into markets. Extensive research on seafood markets in the European Union, for example, shows great variability in preferences by country (Asche and Bjorndal 2011). For Asian markets, Dey et al. (2007) demonstrated widely differing demand elasticities for various types of seafood products that varied by country, species, product form, and income level of consumers.

In the U.S., early surveys documented the development of strong preferences for U.S. farm-raised catfish in the central heartland states of the U.S., as compared to previously-held preferences for wild-caught catfish from the Mississippi River basin (Engle et al. 1990). Other early surveys in the Northeast and Mid-Atlantic regions documented less familiarity with farmed product and growing concerns over the safety of seafood products (Wessells et al. 1994). Foltz et al. (1999) also found that food safety considerations were important in determining consumer preferences for farmed trout. More recent studies have shown that regional differences in preferences have become even more important over time (Singh et al. 2014). Previously contradictory results related to the substitutability of tilapia for catfish in U.S. markets were explained clearly when disaggregated analyses were developed on a regional basis.

While more difficult to study, identification and analysis of new, emerging markets can offer new opportunities for aquaculture and other businesses. In the U.S., contrary to expectations, the market for live fish sales has exhibited rapid growth (Myers et al. 2009; Myers et al. 2010; Puduri et al. 2010; Quagrainie et al. 2011; Thapa et al. 2015). This growth is occurring not only within Asian communities; new, modern supermarket chains in the Northeast, North Central, and West Coast have found that well-designed banks of aquaria in their stores that offer live products can attract new African-American, Hispanic, and Caucasian customers to complement their traditional Asian customer bases.

Overall, the seafood marketing literature clearly shows that the three most important product characteristics are typically taste, quality, and price. Fish has been promoted in recent years for its healthy characteristics, and the emphasis on good nutrition is increasing. Nevertheless, research continues to show that the overriding factor in consumer purchase decisions is the taste of the product. Quality is a complex characteristic that includes freshness, cleanliness, brand identification, brand familiarity, and brand loyalty, as well as other characteristics. If quality standards can be maintained consistently, customers will purchase repeatedly, learn to recognize the brand (brand identification), become familiar with the brand (brand familiarity), and begin to insist on buying only that brand (brand loyalty).

Brand identification has not developed widely in seafood markets. However, as aquaculture companies and industries continue to grow and seafood supplies continue to increase, brand development would be expected to begin to offer some market advantages through differentiating products and developing brand loyalty.

Analysis of business strengths and weaknesses

Careful analysis of the relative strengths and weaknesses of the business should be an integral part of the marketing plan and strategy. These strengths and weaknesses derive from both external and internal factors that can constitute either opportunities or threats to the business.

External threats to seafood businesses can come from a number of sources, but often result from fluctuations in the national economy. For example, economic downturns often result in decreased demand for seafood that causes prices to decline. Unforeseen external shocks to the economy can cause prices to decline. For example, the September 11, 2001, bombing of the World Trade Center in New York City had dramatic effects on seafood sales because restaurant and live fish sales in New York City are dependent on tourism. When tourism falls, demand for aquaculture products sold in these markets also falls. Fluctuating currency exchange rates pose external threats to businesses because a strong currency will attract imports that may compete with domestic production while a weak currency will create profitable export opportunities.

National economic trends that affect income levels can have strong effects on demand for seafood because seafood sales are often related closely to income (Palfreman 1999). Consumers with rising incomes often seek to buy more fish and seafood products. The price of substitute products (i.e., similar types of fish species or products) will also affect demand. Consumers will purchase more of a less expensive type of fish if it is viewed as a good substitute. In a similar manner, national economic policies that result in changes in interest rates can affect demand for fish and seafood products. Interest rate levels will affect decisions to invest in aquaculture businesses and infrastructure. Low interest rates will encourage greater levels of investment. Higher interest rates have the opposite effect. National expectations of higher inflation rates may provide incentives to invest in physical assets such as land, rather than cash-related assets, that may affect the availability of capital for aquaculture investment. Technological changes (computerization, and control and monitoring), political and legal changes (proposals for additional regulations), social and cultural changes (awareness of low-fat characteristics), changes in food consumption habits (fewer set family meals, and more "grazing") are important social changes that are external to the business itself but will affect the demand, and, hence, market price of the product.

There are many other external factors in the marketplace that can affect demand for fish and seafood. One example is how seafood products are handled by buyers. For example, once fish fillets are delivered to a supermarket, the grower and processor no longer have control over how the supermarket treats the product. For example, if the fillets are stacked up high under a light bulb with little ice, the temperature in the middle of the stack may not be adequate to preserve fillet quality. In spite of the fact that high-quality fillets may have been delivered to the supermarket, poor handling by the buyer will result in a poor-quality product.

Other common types of external opportunities and threats include those that involve competitors, customers, distribution channels, and suppliers (Palfreman 1999). Whether competitors have secured cheaper supplies, customers want a different size of box, or if there will be supply shortages in the near future are the types of issues that can represent either a marketing opportunity or a threat to the company. Thus, it is critical that aquaculture owners/managers spend the time to take careful stock of external opportunities and threats at least once a year and adjust overall business goals and objectives to position the business to be successful given external threats and opportunities.

Businesses should also evaluate critically their internal strengths and weaknesses. A small company with a higher cost of production will be better served by developing higher-valued niche markets. A business with expertise to produce certain types of fish that are difficult to spawn may develop a market as a hatchery that supplies scarce and unique fry and fingerlings while another business with access to large amounts of land may concentrate on growout operations of foodfish. Internal strengths may include personnel with detailed knowledge of markets, excellent engineering and maintenance skills, or skill in financial analysis. Examples of internal weaknesses may include assets that have deteriorated, such as ponds that are old, have not been renovated, and may have become too shallow for efficient production. Aging farm personnel may not be able to provide the physical labor required in an efficient manner. Another form of internal weakness is if the business has excessive amounts of labor that may result from down-sizing the farm business.

The analysis of internal strengths and weaknesses must include careful consideration of the financial resources available for market research and any new investment or operating capital requirements. New directions may require re-allocation of company resources, and the company must have a thorough understanding of what the implications will be.

Developing the marketing strategy

Marketing strategy can be thought of as the game plan to achieve the marketing and financial objectives of the business (Palfreman 1999). One strategy may be a low-cost, low-investment model designed to get the most out of previous investments without incurring additional capital outlays before beginning to diversify. Alternatively, if the business sees an opportunity for efficient companies to prosper, it may choose to upgrade. One processor's strategy may be to position the company to be the lowest-cost producer of particular types of value-added products. Such a strategy may depend upon development of the flexibility to make short production runs of more differentiated products that attract higher prices. Other processor strategies may focus on high-volume production of standardized sizes of fillets.

The business's overall marketing strategy should further be developed into a marketing plan of action. The four Ps of the marketing mix (product, place, promotion, and price) can be used to organize the marketing plan of action. Product decisions (i.e., what species of fish, what size of fish to raise, the form of the product) should be based on careful analysis of market conditions and external and internal threats and opportunities. Where (place: geographic market, type of market outlet) to sell fish involves deciding whether to sell fish on the farm, haul to a processing plant, or sell to other farms. Promotion refers to the type of advertising to use to make potential customers aware of the product and its attributes. Pricing strategies are an important part of the marketing plan of action. Careful thought must be paid to appropriate pricing strategies for specific products and markets.

In addition to the above, the timing and seasonality of sales must be analyzed carefully. A baitfish farmer who has borrowed money from a bank with payments scheduled for fall may be in serious financial difficulty if such fish can only be sold in the spring.

The marketing plan of action should include specification of goals and objectives for the short, medium, and long term. Examples of market objectives might be to increase the minimum size of fish purchased by a processing plant to reduce processing costs or to enable the business to compete in a different market segment. Another business may set an objective of increasing market share or penetrating a new market segment. For the above objectives, then, specific, measurable, targets could be specified such as: (1) reduce the percentage of fish less than 0.57 kg (1.25 lb) from 25% to 10% over the next 2 years; (2) increase market share from 20% to 30% over the next 2 years; or (3) generate sales in the new market area equal to 5% of total sales within the next 2 years.

Financial objectives must also be defined clearly. Examples of general financial objectives may be to: (1) survive and avoid bankruptcy; (2) maximize return on investment; (3) increase cash flow; or (4) reduce the debt burden. These may be refined into the following, more specific, targets: (1) within 12 months, reduce overhead expenditures by 20%; (2) undertake capital investment only if it is capable of achieving a rate of return of 15% or above; (3) increase net cash flow from $100,000 per year to $120,000/yr by the end of 3 years; or (4) reduce the debt/equity ratio from 50% to 30% over the next 5 years.

Once marketing and financial objectives have been specified, the strategy or game plan to achieve these objectives can be developed. The following sections will discuss several important considerations and decisions to be made to further develop the marketing strategy.

Developing a retail outlet

Developing a retail outlet for fish requires much advanced planning. It is important to have reliable information on the number of people passing the shop or restaurant each day as well as the proportion of people passing by who might want to buy fish. The amount of money each potential buyer is likely to spend on fish or fish products must be estimated. Gross margins should be estimated from these projections. External factors such as the proximity of supermarkets, the availability of fish suppliers, and relationships with wholesalers must be evaluated. Prospective development of the area, such as road-widening plans, freeway construction, and other possible changes in the locality should be investigated. The business plan should include an estimate of the value of the shop in the event the business should fail.

A successful retail business will pay attention to and follow some common sense guidelines. Employees must be courteous because no one wants to return to a store or restaurant where they have been treated rudely. Prompt service provided to customers is critical to ensure repeat business. The more convenient and easy it is for a customer to purchase from a business, the more sales will be generated. In order to provide service and convenience, it is essential to be flexible. Each individual is different with different tastes and preferences. With a flexible system, it will be easier to meet the needs of each and every customer. Finally, prices charged must be competitive with other businesses.

Market segmentation

Markets can be segmented along many lines. Geographic regions and locations, occupations, special interests, lifestyles, incomes, ages, gender, family size, or certain events are all used to varying degrees by various businesses to segment markets. Market segmentation has become increasingly common throughout the world. The basic concept of market segmentation is to first identify potential segments of a market and then target similar but different products to each segment at different prices. A key criterion for successful market segmentation is that the company understands different preferences and characteristics of the specific buyers in each market segment. For example, a hybrid striped bass grower might segment markets by supplying live hybrid striped bass to ethnic grocery stores in a major urban area at one price, but sales of whole fish on ice to upscale restaurants in the same city at a different price.

The seafood industry historically has relied upon a "mass" or "undifferentiated" marketing approach. Product differentiation has become increasingly common in seafood markets, driven by changing consumer preferences, growing supplies of seafood from aquaculture, and other market conditions. Differentiating and adding value to products often will increase sales but will also increase production, inventory, and promotion costs. Production costs increase because production of two or more products often requires new equipment, separate processing lines, and perhaps separate packaging lines. A factory that specializes in production of one item will be more cost-effective than a factory that manufactures a number of different items. Inventory costs often increase because different products may require different types of storage facilities that can maintain new products at different temperatures. Different distribution systems may be required for different products sold to different markets. Moreover, the greater the number of items marketed, the greater the investment required in safety stocks of inventory carried by companies to guarantee adequate supplies to customers. Differentiated marketing must be accompanied by a range of marketing programs to support the various products sold. Since segmented markets require different promotional programs and messages that appeal to the different types of consumers in each segment, promotional costs will increase. Each advertising program will have a separate cost with overall advertising costs greater the more different products are sold.

Given the potential for increased costs as a company diversifies production, careful analysis is required to identify the most profitable market segments for the company and to target expenditures on new products toward segments with the greatest overall potential for achieving the company's objectives. A segment must be of sufficient size and potential for further growth to justify its development. If over-occupied by competition or if there is no identified need, it may be best for the business to stay with an undifferentiated product. Alternatives to product differentiation may involve concentrating sales of an undifferentiated product in a particular geographic region or to a particular market segment for which the company has a specific strength. Another alternative may be to choose to differentiate its products to capture sales in more than one market segment.

Products and product lines

The identification and selection of products and product lines for the business is an essential component of a successful business and market strategy. Product lines are a series of closely related but somewhat differentiated products. For example, several catfish processing companies have a marinated fillet product line that may include lemon pepper, Cajun, or other seasonings and flavors. The marinated fillet product line is distinct from the nugget, steak, and whole-dressed product lines. Companies with single product lines may have lower costs of production due to production efficiencies, but may also have greater market risk. Differentiated and multiple product lines allow a company to spread risk associated with changing market and economic conditions.

Shrimp, for example, can be processed into many, basic product forms such as: (1) whole, shell-on, raw, frozen; (2) whole, shell-on, cooked, not frozen; (3) whole, shell-on, cooked, frozen; (4) headless, shell-on, raw, frozen; (5) headless, cooked, peeled, frozen; (6) headless, peeled, undeveined, raw, frozen; (7) headless, peeled, deveined, raw, frozen; or (8) headless, cooked, peeled, canned. Primary markets for these different product forms vary considerably and the choice of product forms must be made after careful analysis. A company must establish a unique identity for its product using characteristics or attributes such as price, texture, name, availability, and quality.

The selection of products and product lines must be developed concurrently with the selection of target markets in the company's marketing plan. A product with a high cost of production will need to be of sufficient quality to charge a price sufficiently high to be profitable. Clearly, the target market for such a product would be one in which consumers not only value the particular attributes of that product but also have high enough income levels to be able to pay the price level required. There also need to be enough consumers in that segment to have the volume of sales required to provide an adequate return on any investment incurred in product development.

Product life cycle

The timing of new product development must be considered with regard to the product's projected life cycle (Fig. 9.1). The first goal for most new products (or an existing product being introduced into a new geographic or demographic market) is to penetrate the target market. This phase of a product's life cycle is known as the product introduction phase. The company's objective during this stage will be to generate awareness of the product. Taste tests and sampling opportunities may be important strategies associated with this stage. The product introduction stage is characterized by low sales but high marketing expenditures,

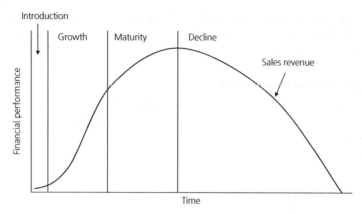

Fig. 9.1 A theoretical diagram of a product life cycle indicating its various stages.

and the product may not generate profits during its introductory stage. The company should seek to generate awareness for the new product quickly and move it into more profitable stages.

As sales increase, the successful product will move into a phase of growth characterized by rapidly increasing sales. The company should begin to generate profit during this stage because, while marketing expenditures are still high, sales begin to grow faster than the increases in marketing expenditures. Key issues during the growth stage involve coordination of the supply chain to ensure timely deliveries and adequate control to guarantee quality throughout the expansion period. A key business objective during the growth stage is to saturate the market with its increased sales.

As the market approaches saturation, the product enters the third stage of maturity. The maturity stage often is characterized by increasing competition from other companies that introduce similar, competing products. Sales continue to increase, but at a slower rate. When a given market segment becomes saturated with that particular product, the business strategy often switches to identification of new markets for the product. Additional promotion and distribution costs are necessary to develop new products, but the costs of production will remain the same.

In the final stage of the product life cycle, sales begin to decrease. It is critical to monitor and manage the stage of decline carefully. When all available markets are saturated, then new products must be developed. Periodic performance review will provide a basis for deciding when to eliminate a product line. The review must consider the hidden costs associated with declining products. Products in the decline phase may take up too much management time, result in short production runs that increase setup time, have unpredictable sales volumes, and may result in less effective advertising expenditures because fewer sales are generated for the same amount of advertising as before.

Product positioning and price-quality considerations

Businesses must make critical decisions related to positioning their product(s) in the marketplace. Consumers' willingness to purchase a product is related to how closely its price matches their perception of its quality. Consumers will pay very high prices for seafood that they view as of the highest quality. This clearly holds true only for markets that include consumers with income levels that allow them to pay these prices. Conversely, they will refuse to pay high prices for a product they view as low quality. Price and quality need to be related for segmentation to be possible.

As a result, the aquaculture business needs to match its price with the quality perceived by consumers for each specific product for that product to be successful. To be financially feasible, the price clearly must exceed production costs for that product. The error committed by many aquaculture businesses has been to set prices based strictly on production costs. Businesses that do not consider the perceptions that prospective customers have of the quality of the product and the consequent implications for its price are doomed from the beginning. Consumers will not pay a high price for a product perceived to be of low quality. Consumers may be suspicious of a product promoted as high quality but with a low price. What is important is to match the price of a product with its quality as perceived by consumers in the market segment being targeted.

Positioning a product as the highest quality with a correspondingly high price, however, may not always be a successful strategy. The quantity demanded for the highest level of quality might not be sufficient for the company to meet its revenue requirements. High-quality products frequently require additional costs related to creating and ensuring the level of quality consumers expect to receive at that price. Careful financial analysis must accompany marketing goals and objectives to be certain that the price consumers are willing to pay exceeds the costs of guaranteeing that level of quality. If the product is not financially feasible at that level of quality, an alternative strategy might be to target a higher-volume, but lower-priced market for which quality standards are not quite as rigid. The lack of comprehensive analysis of price-quality positioning and profitability of alternative price-quality positions has caused many aquaculture businesses to fail.

Techniques that are useful to evaluate alternative product positioning strategies include: (1) a price-quality matrix, and (2) a product space map. These can be developed to consider the position of the company's product or proposed product in relation to other similar or competing products. Pricing strategies should be adopted that match the price-quality positioning of the product.

However, different types of products may be positioned differently even if they are of the same species. For example, small, whole, wild-caught tilapia in Central America is considered a poor-quality, low-priced product. However, fresh and frozen tilapia fillets exported to the U.S. are positioned as medium–high quality and price. Table 9.1 illustrates a potential price-quality matrix for tilapia

Table 9.1 Price-quality matrix, tilapia, Honduras[a].

Product quality	Price		
	High	Average	Low
High			
650 g fresh tilapia fillet	$8.80/kg	$6.60/kg	$5.28/kg
Processed in HACCP-approved	Premium price	Market penetration	Value for money
plant	strategy	strategy	strategy
Average			
350 g whole-dressed tilapia,	$2.64/kg	$2.05/kg	$1.46/kg
constantly on ice			
Processed in HACCP-approved	Market skimming	Average market	Economy strategy
plant	strategy	position strategy	
Low			
250 g whole-dressed tilapia,	$1.91/kg	$1.50/kg	$0.73/kg
occasionally on some ice			
Several days old	Single sale strategy	Inferior goods strategy	Cheap goods strategy

HACCP, hazard analysis of critical control point.
[a] Price data were adapted from Green and Engle (2000); Funez et al. (2003a, b); and Monestime et al. (2003).

produced in Honduras. A fresh fillet from a large 650 g (1.4 lb) tilapia processed in an HACCP-approved plant would constitute a high-quality product. Possible pricing strategies could include a premium price, market penetration, or a value-for-money strategy. A premium price strategy might be pursued for lower-volume sales in a luxury market, while a market penetration pricing strategy to enter a new market for tilapia fillets would be to charge a medium price. If the company has identified a market segment with consumers known to be value-conscious, a lower price might be required as a value-for-money strategy.

An average quality tilapia product in Honduras would be a 350 g (0.77 lb) whole-dressed tilapia on ice. Charging a price at the upper end of the price range for this type of product would constitute a market skimming approach that would be accompanied by low sales volumes. Charging a price at the lower end of the range would be an economy pricing strategy. For low-quality, 250 g (0.55 lb) whole-dressed tilapia that is occasionally held on ice, selling at the upper end of the price range would likely result in only a single sale without repeat sales. To sell additional volumes would require even lower prices in either an inferior good, or cheap goods strategy.

Dover sole has been consistently viewed as a high-quality fish in the Northeastern U.S. Its growing scarcity has further driven its price upwards. Thus, it is considered as a high-quality, high-priced species as viewed in the product space map illustrated in Fig. 9.2. In contrast, buffalofish is considered a low-quality, low-priced product in seafood markets in the Southern U.S.

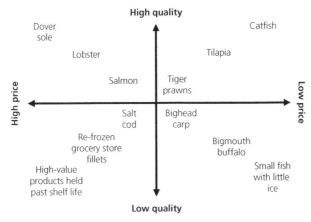

Fig. 9.2 Generalized example of a product-space map with various types of seafood species. The exact position of a product will reflect not only the species but also product form, size, and handling.

Once the company has analyzed carefully the current market situation and understands consumer attitudes and preferences toward its products, the current stage of the product life cycle of its current products, and where these are positioned on the price-quality matrix, broader decisions can be made as to the number of product lines and the size of each product line. The size refers to the number of different products within each product line. These decisions must be based on the supply capacity of the company and the costs associated both with adding new products to existing product lines and adding entirely new product lines. Larger companies that control greater volumes of supply and have larger processing capacity are in a better position to offer a greater degree of product differentiation than are smaller companies.

Fish species with existing demand

Different species of fish are frequently considered to be different products. Asche (2001) indicated that it is easier to market an aquaculture product from species that have traditionally been sold in the area. However, the business should not assume that this is always the case. Roheim et al. (2007) found that product form and other attributes may be more important than the species of fish in terms of customer choices. If a market exists for a particular species, consumers in that market have already developed expectations and perceptions of its quality and the price that they are willing to pay for that quality of product. For high-valued species such as sole, the existing market price for wild-caught species may be high enough to result in profitable sales of aquaculture products. However, there are also cases in which the wild-caught species is offered in a low-quality form (small, whole tilapia with little ice) at a low price (Neira et al. 2003). In these cases, it can be difficult to create a market for a higher-quality, higher-priced aquacultured product. In the case of Nicaragua, it would not be profitable for tilapia farmers to sell farmed product at the price of wild-caught tilapia. Thus, tilapia farmers will either have to seek different markets in which consumers have different perceptions of tilapia or invest in intensive promotional efforts to convince consumers that farm-raised tilapia is a different product from wild-caught tilapia without the negative connotations of wild product. Marketing strategies to overcome these hurdles will need to include educational, promotional, and point-of-sale information for consumers to build a customer base for the product.

The U.S. farm-raised catfish industry faced a similar challenge when it began to develop markets outside the traditional market areas along the Mississippi River. While viewed as a lower-cost fish, catfish was consumed frequently as a major protein source by many in the areas surrounding the Mississippi River and throughout the Southeast. However, consumers outside this area considered catfish to be an undesirable, bottom-feeding scavenger. Years of generic advertising by The Catfish Institute successfully changed these perceptions in regions such as the mid-Atlantic and West Coast regions and increased sales in those areas.

New species

Farmers who raise species for a market in which buyers have no previous experience with that species will have to create and develop the market. This can be a long and sometimes expensive process but is easier than the effort to overcome negative perceptions associated with a species. For example, the companies that export tilapia fillets to the U.S. successfully introduced an entirely new species into the U.S. seafood market. New products offer opportunities for market skimming and market penetration pricing strategies (Table 9.1).

A new species is essentially a new product. Prior to investing in any new product or species, careful research is necessary because the failure rate for new products is extremely high. Businesses must have effective processes in place to screen new ideas to reduce the risk of failure. Surveys can be conducted, but the size and scope of the survey should match the size and scope of the proposed introduction. (See Chapter 10 for details on conducting surveys.) A sales curve should be forecast keeping the product life cycle in mind. The survey data should include some information on consumer attitudes and preferences from which the company can judge the possible price-quality positioning alternatives and select promotion strategies. Care must be taken to ensure that the total cost of the research does not exceed the potential sales value.

Market testing is a critical step in the process of developing a market for a new species or a new product. Key parameters that should be measured in market tests include: (1) actual product trial rate; (2) level and frequency of repeat purchase; (3) relative effectiveness of various marketing plans; (4) consumer acceptance of product benefit claims; (5) reaction of the trade to the new product; and (6) potential distribution problems. The best outcome for the market test is for both trial and repeat sales to be high. This indicates that little effort (and, hence, cost) will be incurred during the product introduction phase and that long-term sales potential is good. If trial sales are low, but repeat sales are high, the company will need to invest more during product introduction to make consumers aware of the product, or to consider alternative product benefit claims and promotion strategies. High repeat sales still indicate favorable longer-term sales. However, high sales during the trial combined with low repeat sales would show that the promotion campaign effectively meets consumer desires, but that the product is not meeting customer expectations. Careful analysis would be required to determine the specific product attributes that would need to be changed and whether or not it is feasible to change them. Low trial and low repeat sales indicate problems both with the image promoted of the product and with product characteristics.

Commodity markets

Chaston (1983) defined commodity markets as "industrial markets" in which products are purchased as an ingredient or element to be used in another product that results in economic return for the buyer. A commodity is a homogeneous product produced by an industry as compared to a series of heterogeneous products with distinctive, smaller niche markets. Many commodities are sold in industrial markets as an input into a supply chain that transforms it one or more times before it reaches the end consumer. Some segments of aquaculture have grown and developed to the point where they can be considered commodities. Salmon futures, for example, are traded by Fish Pool, in a manner similar to futures market exchanges for grain and livestock commodities. An example of a seafood commodity that is sold in an industrial market is the Peruvian ancho-veta that is sold to fishmeal processors. Another example would be shrimp that are sold to a manufacturer for use in a seafood entrée.

Niche markets

Some marketing experts maintain that all markets are niche markets (Palfreman 1999). Nevertheless, niche markets are commonly viewed as low-volume, high-priced, specialty markets. Mass marketing is used to create products that appeal to a broad spectrum of consumers, frequently through development of a brand identity recognized across all consumer segments. Niche markets typically consist of a small segment of a large market. Sales volumes frequently are lower in niche markets than in commodity markets but the strategy is to sell fewer products at a higher price. Smaller companies that successfully identify niche-marketing opportunities may have less competition from larger firms. Typically a niche market is developed through a specific contact, and the grower uniquely supplies a custom product to that one particular market.

Small-scale aquaculture growers are often advised to seek out niche markets, yet there are few specific guidelines for doing so. The key component is the creativity and vision to identify a market opportunity in which a consumer need is not currently being met. Approaching an intermediary in that market line with a new concept is the first step. However, since the product is likely to be new, it is critical that the grower view this as a process of developing a relationship or partnership to develop the market. The grower will need to provide full support in terms of providing material for taste tests, sampling, and point-of-sale materials, as well as guaranteeing consistent product quality.

Niche markets in aquaculture typically have consisted of direct sales from the grower to the end consumer. Thus, the fish farmer performs the wholesaling, distribution, and retail functions of the supply chain. In return, the grower captures the profit margins of each of these phases. However, each of these functions also entails costs in the form of investment in additional holding or processing facilities, utilities, labor, advertising, transportation, and packaging as well as additional time of the grower (Morris 1994).

Niche marketing can be done in a cost-effective manner if basic principles are followed (Gordon 2002). The goal is to meet a unique need of the customer by tailoring the product to meet the customer's needs. It is important to understand and use the jargon of the targeted customer. What is important to a grocery store chain will be different from that of an upscale restaurant. Someone fluent in Spanish would be better positioned to approach Hispanic grocers than non-Spanish-speaking individuals. Direct competitors must be evaluated carefully to identify how to position the new product relative to competing products. It is important to study the advertisements, web sites, logos, and brand names of competitors as well as prices and delivery patterns to identify clues as to needs that can be met with the new product. Do the customers want higher quality, lower price, more convenience, better tasting or safer seafood products? It is important to talk to individual potential customers to identify a currently unmet need for that customer. Test marketing is essential to evaluate how receptive prospective buyers will be to the product. Moving cautiously minimizes risk exposure.

Growers often find it difficult to change their emphasis from production to marketing, but successful niche marketing requires a grower to spend at least 50% of his or her time on marketing. For niche marketing to be successful, value must be added to the product either in terms of convenience, taste, or some other attribute and it takes time and sometimes additional cost to do that. It may be difficult for growers who have made a substantial investment in a particular type of production system to switch to production of something that would move well in a particular niche market or adapt in other ways to meet changing demands of that particular market.

Value-added products

The marketing channel comprises a value-added chain in which some type of value is added each time the product changes hands. Sometimes this value consists of the convenience offered by a large food service distributor that can supply all the food items that a restaurant needs with one telephone call or one visit to a web site. However, the expression "value added" more commonly refers to transformation of the product itself. In many ways, the concept of value added has been discussed under the topic of product differentiation and product lines. For example, a fresh catfish fillet product line may add value to the product and differentiate it from other fresh fillets by adding a Cajun or lemon-pepper marinade to it.

Consumers are demanding ever-greater convenience, nutritional value, and variety while still purchasing based on taste. These consumer trends are creating new opportunities to add value and to differentiate products to capture these emerging market opportunities.

However, developing value-added products alone will rarely solve a particular company's economic problems. A well-developed marketing plan based on sound objectives and carefully analyzed strategy is the answer for struggling companies. For some companies, the move to more extensive and varied product lines may fit the company's business plan whereas such an investment in sales force, processing, and packaging infrastructure would not be feasible for others.

Over 20,000 new products are introduced into U.S. grocery stores each year, and over 90% last less than 3 years. Thus, careful market analysis and testing are required to successfully introduce and grow sales of new products. The reader is referred to the sections on products and product lines, and the product life cycle earlier in this chapter as background material for assessing the feasibility of developing a new value-added product for their company.

Business organization and contracting

Part of a marketing strategy may involve the organizational structure of the business. Many fish farming businesses are organized as sole proprietorships or partnerships, but others are vertically integrated companies. Decisions related to changes in the structure of the business and its impact on the strengths and weaknesses of the business should be analyzed carefully in the marketing and business plan. Vertical integration refers to a single company that has control over several stages of the market channel or supply chain. For example, a shrimp company that owns its own farm, hatchery, and packing plant is vertically integrated. A vertically integrated company controls its own supply chain and, thus, is in a position to be more flexible in terms of meeting customer demand throughout the supply chain. Fish farmers may own shares in processing plants and/or feed mills, but the business is not truly integrated unless it is a single company involved in several levels of input supply, production, processing, and final sales.

While there are a number of examples of vertical integration in aquaculture, contract growing is not as common in aquaculture as it is in some other industries. Contracting companies tend to be market-oriented agribusinesses. A good example of contract farming is found in the U.S. poultry industry. Poultry growers are contracted by poultry processors to supply a certain quantity to the processing plant over a given time period. However, poultry growers bear the yield and financial risk of the growout phase, with no participation in market activities.

Sales

All aquaculture businesses must sell products to generate revenue. While many farmers believe that sales are marketing, this book has demonstrated that sales are only one component of marketing. Selling involves a variety of tasks that can include: (1) taking orders; (2) arranging delivery schedules; (3) delivering the product; (4) building relationships, trust, and goodwill to sustain the relationship; and (5) persuading customers to buy (Shaw 1986). Selling involves communicating the most important information to the prospective customer as to what the product will do for them. In order to communicate the quality of the product the individual handling the sales must be very knowledgeable about the business and able to explain in detail the feeds given, the quality of the water, and the post-harvest handling methods used. Understanding the relative production costs will also provide the seller with some flexibility in terms of negotiating changes in deliveries, packaging, and volumes and whether these changes may adversely impact costs. The seller must learn to listen well and understand the particular needs of the buyer and be prepared to meet those needs.

The marketing plan

Every aquaculture business, regardless of its size, should have an overall business plan, and the marketing plan should be a substantial and integral part of the plan. The marketing plan should focus on answering the question of why a buyer should choose this business's product. Characteristics such as reputation, appearance, delivery times, waiting times, and quality, among others, can be important. There are numerous books and resources available on the Internet and elsewhere on developing business plans.

Table 9.2 presents a typical outline for a marketing plan, and Appendix 9A is a hypothetical marketing plan for an aquaculture business located in the U.S. as an example. A marketing plan typically begins with a situation analysis that includes a descriptive summary of the current market. Important subsections of the current market summary include demographics such as the number of people living in targeted cities or regions. Demographic information typically is divided into potential numbers of customers by outlet types (supermarkets, restaurants) as well as information on relative proportions of the population by age, gender, education levels, household income, lifestyle segments, etc. Consumer needs, likes, and dislikes, and buying trends by geographical area are important components, especially since fish and shellfish markets are dynamic. Each market segment has its own buying patterns, purchase volumes, product forms, price, and delivery needs. Thus, it is important to talk to as many different

Table 9.2 Outline for a marketing plan.

I.   Executive summary
II. Overall market situation analysis
     A. Market summary
         1. Consumer demographics
             a. Geographic areas
             b. Age groups
             c. Family structure
             d. Gender
             e. Income
             f. Education
             g. Lifestyle factors
             h. Spending habits
         2. Supermarket demographics
             a. Geographic areas
             b. Age groups
             c. Family structure
             d. Gender
             e. Income
             f. Education
             g. Lifestyle factors
             h. Spending habits of customers
         3. Restaurant demographics
             a. Geographic areas
             b. Age groups
             c. Family structure
             d. Gender
             e. Income
             f. Education
             g. Lifestyle factors
             h. Spending habits of customers
         4. Market needs
             a. Product(s)
             b. Convenience/service
             c. Pricing
         5. Market trends
             a. Supply
             b. Packaging
             c. Health consciousness
         6. Market growth
     B. Analysis of strengths and weaknesses of business
         1. Strengths
         2. Weaknesses
         3. Opportunities
         4. Threats
     C. Competition
     D. Product offering
     E. Keys to success
     F. Critical issues

(Continued)

Table 9.2 (*Continued*)

VII. Marketing strategy
    A.   Mission and strategy
    B.   Marketing objectives
    C.   Financial objectives
    D.   Target markets
    E.   Distribution channels
    F.   Marketing mix
    G.   Positioning and promotion
    H.   Marketing research
IV. Financial analysis
    A.   Planned expenses
        1.   Sales force requirements
        2.   Advertising expenditures
    B.   Sales forecast
    C.   Break-even analysis
IV. Controls
    A.   Implementation
    B.   Marketing organization
    C.   Contingency planning

prospective buyers as possible in the markets targeted, to determine their needs. Useful insights can be gleaned from conversations with aquaculturists and buyers in regions where the product is being sold.

After describing the characteristics of consumers in the target market, the plan should move to an analysis of the position of the product types already being sold. Substitute products sold locally should be identified and market inquiries made. The recent history of sales and revenue for current products should be described in terms of market share, product quality, and promotional strategies. Personal visits to retail markets in the target market area can provide insight into important competitive attributes such as price, product form, product quality, species availability, sources of competing supply, and buyer preferences. Distribution patterns for competitive products should be described in detail in terms of sales through brokers, wholesalers, and retailers. Finally, the macroeconomic environment of population, economic climate, and technology, legal, and social issues should be addressed.

If the target market is a processing plant, it is still important to visit the plant and identify delivery requirements. Some important types of information to obtain from a processor include: contracts; delivery volume requirements; delivery quotas and scheduling; seasonality trends as these affect fish deliveries at the plant; fish size requirements; quality standards and quality control procedures; transportation charges, if any; historical prices paid; dockage rates; frequency of payment to growers; and bonding requirements.

The plan should include a description of overall market trends that are relevant to the business along with an assessment of the potential for market growth. The description of market trends related to the products that the business intends to sell should include discussion of supply and demand characteristics, market size, and past growth by geographic area and demographic segment. The potential for growth should be based on past historical trends in the context of projected changes in consumer preferences, economic conditions, and patterns of international trade.

When the market summary is completed, the next step of the situation analysis is to assess the strengths and weaknesses of the business in relation to opportunities and threats facing the company from both external and internal factors and conditions. Analysis of internal strengths and weaknesses should include: (1) relationships (with buyers, suppliers, people who work in the business, and other businesses); (2) reputation; (3) innovation; and (4) strategic assets. Relationships are key to the success of any business. Establishing and maintaining good relationships with buyers will give a business an advantage over the competition (Palfreman 1999). Special relationships with suppliers and repeated transactions may enable a business to benefit from improved services, short-term credit, improved quality, or even better prices. Within the business, a higher degree of commitment or team spirit may result in greater productivity or efficiency. Good relationships with other businesses may offer opportunities to share information or contracts, or to purchase supplies at bulk prices. The reputation of the business may provide a competitive advantage or disadvantage. Companies with excellent reputations will attract more business and have greater ease of attracting sources of supply. Innovation is required to improve productivity and profits. While innovation can be copied, it cannot be avoided if the business is to be successful. An entrepreneur needs to look deeply within his or her own business and ask what special abilities exist and whether these can provide the business with a competitive advantage in the marketplace (Palfreman 1999).

Following the analysis of strengths and weaknesses in the situation analysis is a discussion of the competition in the market, relative to the proposed product offerings of the business. Competing products should be described in as much detail as possible in terms of product offerings, pricing, and volumes sold in various markets. Distribution patterns of potential customers and current level of customer service should be assessed with the goal of identifying unmet customer needs and gaps in the market.

The key to success for businesses is to identify and provide either a product or service that is not currently offered but would be preferred by customers. The plan should describe these opportunities in detail and also discuss the issues that will be critical to the business's success.

The second major segment of the marketing plan is the description of the marketing strategy itself. The strategy first needs to be articulated succinctly in a paragraph or two. Sometimes a mission statement is included and then specific marketing and financial objectives are listed.

The strategy section of the plan should then list the key markets to be targeted, beginning with a description of the serviceable geographic market area, taking into consideration the travel distance and time. Specific market segments to be targeted within that geographic area should then be described. Since any given market area includes a variety of different types of customers, the plan must determine whether or not there are enough potential buyers of the product to support the specific products proposed by the business. Consumer census data and business or economic development data can be used to estimate the number of potential buyers in the targeted market area. Distribution channels then need to be planned according to the volumes expected and geographic areas. The desired marketing mix is described and divided into key categories. Decisions related to selling to processors or wholesalers as compared to selling directly to retail outlets are important considerations.

The strategy should include a thoughtful analysis of the position of each proposed product in the market. The positioning decisions should be accompanied by a detailed plan for promotion and advertising. This plan should highlight the characteristics of the product that fill unmet customer wants and needs.

The third major section of the marketing plan is the financial analysis. A break-even analysis is developed for the business's marketing strategy. The market potential is estimated through sales forecasts, typically on a monthly basis, by type of market outlet and target market. The forecast establishes goals for annual sales. Costs are projected in the pro forma income statements (also called profit and loss statements), balance sheets, and cash flow budgets. From the financial analysis, specific financial goals can be established. Specific goals for the upcoming year should be based upon improvement in the weakest part of the projected financial performance. A detailed discussion of analyzing and monitoring financial performance can be found in Engle (2010). For example, a business may set a profitability goal of achieving a return on the investment of 15% per year. Alternatively, a company could set a business goal of increasing sales by 40% over the previous year. The business plan will also specify the size, type, and quality of the sales force. The level and quality of customer service should be described. The amount of advertising and sales promotion will be specified along with the amount, types, timing, and projected success of research and development needed.

The final segment of the marketing strategy describes what type of market research will be developed. Even small businesses must have a plan to obtain information on changing market conditions and consumer preferences to be able to make adjustments and adapt to changing markets in a timely fashion.

The plan must also include a detailed methodology for monitoring and evaluating the company's performance in following the marketing plan. Typically revenue, expenses, repeat business, and customer satisfaction are categories that would be monitored to gauge performance. Contingency planning in the event of performance that does not meet expectations is a critical component of the plan.

Aquaculture market synopsis: mussels

Mussels have been raised, captured, eaten, and sold for many centuries in various parts of the world (Avault 1996). In Europe, France, Spain, The Netherlands, and Sweden have long histories of mussel culture (Girard and Mariojouls 2003). For example, the earliest reports of aquaculture in France were of mussel production dating back to 1235 (Bardach et al. 1972). In Asia, the Philippines and Thailand similarly have long histories of mussel production.

Global mussel production worldwide has generally increased over time to reach 2.92 million metric tons in 2012 (Fig. 9.3). Wild-caught mussel production reached a peak of 317,852 metric tons in 1971 and has generally declined since 2003 to just under 100,000 metric tons in 2012. Clearly, the world's supply of mussels comes primarily (95%) from aquaculture production.

The primary mussel species raised in the early years of aquaculture production was the blue mussel (*Mytilus edulis*) with some production of green mussels (*Perna viridis*) and Chilean mussels (*Mytilus chilensis*). By 1980, the FAO records show that the category "sea mussels" became the dominant type of mussel culture. By 1998, production of sea mussels composed 50% of the total farmed supply, followed by Chilean mussels (13%), blue mussels (10%), green mussels (8%), Mediterranean mussels (6%), New Zealand mussels (*Perna canaliculus*) (4%), Swan mussels (*Anodonta cygnea*) (5%), and Korean mussels (*Mytilus coruscus*) (3%) (Fig. 9.4). "Sea mussels" includes various species of mussels, including blue, green, and possibly other species, but the available data from FAO do not allow disaggregation of the "sea mussel" category reported. Production identified exclusively as blue mussels has remained stable over the years, but that of green mussels has declined since 2002 (Fig. 9.5). In contrast, production of Chilean mussels has increased since 2002.

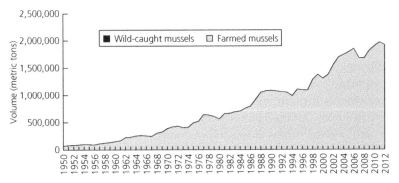

Fig. 9.3 Global production of farmed and wild-caught mussels, 1950–2012. Source: FAO FishStatJ (2014).

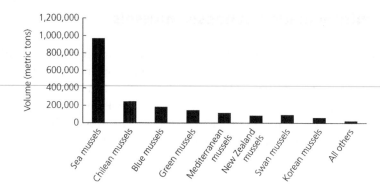

Fig. 9.4 Global production of farmed mussels by species, 2012. Source: FAO FishStatJ (2014).

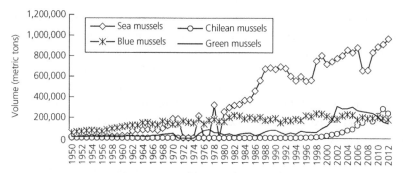

Fig. 9.5 Global production of mussels by species, 1950–2012. Source: FAO FishStatJ (2014).

China is by far the world's largest producer of mussels, producing 45% of all farmed mussels in 2012 (Fig. 9.6). The next largest producers are Chile, Spain, Thailand, New Zealand, Italy, France, and The Republic of Korea.

However, the various species cultured vary by geographic region. The blue mussel is the most commonly cultured mussel in Europe. Spain is the leading producer, producing 50% of the blue mussels cultured in 2012; France is the next largest, with 14% of total production, followed by The Netherlands (11%), Ireland (8%), and Canada (5%). The marketable size of mussels is about 8–15 cm (3 inches).

Traditional on-bottom culture methods were expanded in the 1970s to include new rope culture, or longline culture methods. More recent technological developments include improved spat collecting techniques that improved reliability of supply. Additional hatchery innovations have developed polypoid, hybrids, and selected strains of mussels. Advances in conditioning adult mussels by using algal food and temperature control further contributed to improved reliability of seed supply. Growout culture techniques that are still practiced include on-bottom culture, bouchot culture (pole), raft culture, and longline culture. Mussels, like other types of shellfish and seafoods, have potential for

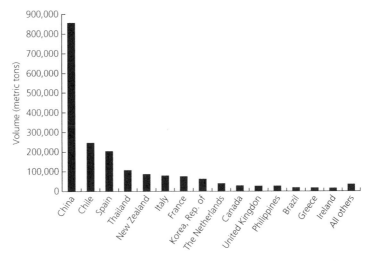

Fig. 9.6 Global production of farmed mussels by country, 2012. Source: FAO FishStatJ (2014).

culture in emerging production systems such as integrated multi-trophic aqua-culture, IMTA (Ridler et al. 2007).

The primary market for mussels continues to be as a live product. However, mussels do not transport well as a live product over long distances because they do not close their shells when out of water as do some other types of shellfish. Interest in mussels has grown in the restaurant trade as away-from-home sales of mussels have grown. Due to the difficulties involved with long-distance shipping of mussels, most export products include processed products in canned, cooked, and frozen forms.

Trade in mussels has generally increased within the European market (Girard and Mariojouls 2003). The leading importing nations in Europe are France, Belgium, and Italy, and the main exporters are The Netherlands, Denmark, and Spain. Mussels are traded primarily as a fresh product, 80% of the total volume traded. Since mussels are primarily consumed as a fresh, whole product, the proximity of the main production areas to the major markets has greatly facili-tated this exchange.

In France, 60% of the supply of mussels is from domestic production (Girard and Mariojouls 2003). The market is segmented based on the culture method (rope-cultured mussels, "bouchot" mussels that are cultured on fixed, wooden poles, and wild mussels) and by species (between the blue and Mediterranean mussels). French "bouchot" mussels are considered a premium product with the highest market price. Mediterranean mussels grown in France and imported from Spain are intermediate-priced products. Mussels imported from The Netherlands ("Dutch" mussels) are the lowest-priced mussel product in France. Dutch mussels are sold primarily in supermarkets for lower prices (Paquotte 1998). However, the price of Dutch mussels increased in 2001, likely due to an increased supply of washed, debyssed, and ready-to-cook mussel products. Generally, wild-caught mussels are the lowest-priced products, with the exception of those harvested from the Basin of Marennes-Oléron (Girard and Mariojouls 2003). This region of France created a regional trademark in 1974 that has successfully resulted in higher prices for its products.

Imports of mussels into France occur mainly during the period of February to April (Paquotte 1996). This is the season of the year when French production is low, and the supply shifts to imports from the United Kingdom, Ireland, and The Netherlands. Mussels are distributed primarily by large retailers in France (Girard and Mariojouls 2003). Most mussels are consumed at home in France, and are prepared as a cooked appetizer or as a main dish. However, away-from-home sales are increasing, including the popular "mussels and chip" dishes which are gaining popularity in many restaurants.

The first companies to market value-added, convenience packs of mussels were Dutch (Girard and Mariojouls 2003). These companies developed ready-to-cook family packs of washed mussels in package sizes of 1–2 kg each. Dutch companies have continued to develop new products, and other companies, notably in Ireland and France, have followed suit. Fresh cooked dishes, pre-cooked, vacuum-packed mussels, and intermediate products have been developed in recent years. Modified atmospheric and vacuum packaging technologies provide opportunities for adding further value to mussel products to preserve freshness, safety, and quality of products. The new packaging technologies provide additional opportunities to add consumer convenience to mussel products.

Economies of scale in mussel farming have been documented on Mediterranean mussel farms in Greece (Theodorou et al. 2010). On the local scale, mussel farming can be quite important. For example, Prince Edward Island produces 80% of all mussel production in Canada and 71% of all production in North America (Department of Fisheries and Oceans 2006). Production of blue mussels on Prince Edward Island represented approximately 1% of the province's total gross domestic product in 2004.

Filter-feeding animals like mussels will grow faster in waters that are more "productive", that is, those that have more nutrients and appropriate water chemistry to support growth of phytoplankton (algae), the base of the food chain. Conversely, growth of mussels and other filter-feeding animals will be slower in waters that have fewer nutrients and lower primary productivity (production at the base of the food chain). However, location decisions of mussel growers must also be balanced against the risk of losses due to predation by birds (Mongruel and Thébaud 2006).

The mussel industry is challenged with food safety concerns relating to what may be filtered from the water by shellfish such as mussels. Shellfish beds may be closed due to contamination of public waters where beds are located.

Contamination may result from a variety of sources including pathogens, harmful compounds released into the waters, and toxic algal blooms. Wessells et al. (1995) chronicles a case study of the impact on demand for mussels in Montreal following reports of harmful algal blooms in mussel-growing areas. The study documented the economic losses during and after domoic acid contamination of Prince Edward Island mussels. The effect of decreased demand on sales of mussels was calculated. In this case, losses consisted of the direct losses during a 4-week ban on all mussel sales. However, loss of sales continued after the ban was lifted as media reports of the contamination event continued in the press. Those farms located outside the contamination area that had clear labels of product origin and location of the farms experienced fewer losses than farms with unlabeled product.

There have also been a few international trade conflicts involving mussels. The Great Eastern Mussel Farms of Tenants Harbor, Maine, filed an antidumping petition against Prince Edward Island mussel producers in 2001. Tariffs were imposed initially, but when the Prince Edward Island mussel growers raised prices the antidumping lawsuit was withdrawn.

Mussel farmers face a wide array of regulations that are complicated by a variety of property rights issues in coastal and marine environments. The regulations vary by country but often include federal, state/provincial, and local laws and ordinances. These laws may variously regulate access or leases to coastal areas, water quality, and endangered species laws (Engle and Stone 2013). If the mussel that a farmer wishes to raise is not a native species, additional laws and restrictions can apply.

Summary

This chapter presents specific details on the process and components of market plans and development of associated marketing strategies. Techniques and information sources for developing an analysis of the current market situation provide a means to identify some potential market opportunities. Understanding the competition and consumer attitudes and preferences is key to uncovering unmet consumer needs and wants. Analysis of the strengths and weaknesses of the business to meet those unmet needs is a critical step in developing the plan. Analysis of the external and internal strengths and weaknesses should result in the identification of competitive advantages for the business. The marketing strategy and plan is then developed based on the business's answer to the question of what unmet consumer need this business can fulfill better than any other business. The strategy is developed, then, to specify the sales goals of the products, the associated costs to supply the markets identified, and the overall feasibility.

Study and discussion questions

1 What is market segmentation? Give an aquaculture example.
2 Explain the product life cycle, using seafood examples.
3 Explain how to successfully develop new markets and use a recent seafood example.
4 Explain the costs associated with product diversification.
5 Explain and give seafood examples of a product-space map and a price-quality matrix.
6 What is the difference between an industrial market and a consumer market? Give an aquaculture example of an industrial market.
7 How does one determine what scale of market research should be undertaken and whether the emphasis should be on collecting primary or secondary data?
8 What are the four Ps of the marketing mix? Explain and describe aquaculture examples of each.
9 Think of examples of business strengths and weaknesses and how these can be used to develop a marketing strategy.
10 What are some differences in developing market strategies for species with existing demand as compared to new species?

Appendix 9A: A sample market plan (hypothetical)

An enterprising family would like to start an aquaculture business. They live in a small city in the southern part of the U.S. They need to develop a market plan to start the business off in a well-organized and well-thought-out business direction.

Executive summary
This family-owned and operated aquaculture business will meet an unmet demand for live, fresh catfish in a small city in the southern U.S. The advantage of this business is the family's love of and enthusiasm for quality fish. Their marketing challenge will be to tap into word-of-mouth advertising to be the supplier of choice to market segments that prefer very fresh fish at a reasonable price.

Vision
This family-owned farm business is based on the assumption that people will prefer to purchase live catfish due to its guaranteed and obvious freshness. The farm business will serve its clients by providing consistently on-flavor fish delivered as and when ordered to provide for their fish supply needs.

Overall market situation analysis
Market summary
Small City USA is located in the southern part of the U.S. People in the area are accustomed to eating freshwater fish such as catfish, buffalofish, largemouth bass, and crappie that they have caught while fishing in the rivers and ponds in the region. Small City USA has tightly knit family groups and is a conservative town that revolves around church and family. Incomes are not high. Many rural poor looking for a better life cycle have out-migrated from depressed farming communities to Small City USA. Educational levels generally are lower than the national average. There is a higher-income segment in the city, particularly in the areas surrounding the hospital and federal facilities located in Small City USA. However, many of the higher-income residents often travel to larger cities within a few hours drive for entertainment and recreation. The city has a population of approximately 50,000 people and is roughly half white and half African-American. However, the Hispanic population is growing rapidly and there are a few Asian families in the area.

Supermarkets cater to the southern lifestyle and feature the main ingredients of southern cooking. Supermarkets located in closer proximity to the hospital carry a wider variety of specialty foods and spices, but the majority of supermarkets are discount types of supermarkets that compete primarily on offering lower-priced foods.

Restaurants in Small City USA include many fast food chains, a number of Mexican and Chinese establishments, and a few barbecue houses. There are several catfish restaurants in the city that are popular, particularly on weekends and in the evenings. Other restaurants advertise plate lunches and southern cooking, and there are a few steak restaurants in the city. One or two restaurants offer some Cajun dishes and a few more innovative dishes, but these are few.

Given the lower-than-average income levels of residents of Small City USA, pricing of products is extremely important. Restaurants that offer specialized cuisine at menu prices in excess of $10 generally do not fare well in the city. Adherence to southern lifestyles also is important and a large percentage of the population does not have a strong sense of adventure with foods.

There has been a marked increase in foods catering to the growing Hispanic population in the city. There has also been an increase in the percentage of African-Americans in the city.

Prices of catfish in local supermarkets are at levels that restrict purchases. Fresh fillets on ice also lack the freshness of live products. Live catfish that guarantee freshness and can be sold at lower per-unit prices may have potential in this market. Accessibility, customer service, and competitive pricing will be important.

Analysis of strengths and weaknesses of business

This family has experience raising freshwater fish, has a strong work ethic, and owns some land within 10 miles of Small City USA. The family owns 25 acres of land with plentiful groundwater supplies. A well that pumps 350 gal/min is already in place.

The family is not from the local area and does not have strong personal ties through the family-church network of relationships. The family is also Caucasian. Since Small City USA has an increasing percentage of African-American and Hispanic residents, it may be difficult for Caucasians to develop strong market relationships with individuals of other races, given continued racial divisions in the community.

Catfish is a well-known and desired product in the community. Prices of catfish offered in the supermarkets and restaurants are medium-high as compared to chicken, beef, and other protein sources. The growing African-American and Hispanic populations also offer market segments that, per capita, tend to eat more fish than do other population segments. However, prices must be reasonable and present greater value than fish products sold in supermarkets and in restaurants.

The primary competition will come from catfish restaurants, supermarkets that carry catfish (a major discount chain has begun to sell catfish in its superstore), and fish markets that carry wild-caught catfish, buffalofish, and other freshwater species. Other competition might come from larger catfish farms that might choose to sell directly to the public.

The product offered will be live catfish. The emphasis will be on the freshness and quality of the product and exceptional service in delivering product to customers.

The keys to success will be to satisfy customers who will be carrying live fish home from the farm or from a truck parked at strategic locations. Critical issues may include the willingness of customers to drive to the farm to purchase fish or the identification of locations in the city where a truck could sell fish successfully. Establishment of effective delivery routes to maximize convenience may be important to the success of the business. The willingness of individuals to clean the fish purchased may be a constraint.

Marketing strategy

The mission of the business is to be the most preferred source of quality live fish for Small City USA. The marketing objectives are:

1 Sell 54,000 lb of live catfish a year.
2 Develop effective word-of-mouth advertising.
3 Become the preferred supplier of catfish for church and family reunion fish frys.

Financial objectives are to:

1 Develop sufficient cash flow for the business to survive in year 1.
2 Beginning in year 2, reduce the debt-asset ratio by 10% a year.
3 Begin to show normal profit in year 3 as markets are developed and sales stabilize.

Target markets will be the African-American and Hispanic populations that prefer quality, very fresh fish. Church and family reunion fish frys will be targeted. Sales will be direct to the public with no intermediaries. The live catfish will be positioned as a higher-quality but lower-priced alternative to fillets sold in supermarkets. The strategy will be to advertise to church ministers, invite church groups to visit the farm, organize youth fishing activities, and provide samples, radio advertisements, and flyers.

Planned expenses are as outlined in Engle and Stone (2014). Ponds will need to be built and equipment purchased that will include an all-terrain vehicle, electric paddlewheel aerators, oxygen meter, mowers, a tractor, waders, nets, a feed bin, and a live car for holding fish. Operating costs will include fingerlings, feed, some part-time labor, fuel, electricity and other utilities, and insurance. Total annual costs (including non-cash costs such as depreciation) are estimated to be $71,186 per year. The break-even price of fish is estimated to be $0.87/lb above operating cost and $1.32/lb above total cost.

The family will serve as the sales force. The part-time labor will also be asked to help spread the word about the farm. Sales the first year are expected to be 12,000 lb, increasing to 54,000 lb by the end of the second year. Anticipated sales price is $1.50/lb, to generate a profit of $0.18/lb, or net returns above all costs of $9,720 per year.

References

Asche, F. 2001. Testing the effect of an anti-dumping duty: the US salmon market. *Empirical Economics* 26:343–355.

Asche, F. and T. Bjorndal. 2011. *The Economics of Salmon Aquaculture*. John Wiley & Sons Ltd., Chichester, UK.

Avault, J.W. 1996. *Fundamentals of Aquaculture*. AVA Publishing Company, Inc., Baton Rouge, Louisiana.

Bardach, J.E., J.H. Ryther, and W.O. McLarney. 1972. *Aquaculture: The Farming and Husbandry of Freshwater and Marine Organisms*. Wiley-Interscience, New York.

Chaston, I. 1983. *Marketing in Fisheries and Aquaculture*. Fishing News Books, London, UK.

Department of Fisheries and Oceans. 2006. An economic analysis of the mussel industry in Prince Edward Island. Policy and Economics Branch, Gulf Region, Department of Fisheries and Oceans, Moncton, New Brunswick, Canada. 25 pp.

Dey, M.M., R.M. Briones, and M. Ahmed. 2007. Disaggregated analysis of fish supply, demand, and trade in Asia: baseline model and estimation strategy. *Aquaculture Economics & Management* 11:1–2.

Engle, C.R. 2010. *Aquaculture Economics and Financing: Management and Analysis*. Blackwell Scientific, Ames, Iowa.

Engle, C.R. and N.M. Stone. 2013. Competitiveness of U.S. aquaculture within the current U.S. regulatory framework. *Aquaculture Economics & Management* 17:251–280.

Engle, C.R. and N.M. Stone. 2014. Costs of small-scale catfish production for direct sales. Southern Regional Aquaculture Center Publication No. 1800, Stoneville, Mississippi.

Engle, C.R., O. Capps, Jr., L. Dellenbarger, J. Dillard, U. Hatch, H. Kinnucan, and R. Pomeroy. 1990. The U.S. market for farm-raised catfish: an overview of consumer, supermarket and restaurant surveys. University of Arkansas Division of Agriculture Bulletin 925, Fayetteville, Arkansas.

Foltz, J., S. Dasgupta, and S. Devadoss. 1999. Consumer perceptions of trout as a food item. *International Food and Agribusiness Management Review* 2:83–101.

Food and Agriculture Organization of the United Nations (FAO). 2014. FishStatJ. FAO, Rome. Available at www.fao.org.

Funez, O., I. Neira, and C. Engle. 2003a. Potential for open-air fish markets outlets for tilapia in Honduras. CRSP Research Report 03-193. Oregon State University, Corvallis, Oregon.

Funez, O., I. Neira, and C. Engle. 2003b. Potential for supermarket outlets for tilapia in Honduras. CRSP Research Report 03-189. Oregon State University, Corvallis, Oregon.

Girard, S. and C. Mariojouls. 2003. French consumption of oysters and mussels analyzed within the European market. *Aquaculture Economics & Management* 7:319–333.

Gordon, K.T. 2002. 3 Rules for niche marketing. Entrepreneur.com. www.entrepreneur.com/article/49608.

Green, B.W. and C.R. Engle. 2000. Commercial tilapia aquaculture in Honduras. In: Costa-Pierce, B.A. and J.E. Rakocy, eds. *Tilapia Aquaculture in the Americas*, vol. 2, pp. 151–170. World Aquaculture Society, Baton Rouge, Louisiana.

Hanson, G.D., J.W. Dunn, and G.P. Rauniyar. 1996. Marketing characteristics associated with seafood counters in grocery stores. *Marine Resource Economics* 11:11–22.

Monestime, D., I. Neira, O. Funez, and C.R. Engle. 2003. Potential for restaurant markets for tilapia in Honduras. CRSP Research Report 03-191. Oregon State University, Corvallis, Oregon.

Mongruel, R. and O. Thébaud. 2006. Externalities, institutions, and the location choices of shellfish producers: the case of blue mussel farming in the Mont-Saint-Michel Bay (France). *Aquaculture Economics & Management* 10:163–181.

Morris, J. 1994. Niche marketing your aquaculture products. North Central Regional Aquaculture Center. Iowa State University, Ames, Iowa.

Myers, J.M., R. Govindasamy, J.W. Ewart, B. Liu, Y. You, V.S. Puduri, and L.J. O'Dierno. 2009. Survey of ethnic live seafood market operators in the Northeastern USA. *Aquaculture Economics & Management* 13:222–234.

Myers, J.J., R. Govindasamy, J. Ewart, B. Liu, Y. You, V.S. Puduri, and L.J. O'Dierno. 2010. Consumer analysis in ethnic live seafood markets in the Northeast Region of the United States. *Journal of Food Products Marketing* 16:147–165.

Nash, B. and N. Sharpless. 2011. Understanding the requirements of Carteret County fishermen and dealers to meet the rising demand for local seafood within North Carolina: a situation assessment. Report prepared for the Carteret County Commercial Seafood Industry and the North Carolina Seafood Coalition, Moorehead City, North Carolina. 16 pp.

National Agricultural Statistics Services (NASS). 2004. Aquaculture Situation and Outlook Report. United States Department of Agriculture, Washington, D.C. Can be found at: www.nass.usda.gov.

National Marine Fisheries Service (NMFS). 2004. Capture species. National Marine Fisheries Service, Washington, D.C. Can be found at www.st.nmfs.noaa.gov.

Neira, I., C.R. Engle, and K. Quagrainie. 2003. Potential restaurant markets for farm-raised tilapia in Nicaragua. *Aquaculture Economics & Management* 7:231–247.

Olowolayemo, S.O., U. Hatch, and W. Zidack. 1993. Potential U.S. retail grocery markets for farm-raised catfish. *Journal of Applied Aquaculture* 1:51–72.

Palfreman, A. 1999. *Fish Business Management*. Fishing News Books, Blackwell Science, Oxford, UK.

Paquotte, P. 1996. The French mussel market: evolution of supply and demand. Actas des VIII'emes Rencontres interrégionales de l'AGLIA, Galmont, St. Hilaire (Vendée), Septembre 1995, Publication ADA 49:43–50.

Paquotte, P. 1998. Le marché de la moule en France: evolution de l'offre et de la demande. VIII'emes Rencontres interrégionales de l'AGLIA, 14–15 septembre 1995. *Globefish* 55:65–80.

Puduri, V.S., R. Govindasamy, J.J. Myers, and L.J. O'Dierno. 2010. Demand for live aquatic products in the mid-Atlantic states: an ordered probit analysis towards consumers' preferences. *Aquaculture Economics & Management* 14:30–42.

Quagrainie, K.K., A. Xing, and K.G. Hughes. 2011. Factors influencing the purchase of live seafood in the North Central Region of the United States. *Marine Resource Economics* 26:59–74.

Ridler, N., M. Wowchuk, B. Robinson, K. Barrington, T. Chopin, S. Robinson, F. Page, G. Reid, and K. Haya. 2007. Integrated multi-trophic aquaculture (IMTA): a potential strategic choice for farmers. *Aquaculture Economics & Management* 11:99–110.

Roheim, C.A., L. Gardner, and F. Asche. 2007. Value of brands and other attributes: hedonic analysis of retail frozen fish in the UK. *Marine Resource Economics* 22:239–253.

Shaw, S.A. 1986. Marketing the products of aquaculture. FAO Fisheries Technical Paper 276, Food and Agricultural Organization of the United Nations, Rome.

Singh, K., M.M. Dey, and P. Surathkal. 2014. Seasonal and spatial variations in demand for and elasticities of fish products in the United States: an analysis based on market-level scanner data. *Canadian Journal of Agricultural Economics* 62:343–363.

Thapa, G., M.M. Dey, C. Engle, and K. Singh. 2015. Consumer preferences for live seafood in the Northeastern Region of USA: results from Asian ethnic fish market survey. *Aquaculture Economics & Management* 19:210–225.

Theodorou, J.A., P. Sogeloos, C.M. Adams, J. Viaene, and I. Tzovenis. 2010. Optimal farm size for the production of the Mediterranean mussel (*Mytilus galloprovincialis*) in Greece. Proceedings of the International Institute for Fisheries Economics and Trade, Montpellier, France.

Wessells, C.R., S.F. Morse, A. Manalo, and C.M. Gempesaw II. 1994. Consumer preferences for Northeastern aquaculture products; report on the results from a survey of Northeastern and Mid-Atlantic consumers. Rhode Island Experiment Station Publication No. 3100, University of Rhode Island, Kingston, RI 02881.

Wessells, C.R., C.J. Miller, and P. Brooks. 1995. Toxic algae contamination and demand for shellfish: a case study of demand for mussels in Montreal. *Marine Resource Economics* 10:143–159.

CHAPTER 10
Marketing research methodologies

Marketing research is essential to the overall success of any business because the major objectives of any seafood business are to meet consumer demand and operate efficiently at a profit. To stay in business and remain competitive, companies rely on various types of marketing research information to formulate marketing strategies, make marketing decisions, or implement marketing concepts. For example, marketing research will help the business manager to find answers to questions such as: "What are the attitudes and desires of consumers?"; "Is there a demand for our product?"; "What is our volume of sales compared to our competitors, or what is our share of the market for the product?"; and "What products will consumers demand in the future?" Answers to such questions are important for business planning as they allow a business to find out more about the current market situation relating to a product of interest as well as to predict future market situations. Market research can also be used to find solutions to specific marketing problems that a company might have.

The American Marketing Association defines marketing research as the function that links the consumer and the public to the marketer through information that is used to: identify and define marketing opportunities and problems; generate, refine, and evaluate marketing performance; and improve understanding of marketing as a process (Bennet 1988). This definition elaborates on the several functions and uses of marketing research. A seafood company that is not doing very well in sales may conduct a marketing research study to obtain information about why their product is not selling and what can be done to improve sales. A new company that wants to introduce a seafood product to the market will first have to find an answer to the question, "Will there be a market for this product or will this product meet a need on the market that has not been satisfied?" Marketing research is therefore conducted for various reasons and it is essential that the research be conducted appropriately.

Seafood and Aquaculture Marketing Handbook, Second Edition. Carole R. Engle,
Kwamena K. Quagrainie and Madan M. Dey.
© 2017 John Wiley & Sons, Ltd. Published 2017 by John Wiley & Sons, Ltd.

An effective market research process can be financially rewarding for a company. If done poorly, however, it could result in the failure of the business. Before embarking on marketing research, the business owner should know the purpose for which the research is to be conducted. Any company embarking on market research should know the type of information it needs and the cost of obtaining that information. Sometimes, marketing research is needed to obtain some general information or market outlook while at other times it is required to solve specific problems.

Types of research and design

The process of conducting marketing research consists of gathering, sorting, analyzing, evaluating and disseminating information for timely and accurate market decision-making. There is so much information in the marketplace that the focus of the process should be to target information necessary to make informed decisions. Market research can be designed in one of three forms: (1) exploratory research; (2) qualitative research; or (3) quantitative research. The type of research that is most appropriate depends upon the objectives. For example, exploratory research would be most appropriate for a new startup business that is taking its first steps in identifying potential markets. Qualitative research would be appropriate for a company attempting to decide whether to change its brand or whether its advertising program should focus more on emphasizing the color, taste, or safety of its fish fillets. Quantitative research could help a company estimate the size of a prospective new market. Each of these types of research is discussed below.

Exploratory research
Through exploratory research, information can be obtained that allows seafood companies to identify and clarify some problems or issues confronting them. It may also provide information that helps a seafood company to identify potential challenges and opportunities and to establish research priorities. Exploratory research raises awareness and provides insights.

In exploratory research, there are no specified objectives; neither is there any structure to the process of gathering market data and information. It is a very informal approach to research which may involve mere observations of things of interest, such as: observing customers as they shop, consumers' buying patterns, clients as sales personnel interact with them, sales or revenue figures; reading periodicals; surfing the Internet; visiting the library; or enquiring about certain products, services, prices, market situations, and current trends and issues. There is no structure to this form of research and it can therefore be used in a number of situations. Related to exploratory research is what is often referred to as market intelligence, or the art of obtaining updates about relevant developments in the market. The market intelligence system can also be informal or formal with a focus on searching for information and anything that may be of interest to the company.

Sometimes, the process of gathering information could consist of purchasing and tasting products of competitors, or scanning periodicals for specific information about the seafood market or seafood products of interest. Many companies subscribe to newspapers, magazines, and industry and trade publications for the purpose of keeping up with industry affairs.

The case of Ippolito's Seafood, Philadelphia

Ippolito's is a wholesale seafood company that sells frozen and fresh seafood to hotels and restaurants across the Philadelphia region. The company began as a seafood retailer selling frozen shrimp, lobster tails, and fish fillets. However, the company was struggling to stay in business because of increased competition from grocery outlets and supermarkets. In the early1980s, through exploratory research, the company realized the need for niche wholesaling in the region that would target the food service industry. Large general-line food distributors performed the seafood wholesale functions in the region at the time. Ippolito's also realized that the traditional wholesalers did not deliver seafood on Saturdays. The company therefore launched into seafood wholesaling offering its traditional products of frozen shrimp, lobster tails, and fish fillets as well as imported Chilean sea bass, New Zealand orange roughy, and fresh octopus and *loup de mer* from the Mediterranean Sea. In 2001, total sales revenue for the company was $47.3 million. Ippolito's clients included the Four Seasons Hotel, Rittenhouse Hotel, and the Park Hyatt as well as neighborhood taverns and restaurants in the Philadelphia area (Bennett 2001).

Qualitative research

Qualitative research also raises awareness and increases insights. However, in qualitative research, theoretical concepts can be tested to provide some definitive explanations. Other textbooks refer to qualitative research by different names such as descriptive research, subjective research, inductive research, and case studies. You can readily see that the name depends on the purpose of the research.

Qualitative research is a more structured and formal type of research that is concerned with obtaining explanations of certain issues or subjects of interest. It deals in words, images, and subjective assessments. For example, qualitative research may be used to describe a purchase behavior or pattern, event or concept, or to understand a market situation from a holistic perspective. This approach is well suited for a store that wants to examine its own brand of products and compare them to national brands of similar products. Qualitative research can also be used where there are concerns about customer opinions, experiences, and feelings. In effect, qualitative research is concerned with finding answers to questions that relate to why?, how?, and what? Data for this type of research can be collected through direct observations, interviews, or surveys. The data and information are then used to develop concepts that help to understand the marketplace.

Quantitative research

Quantitative research deals in numbers, logic or theory, and objective measures to provide measurement and statistical predictability of results to the total target population (customers, consumers, etc.). Some level of certainty is required in quantitative research, for example, if the business owner wants to know the size of a target group for a certain product on the market, or the extent of customer satisfaction with a product or service. Quantitative research methods include the use of questionnaire surveys or telephone interviews, and subsequent statistical analyses.

Decisions regarding planning and implementing marketing measures and for making organizational changes can be made with a relatively high level of certainty from quantitative research compared to the other approaches. Good quantitative research requires three elements: a well-designed questionnaire, a randomly selected sample, and a sufficiently large sample. These will be discussed in detail below.

Data collection

Whether the research effort is exploratory, qualitative or quantitative, data need to be collected. Data to be collected should relate directly to the research objectives, research questions, and research hypotheses. There are two basic types of data that can be gathered: primary and secondary data. Primary data collection is very expensive. Companies should carefully weigh the anticipated value of new sales generated as a result of investing in primary data collection with the cost of doing the research. For example, a tilapia company that likely would increase sales by $150,000 would not be wise to invest $500,000 to generate primary data. Even in cases where the results would be worth the cost of research, spending some time gathering secondary data should be the first step.

Secondary data

Since primary data can be expensive to collect, it is often worthwhile to access data and information previously gathered by others. This is referred to as secondary data. The benefits of using secondary data are that significant time and financial investment are not required for gathering the data. Moreover, it is always useful to ascertain that the data needed for a research study are not already available.

The major disadvantage of secondary data is that the researcher does not have control over the design of the data-gathering process, the data collection process, or any manipulations of the data. Data may be available only in forms that are not suitable for your purpose and therefore require some manipulation in order to be useful. Secondary data are generally published data and can be obtained from a number of sources that include: established archives, government and state agencies, private companies, or directly from principal investigators and researchers (see the Annotated Webliography for summaries of various sources of secondary data for marketing aquaculture products).

Primary data

Primary data are gathered by researchers. Any systematic documentation of personal observations, interviews, surveys, focus groups, or personal experience constitutes primary data. The most common primary data collected in marketing research is the documentation of consumer attitudes and behavior using focus groups, interviews, or surveys. Each of these will be discussed below. More information on primary data collection methods can be obtained from the American Statistical Association's (1997) series on *Survey Research Methods* (see the Annotated Webliography) and from marketing research textbooks, such as Blankenship et al. (1998).

Focus groups

Focus groups are informal techniques to assess consumer preferences and needs, new product concepts, and purchase behavior for a good or service. Focus groups consist of 6–12 carefully selected participants with some common characteristics that relate to the objective of the study. The homogeneity of group participants is vital to generating important data and information from the sessions. With consumer preference studies, the most important characteristics of participants often include income, age, and ethnicity. It is useful to use different groups to obtain a diversity of responses.

Focus group sessions are conducted in the form of a discussion with a moderator who maintains the group's focus. The moderator should promote free-flowing individual participation in the discussion. However, the moderator must follow an agenda on specific issues and goals that relate to the type of information to be gathered and ensure that all group members contribute to the discussion. Discussion questions should be open-ended to allow all possible responses. As much as possible, the moderator should promote give-and-take discussion among participants. These group sessions can last from an hour and a half to two hours.

A well-moderated focus group session can generate new product ideas or concepts, reveal consumer reactions to potential new products, and discover potential market prices for a product. The session can also reveal information about competing products, product usage, preferred packaging, and effective advertising strategies. Some group dynamics and organizational issues can also be observed during a focus group session.

The major disadvantage of focus group research is that the responses cannot be analyzed statistically or quantitatively. Information obtained from focus group sessions relates more to words and behaviors of the participants who are not representative of a target population. Focus group research is, therefore, qualitative research.

Surveys

Surveys are methods of gathering systematic information from a sample of a target population. In market research, surveys provide a speedy and economical means of determining consumer attitudes, beliefs, expectations, and behaviors about products and services. For example, a seafood product manufacturer might do a survey of the potential market before introducing a new product. Surveys can be conducted in a variety of ways that include telephone, mail, or face-to-face, and in-person. Surveys can also be self-administered. Some surveys may combine several of these methods. For example, a telephone survey can be employed to select eligible respondents and make appointments for in-person interviews.

Personnel involved in market research surveys must have some training in interviewing. Interviewers should possess the ability to approach people in person or on the phone, persuade them to participate in a survey, and collect the needed data. The whole survey process requires skills in survey planning, sample selection, questionnaire development, data processing, data analysis, and reporting. Survey results should always be presented in broad categories such that individual respondents cannot be identified (see Blankenship et al. 1998 for an overview of survey methods).

Mail surveys

Mail surveys have the advantage of being a relatively lower cost method as compared to the other survey methods. When respondents cooperate, mail surveys can also provide more thoughtful responses to the survey questionnaire. Moreover, there is no potential for interviewer bias with mail surveys. Visualization may be required for respondents to answer survey questions. For example, the use of a color chart, or a series of advertisements may make it easier for respondents to understand the questions. Some surveys require respondents to refer to and provide data from records they keep. In these instances, mail surveys can be an effective data collection method.

The major disadvantage of mail surveys is often a low response rate due to lack of cooperation from respondents. Also, if timing is important for the completion of the research problem at hand, mail surveys may not be appropriate since they require more time. Mail survey questions must be clear and simple to understand; otherwise, respondents will give different interpretations and meanings to the same question. This results in unreliable data that are difficult to interpret. Other potential problems with mail surveys are non-responses to certain questions and inaccurate responses to particular questions. Respondents may also skip questions, answer questions incompletely, or record illegible responses.

Various techniques have been developed for improving the efficiency and response rate of mail surveys. These include:

1 Notification of recipients well in advance of their participation in the impending survey. This can be done through a letter or postcard. This is very common with surveys conducted by the government.
2 Addressing all correspondence using recipient names and not "current occupant," if it is a consumer or household survey.
3 Including a cover letter with the survey questionnaire that outlines the purpose of the survey, the importance of the respondent's response and participation, and the benefits of the study to them. An estimated time for completion of the questionnaire should be included in the cover letter because recipients are likelier to cooperate and respond to the survey if the time required is short. For household surveys, open-ended and lengthy questionnaires should be avoided. Generally, a mail questionnaire should be short and require straight answers such as questions that have response categories that can be checked off quickly.
4 Providing incentives for participation. This is a good idea and could be in the form of offering each participant some cash or coupon for participation. This type of incentive should be a token amount due to the total survey expense, particularly with a large sample size. Alternatively, cash or coupons can be offered as prizes for drawings, where respondents have a chance of winning a prize for participation.
5 Providing postage-paid return envelopes with the survey questionnaires.
6 Sending follow-up postcards or letters to remind recipients about responding to the survey questionnaires. The message should be a shortened version of the cover letter and should include an expression of appreciation to those who have completed and returned the questionnaire. It should also include the willingness to send another questionnaire and a return envelope to the recipient if the first has been misplaced. It is recommended that this be done after the second week of mailing the survey questionnaire because the majority of responses to mail surveys are returned within two weeks. This helps with the response rate, especially by getting the attention of recipients who did not respond to the first mailing.

Telephone surveys
Telephone interviews are efficient methods of collecting some types of data and are being used increasingly in marketing research. Compared to mail surveys, telephone surveys are relatively expensive but quicker to administer. Depending

on the type of data required, telephone surveys can generate a great deal of quality information. Interviewers exert control over the entire process and can probe for additional information on open-ended questions when a respondent provides an answer that is incomplete or unclear. During the interview process, the respondent does not know what the next question will be, which allows substantially greater flexibility in questionnaire design. Another advantage of telephone surveys is that they lend themselves to proper sampling techniques because almost every household and all businesses have telephones, and when conducted at the appropriate time, the response rate can be very high. Telephone surveys are most suitable when time is of the essence and the length of the survey is limited. However, they cannot be used for elaborate and detailed surveys that require respondents to consult records to provide an accurate response or where visual aids and display materials are associated with survey questions.

Trained interviewers normally conduct telephone surveys using a computer-aided telephone interviewing (CATI) system, with which responses are entered directly into a computer database while the interview is taking place. This approach reduces the setup time and costs. During the telephone interview process, supervisors usually monitor the process and the interviewers to assure the accuracy and integrity of the collected data. The supervisors have facilities that allow them to listen in while the interviewing is proceeding. The telephone interviewing facility usually contains interviewing stations or booths, high-speed modem autodialing, and, in some cases, a visual and audio monitoring system.

Direct, in-person surveys

Direct, in-person interviews can be conducted in homes, offices, shopping outlets and shopping malls, or other locations where the interviewer and the respondent can meet face to face. The most common form of this type of survey is that in which interviewers intercept shoppers at shopping outlets or malls (mall-intercept method). In-person surveys are much more expensive than mail or telephone surveys in terms of the cost per interview. The cost can be extremely high if the survey involves travel by interviewers. Another limitation of in-person surveys is bias that may result from the interviewer during the interview process. Interviewers can have their own biases that may affect the responses. This is especially the case with open-ended questions. Interviewers should be as neutral as possible and should not, in any way, influence the answer provided by the respondent. Bias can also be introduced when selecting the individuals to approach for interviewing. In today's society where there is always suspicion and mistrust, interviewers may tend to approach neat, safe-looking people to interview from a stream of shoppers at a shopping mall. This is what is referred to as convenience sampling. The primary questions that arise for the researcher are: (1) how are people or homes selected to approach for interview?

(2) will every person/home or every other person/home be selected, or will some other sampling technique be used? Sampling techniques will be discussed in a later section of this chapter.

Despite the limitations of in-person surveys, the method is convenient when it is necessary to display advertisements, products, packaging, and other materials associated with the survey. In some instances, intercept surveys at shopping outlets and malls can be low cost with no travel cost. Intercept surveys can also provide a good demographic spread and diversity of respondents.

Interviewers require training to be able to effectively solicit and gain cooperation from respondents. The following factors about the interviewer are important for obtaining accurate data and information from face-to-face interviewing.

1 *Appearance*: The first impression of the interviewer is very important in determining how cooperative a respondent can be. While there are no dress codes for the interviewer, he or she should not be over-dressed or under-dressed. Many market research firms provide jackets with signs to indicate what the interviews are about.

2 *Good interpersonal skills*: This requires interviewers to have the ability to approach strangers and secure their cooperation for the survey with little opposition. Interviewers should be able to quickly establish a rapport with potential respondents and interest them in the survey.

3 *Good judgment*: Interviewers should have the skill to make judgments relating to cooperation and non-cooperation from respondents. For example, in an intercept survey, it can be difficult to get cooperation from people who are in a hurry. In the case of a home survey, it will be difficult to get cooperation from households during their meal times or during the Superbowl.

Self-administered surveys

Self-administered surveys are administered entirely by the respondent. Potential respondents pick up survey materials, complete them, and return them at their convenience. This kind of survey is common with questionnaires relating to customer satisfaction. Most service firms or even shopping outlets place survey questionnaires at entrances to their facilities for their clients or customers to complete and drop in a box. Others can be taken and return-mailed with prepaid postage. Internet surveys are becoming popular as self-administered surveys, particularly for consumer opinion research. With Internet surveys, potential respondents are directed to a website through electronic mail lists or user groups. The website includes the questionnaire to be completed.

Self-administered surveys are convenient and less costly, and can provide well-thought-out responses to survey questions. However, respondents might not constitute a representative sample. Some respondents may not be within the intended target population, and responses provided by them are not appropriate. Respondents of self-administered surveys can be people with some strong opinions.

Sampling

The basis of quantitative marketing research is to gain information about an entire group of people (i.e., population), such as consumers, households, and clients. If information about all seafood buyers in the nation is desired, that group is the population. Obtaining information about the entire population is ideal. Depending on the size of the target population, it may be possible to survey the entire population. For example, a teacher may want to survey his or her class about their interest in a particular teaching style. In this case the entire class will be the population. In marketing research, however, it frequently is not practicable to survey the entire population of potential consumers, or clients. It is usually necessary to draw a sample (portion of the population) in order to obtain information about the entire population. The selection of a valid and efficient sample is crucial to the success of applying information obtained about the sample to the entire population. Consequently, an efficient method to choose the sample from the population is needed. This is referred to as sample design. The accuracy of the survey results will depend on the quality of sampling information available at the design stage, and particularly on the implementation of the sampling procedure.

Sample design involves the following steps:

1 Define the population or group of people to be studied. This is the intended target group, from which you wish to obtain information. For example, in a study of tilapia consumption, the target population could be grocery shoppers, seafood consumers in general, or only those who consume seafood in restaurants. Defining the target population is important, especially in a study for which the results of the survey will be used in decisions relating to marketing management and strategy development.

2 Determine how the potential respondents will be identified. For in-person surveys, potential respondents are the people who will be contacted in person. These would be shoppers in the case of an intercept survey, heads of households in the case of a home survey, or restaurant managers in a particular geographic area. For telephone and mail surveys, potential names and contact information of respondents are needed. These names and contact information can be obtained from telephone books or can be purchased from market research or communications companies.

3 Determine sample size. There is no simple (one-size-fits-all) formula for the selection of a sample size to be used in a survey. For large populations, the sample size to use depends on the level of statistical accuracy and reliability necessary to associate with the survey results. It requires establishing a statistical level of confidence and a margin of error. A high confidence of 95% and a small margin of error of 1% can be obtained with a large sample. In general, the larger the sample, the better the sample results will reflect the population. The most common approach to determining the sample size for large

populations is to assume a normal distribution of the target population and a random sampling procedure. Thus, the sample size can be calculated as:

$$n = \left[\frac{z_\alpha * s}{m} \right]^2$$

where n is the sample size, z_α is the critical value from a standard normal curve based on the desired confidence level α (commonly set 95% or 99% level of confidence for z_α values of 1.96 or 2.575, respectively), s is the sample standard deviation (commonly set at 0.5), and m is the desired margin of error (commonly set not to exceed $\alpha = 5\%$). The above formula can only be used under the assumptions that the population to be studied is normally distributed, the sample is generated randomly from the population, and the sample is sufficiently large that the sample standard deviation is close to the population standard deviation.

For small populations, selecting a sample size can be calculated with the standard error computed with a finite population N correction included.

$$n = \left[\frac{z_\alpha * s}{m} * \sqrt{\frac{N-n}{N-1}} \right]^2$$

Solving for n becomes:

$$n = \left[\frac{\left(z_\alpha * s\right)^2 N}{\left(z_\alpha * s\right)^2 + (N-1) \, m^2} \right]$$

In practice, the use of the above formulas can yield a large sample size that will be too expensive to survey. In practice, the choice of sample size is often based on professional experience, available resources, and the purpose of the study, when the calculated sample size is high.

4 Choose a sampling method. The choice of the sampling method is often determined by the study objectives, population characteristics, time, cost, and sometimes convenience. The various methods available for selecting samples include:

(a) Simple random sampling: The sample consists of individuals from the population chosen in such a way that each person in the population has an equal chance of selection. This allows the results to be projected reliably from the sample to the larger population.

(b) Systematic random sampling: The selection procedure consists of selecting every n^{th} individual in the population. If the population is in a random order, systematic random sampling approximates the simple random sampling procedure.

(c) Stratified random sampling: This procedure is applicable where there is a particular interest in a specific group or subdivision of the population. For example, if individuals of the same age or race were believed to have similar preferences for fish, a stratified random sample would allow the researcher to test for this. The population is first divided into subdivisions, called strata, and random samples are selected from each stratum. The information collected from each stratum is then combined. This procedure is useful to capture the variability within various strata.

(d) Multi-stage or cluster sampling: A cluster is a random selection of individuals, but the sample is chosen in stages. For example, in a survey of households for grocery coupon use, a researcher will first divide the nation into clusters, perhaps counties. A random sample of counties is then selected. From the selected counties, a list of cities and towns is then selected, and from this, a sample of households is selected.

(e) Ad-hoc sampling: The general framework for the above four sampling methods is probability, in which each individual of the population has a known chance to be selected. In contrast, ad-hoc sampling is arbitrary and not based on any known probability or chance. Examples include convenience sampling that is often used in intercept surveys at shopping outlets and malls. The sampling is based on those shoppers who pass by. Ad-hoc sampling methods also include sampling based on personal preferences and judgment.

Questionnaire design

It is important to design questionnaires (survey instruments) carefully. Poorly worded questions, poorly structured questionnaires, and inappropriate questions can result in erroneous and misleading information. There are many good reference sources available that provide detailed instructions on proper questionnaire design. Several marketing research college texts have sections on questionnaire design and issues.

A questionnaire consists of several components including words, questions, formats, and hypothesis. Word selection can influence the response to a question; therefore the researcher should carefully choose the words for formulating the questions or scales. There should be no ambiguity or abstraction in the wording. There are two question formats: unstructured questions and structured questions. Unstructured questions are open-ended questions that allow respondents to write in their response; structured questions are closed-ended and require the respondent to choose from a predetermined set of responses or scale points.

A questionnaire should begin with easy-to-answer questions. Subsequent questions should flow naturally and in a logical fashion. More sensitive questions that relate to demographic information on age or income, for example, should be at the end. Questionnaires should be checked carefully to eliminate wording that is considered unanswerable, leading (or loaded), double-barreled, or incomprehensible to the respondent (see Glossary for details). Validity and reliability tests should also be conducted (see Glossary for details). Moreover, pre-testing questionnaires is essential and will allow the researcher to correct problems with vague and imprecise wording and misunderstanding.

Different types of questions will generate different levels of information. "Yes/No" questions give some indication of consumer attitudes but a multiple-choice question will allow for assessment of finer distinctions in attitudes. Other types of questions are the Likert scale, rank order, rating scale, true/false, and semantic differential scale (see Glossary for details).

Response rate

How many responses are enough? The answer depends on how representative the sample was and the survey method. An attempt should be made to re-contact a small sample of the non-respondents to be certain that the survey was not biased toward certain groups of people with certain characteristics. Generally, response rates of 35–50% are considered acceptable as long as no non-respondent bias was observed.

Research on attitudes and preferences

In marketing, one of the fundamental axioms that are stressed repeatedly is "know your customers." People are complex biological organisms and the information needed about consumers depends to a large extent on the intended uses. Those in turn depend on the market conditions and on the nature of the products being sold. There are several different forms of information collection related to consumer attitudes and preferences, and each type requires a different approach.

Suppose a company is considering adjusting the price of its product. It would be important to obtain information on how the quantity demanded by customers is likely to respond to the price change. If there are other competing firms selling similar products, it will also be necessary to know how the competitors will respond to its price decisions. More importantly, the company needs to know the extent to which consumers will substitute one product for another. This is an example of understanding behavioral responses by gauging the effects of a price change.

If customers are not aware of a new product being sold, or of the new lower price of a familiar product, then advertising could affect the sales of the new product or the old product at the new price. This advertising is commonly practiced by grocery outlets and involves issuing store flyers periodically or doing in-store advertising. Information on how consumers react to advertising strategies will also be needed. For an entirely new product or a change in an existing product, information about how consumers will respond to it is required.

Gathering information about consumers is not easy. It can be obtained through the marketing research process but is expensive. However, an effective marketing research process can provide good knowledge of consumers. Gathering information that is relevant for some intended uses implies knowing the customer well enough to formulate effective marketing strategies for your product or services.

Theories of choice behavior

Understanding behavior and preferences as these relate to choice by individuals or a group of people can be complex. People's choices manifest themselves in many ways, but are particularly expressed through active or passive responses, such as through purchasing specific products or services. Individual choices are influenced by factors such as income, habit, experience, advertising, peer pressure, family, and accumulated beliefs. These factors reflect the dynamic nature of human attitudes and preferences. Several theoretical frameworks have been proposed for examining consumer behavior but there are three basic theories of choice behavior:

Neoclassical preference theory

Preferences are expressed as utility, which is a generalized term for the satisfaction obtained by an individual from the choice of a product (good or service). Preference is measured by the price the individual is willing to pay for the product. Total satisfaction obtained from the product is termed total utility, and the additional utility obtained from the use of an additional unit of the product is called marginal utility. The classical theory assumes that a rational individual purchases a combination of quantities of products that yields the maximum utility subject to constraints of the level of income and prevailing prices (see Pindyck and Rubinfeld 2001). The behavioral assumptions are that a representative consumer chooses between alternative commodity combinations to maximize utility, has perfect knowledge of all alternative commodity combinations and their prices, and is capable of evaluating the alternatives. The utility is ordinal, that is, a consumer is able to order commodity combinations by level of utility (first, second, third). The utility does not require cardinality (the ability to specify the actual numeric level of utility). A demand schedule for a representative consumer is derived from the behavioral assumption of utility maximization.

Revealed preference theory

The neoclassical ordinal utility theory is based upon a set of psychological assumptions, but revealed preference is based on actual behavior. By changing the budget allocation of a consumer and observing the consumer's purchasing pattern, that is, which commodity combinations are actually purchased, we can derive a preference schedule for the consumer through "revealed preference" theory (see Deaton and Muellbauer 1980). Revealed preference utilizes actual behavior of consumers to derive preference curves and consequently a demand schedule.

Hedonic theory

The classical preference theory assumes consumer preferences are for quantities of products. With hedonic theory, consumer preferences relate to the bundle of attributes or qualities contained in that product and not the quantities of the product. Hedonic theory assumes that the qualities of the product are the ultimate source of utility for consumers and that a product is described solely by its characteristics (see Lancaster 1971). These characteristics refer to price, flavor, texture, color, and others. This assumption of consumer preference for products provides the ability to derive implicit relative prices of product attributes or qualities and how much consumers are willing to pay for each of the attributes. The relative implicit prices of the attributes, as valued by consumers, differentiate similar products in the marketplace. Closely related to hedonic theory is the conjoint analysis, which is used in new product research.

Product research

Product ideas

Research in market products begins with a testable product concept. Examples of product concepts that have been translated into successful marketing products over the years may include ready-to-eat products, re-sealable retail packs, reduced fat products, or the elimination of preservatives in the products. The concept should try to address some current or emerging consumer need. In particular, product attributes, packaging, positioning, and pricing play a vital role in the development of any product concept.

The development of a new product often depends on the type of product. Whether the product is evolving from a known or an existing product or if it is an entirely new product will affect its market development path. For example, repackaging a known food product or adding new flavors to an existing product is an evolutionary concept. An existing product can be modified to suit particular needs or improve on particular experiences of customers in the use of the product.

An entirely new product concept can be termed a revolutionary concept. Revolutionary product concepts involve discovery of consumer needs that have not been met by existing products. For example, the microwave oven was a revolutionary concept that allowed for the preparation of many ready-to-eat food products to meet the increasingly busy schedules of the working population.

Product testing

Testing a concept identifies potentially successful new products and determines the probability that consumers will accept the product. Evolutionary product concepts are best tested using qualitative research such as in-depth interviews and focus groups in combination with quantitative survey-based research methods. The use of the qualitative phase allows for fine-tuning the product concept and formulating hypotheses for the quantitative phase. The quantitative research provides specific measurements to assist in marketing decisions.

Revolutionary concepts can best be tested with qualitative methods such as focus groups and in-depth interviews. New products are usually developed based on some perception of market need. However, there may be many possible ways to meet that perceived need. The relative strengths and weaknesses of the product concept can be better evaluated through qualitative research. Chances are that there may even be some prototypes available on the market. Qualitative market research will provide information on any product flaws, flavor preferences, size, packaging, and a host of other modifiable attributes before going into mass production.

Product testing could also involve testing the name to avoid confusion with similar products, or problems with pronouncing and writing the name. Evaluation of the packaging is also important. A test helps to identify how readily customers would identify the product on the shelf, open the package, and follow the cooking and preparation instructions. A product test involving sensory evaluation will provide understanding of consumer preferences regarding texture, flavor, color, and other basic product attributes.

Generally, product testing allows the prediction of consumer acceptance of new products. With product testing, there is the possibility of achieving product superiority over the competition. Companies that do frequent testing can continuously improve product quality and customer satisfaction, especially as consumer tastes evolve over time, and will also be able to monitor the potential threat levels posed by competing products. Product testing will provide some understanding of competitive strengths and weaknesses and can also allow the implicit measurement of the effects of price, brand name, or packaging on perceived product quality. It is often recommended that tests be conducted in real environment situations. For example, for food products, an in-home usage test is recommended because it provides a more accurate and predictive response.

During the testing process, the critical variables to examine should be the quality attributes of the product from the consumer's perspective. It is important to determine what product attributes are truly important to consumers and what factors determine consumer satisfaction. Once consumer acceptance is ascertained for the product, the product can be introduced into a limited geographic area for a period of time (to observe product repeat purchase patterns) before venturing into general markets.

Market share research

Market share is among the important parameters in the marketing research process. It is a critical indicator of relative performance compared to the competition and shows which company's products or services are bought the most and who are the competitors in the market. Therefore, market share research provides measurements of the proportion of the market supplied by the company's specific product. It is the percentage of market unit volume or dollar value held by a company as a proportion of total market size. Market share may be expressed either in unit sales or dollar values, as follows:

$$\text{Market Share} = \frac{\text{Total company sales (units or dollars)}}{\text{Total market or industry sales (units or dollars)}}$$

In the business world, attaining the highest market share is the objective for most companies. It is believed that, regardless of the price of the product or service, a company with a high market share will remain more profitable than the competitors. However, some small companies with small market shares can function profitably in large marketplaces. This is because they develop and service a large share of a small segment of the total market.

Business mergers and acquisitions are common in the marketplace. Therefore, the level of market share also suggests the safety and stability of the position occupied by the company in the market. Large competitors have frequently absorbed smaller competitors to increase market share.

Because of the competitive nature of the business world, it is always important for companies to monitor how their share of the market changes over time. A company's sales may be growing at the same time that market share is decreasing. Monitoring market share over time should be a vital part of a company's overall strategic business, marketing, and sales plan. With good market share information, a company can adjust its marketing strategies and improve its revenue, customer base, and brand value.

To establish the market share of a company's product or service, interviews and surveys can be used to obtain primary sales information from manufacturers, vendors, and customers. In general, market share research has primarily been concerned with the top players in the market for a single product or an entire product line within a single or a segmented market.

Advertising research

Advertising is a major part of marketing products. Advertising may be used to convey information about a product or service to a target audience or it may be used to create awareness or change perceptions about a product or service. Generally speaking, the ultimate purpose of advertising is to influence consumer behavior through either a change in behavior or reinforcement of an impression or perception for the benefit of the advertiser. The change in consumer behavior is a change in purchase of the product or service being advertised that will result in an increase in sales. Thus, in food marketing, changes in sales or similar measures are used to assess the effectiveness of advertising programs.

Companies recognize that consumer preferences are not static but are subject to some degree of randomness and systematic change. Humans, by nature, are dynamic in terms of their preferences. New products continue to be developed and introduced to the marketplace from time to time to meet the changing tastes and needs of consumers. Technological advances also affect new product development and influence food preferences and demand.

Advertising can lead to changes in consumer behavior, but the degree of change will differ by the type of commodity and potentially by the nature and quality of the advertising. There are advertisements that can readily alter the preferences of consumers. Others rarely alter consumer preferences such as those for products that may be well defined and stable. Food is considered a typical example of a product with a stable preference because of the inelastic nature of demand for food. However, demand for food depends on the prevailing price, price of substitutes, and perhaps the attributes of the product. Thus, consumption depends on consumer knowledge and perceptions of product attributes. Advertising plays the role of influencing knowledge and perceptions of food products.

The marketing literature includes a number of examples of different measures used to determine the effectiveness of advertising. Copy testing is the most common measurement approach, assessing the effectiveness of advertising within minutes or hours of exposure to the advertisement. Some of the measurements used in copy testing can be classified into the following (Haley and Baldinger 1991):

1 Persuasion measures: Using a survey instrument, one can solicit choice of a brand among a product category, overall brand ratings, and purchase interest and intentions of particular products or brands.

2 Salience measures: With this type of measure, respondents are examined for high brand awareness, such as top-of-mind awareness, unaided awareness, and total awareness (unaided and aided) of particular products or brands.

3 Recall measures: These measure the ability of respondents to recall brand from cues of product category and brand category.

4 Communication measures: Sometimes advertising research focuses on the main point of communication (TV, print media, sales point), and nature of advertisement (ad situation, visual characteristics).

5 Diagnostics measures: These relate to reaction to the advertisement. A positive reaction invites responses such as "I learned a lot from this advertisement," "Ad tells me a lot about the product," and "I learned something new about the product." A negative response could be "The product does not taste as good as the ad claims."

In practical work, persuasion and recall measures have been found to perform better than others in terms of predicting the effectiveness of advertising.

In farm commodity marketing, advertising has usually been generic. In evaluating generic advertising programs, researchers have typically used (1) advertising expenditure and (2) gross rating point as the primary indicators of effectiveness. Advertising expenditure is used as a proxy for advertising intensity and assumes that there is a positive relationship between the amount spent on advertising and sales. The gross rating point is a product of the reach of the advertisement and the average of its distribution of exposures delivered to a target audience.

Sales control research

Companies are always looking for ways to have a competitive advantage in the marketplace. One of the surest ways of gaining that edge over the competition is to research the market and monitor sales performance of various products. In addition to research on consumer attitudes and preferences, products and services, market share and advertising, forecasting sales is also an important aspect of market research. Sales forecasting helps to determine trends in the marketplace and how to benefit from such trends.

The process of sales forecasting involves organizing and analyzing information in a way to estimate future sales. Sales are generally affected by several factors that include season, holidays, special events, direct and indirect competition, labor events, productivity changes, demographic trends, fashions or styles, political events, and weather. These can be considered as external factors. Within the company, factors that can potentially affect sales include changes in product form, product quality, production capacity, advertising and promotion, sales efforts and strategies, price changes, inventory, distribution methods, and credit policy changes, among others.

A qualitative type of research is required for developing a good sales forecast with internal and external information including information from competitors, neighboring businesses, trade suppliers, business associations, trade associations, and trade publications. The information should be useful in describing a purchase pattern and event, and to understand market situations and trends from a holistic perspective.

There are a number of indicators that can be followed to develop sales forecasts:

1 Sales revenues from the same month or quarter in the previous year are good predictors of sales for the same period in succeeding years, but trends and forecasts in the economy and the industry must be accounted for.

2 Actual customer contacts and salespersons closely associated with customers and particular products, services, market, or territory can provide some good estimates.

New businesses should begin with the following to develop sales forecasts:

1 Developing customer profiles and determining industry trends.
2 Making some basic assumptions about the customers in the target market by developing a profile of the principal market. For example, assess the business (is it a small to medium sized grocery outlet and what are the sales volumes?). Determine the profile of, say, 20% of the target market (males, aged 20–34, professional, middle income, fitness conscious, or young families, with parents aged 25–39, middle income, home owners).
3 Determining trends by talking to trade suppliers about what is selling well and what is not, reading the industry's trade magazines and business periodicals.
4 Establishing the approximate size and location of the business area, using available statistics to determine the general characteristics of the area, unique characteristics, how far the average customer travels to buy from the outlet, or how far to go to distribute or promote the product. Government statistics can be used to estimate the number of individuals, households, or businesses.
5 Listing and profiling competitors in the business area. Study the competitors, visit their stores or locations, analyze the location, customer volumes, traffic patterns, hours of operation, busy periods, prices, quality of goods and services, product lines carried, promotional techniques, positioning, product catalogues and other handouts, and talk to customers and sales staff.
6 Estimating sales on a periodic basis (monthly or quarterly). The basis for the sales forecast can be the average monthly or quarterly sales of similar-sized competitors operating in a similar market, making adjustments for predicted trends in the industry.
7 Considering how well competitors satisfy the needs of potential customers in that trading area and determining how the new business' products fit in and what niche can be filled (a better location with more convenience, a better price, better quality, or better service).
8 Considering population and economic growth in the trading area and estimating market share.
9 Reviewing forecasts periodically using actual sales figures, and revising the forecast accordingly.

Value chain research

Value chain research involves analyzing a firm's or an industry's activities relating to the design, production, marketing, delivery, and support of its products or services (Porter 1980; 1985). These activities are outlined in the form of value-adding activities in the firm or industry. For a firm, the value chain analysis should be approached as part of the larger "value system" of related value chains because the analysis can be used as a way of investigating relationships among activities within the firm and between the firm and its customers and suppliers.

A firm or an industry requires a good understanding of its own value chain to compare to the competition. Good value chain research should capture, analyze, and develop appropriate strategies to enhance competitiveness.

Porter (1980; 1985) suggested five useful strategic frameworks for value chain research to analyze the forces that influence the profitability of a firm. The framework is commonly called Porter's five forces and includes an assessment of: (1) the bargaining power of suppliers; (2) the bargaining power of buyers; (3) the threat of substitute products or services; (4) the threat of new entrants; and (5) the intensity of competition. An assessment of the bargaining power of suppliers could involve differentiating inputs used by the firm, analysis of its supplier concentration, assessment of transaction volume, and assessment of cost effectiveness. Similarly, issues relating to buyer concentration that can be examined include volume of transactions and buyer integration as part of an assessment of bargaining power of buyers. An analysis of threats and intensity of competition involves research into factors such as economies of scale, product differences, brand identity, access to distribution, cost advantages, and government policy.

Internal cost analysis is a key component of value chain analysis. The value-adding activities involve processes. Thus, all processes along the chain need to be identified. The contribution of each stage of the process to total product cost must be determined, the cost drivers identified for each process, the links between the processes identified, and the opportunities for achieving relative cost advantage evaluated. The bottom line is to ensure cost effectiveness that will enhance overall profitability.

In addition, the firm or industry needs a better understanding of how its products or services are differentiated from the competition. This requires identifying and analyzing processes that relate to product or service features, marketing channels, support/service, brand or image positioning, and price. A competitive firm or industry should strive to have an edge over competitors in all of these factors. Gaining and sustaining a competitive advantage in the entire value delivery system also requires knowledge of activities of all participants in the delivery system to understand a firm's or industry's cost and differentiation positioning, because the end-use customers ultimately pay for all the profit margins along the entire value chain (Shank and Govindarajan 1993).

Value chain analysis has been conducted for a number of aquaculture industries around the world (see for example Sankaran and Suchitra 2006; Ardjosoediro and Goetz 2007; Veliu et al. 2009; Christensen et al. 2011; Jacinto and Pomeroy 2011; Macfadyen et al. 2012; Tran et al. 2013). Tran et al. (2013) reported that in Vietnam, traders visiting remote shrimp farms provide the first linkage between producers and market, and the shrimp industry in Vietnam is buyer-driven. The authors also reported that standards and certifications have had limited impact on the shrimp value chain in Vietnam though certifications are necessary to assure access to lucrative markets in the U.S. and the EU. They concluded that the fragmented nature of production and initial marketing make certification difficult for small-scale producers in Vietnam. In Egypt, Macfadyen et al. (2012) found that the tilapia industry generated a combined $775 of value added (i.e., profits plus wages/earnings) for farmers, traders, and retailers for each ton of fish produced, and the industry also generated 14 full-time equivalent jobs for every 100 tons of fish produced. The critical factors the study found impacting aquaculture value-chain performance in Egypt related to: (1) inputs, mainly rising feed costs and poor quality fry; (2) production challenges in the form of poor feed management, farm design/construction, fish health management, and stocking densities; and (3) marketing, transportation, and sale of products.

Data analysis

From the above, it is clear that much information and data can be collected in market research. What can be done with all the data and information gathered? Comprehensive analysis of the data is necessary to fully understand the market implications.

The large amount of information and data gathered during the market research process can only be useful if it is presented in a form that makes the information meaningful. There are several software applications that can present the data and information in desired forms. The common statistical software application used in market research is Statistical Package for the Social Sciences (SPSS) owned by IBM. There are even some integrated questionnaire design and analysis software programs that allow you to design the interview or survey material and analyze the responses, such as SurveyPro by Apian Software of Berkeley, California. Whatever data and information are gathered, the analysis needs to relate directly to the nature of the data gathered, and the nature of the research objectives, questions, or hypotheses.

To perform any analysis with data from a survey instrument, the responses need to be converted into numbers for analysis. This is commonly called "coding." Code numbers are assigned to particular responses in survey questionnaires. This allows the presentation of market research data in the form of statistical summaries and inferences, as well as relationships among variables.

Statistical summaries
Statistical summaries can be presented in graphical forms such as charts (e.g., line graphs, bar charts, histograms, stem-plots, etc,) and/or in tabular forms. These give a snapshot of all the data gathered. The following are some examples of useful statistical measures:

Proportions

Determining the proportion of all respondents that responded in a particular way to specific questions may be useful. For example, after a survey of grocery retailers, one might be interested in what proportion of respondents have fish counters, or what proportion of respondents answered "Yes" to a particular question. This proportion is simply the number of responses of interest over the total number of responses. For example, if 817 out of 950 respondents have fish counters, the proportion is 817/950=0.86 or 86%.

Central measures

The mean or the arithmetic average is commonly used to summarize survey response data. Another measure of the center of the distribution of responses is the median, which is the middle value of all the responses of interest from the lower value to the highest value. A third measure of the center is the mode, which is the value that has the highest number of occurrences. For example, if data were obtained on sales volumes, we might not only be interested in the lowest and highest, we might also be interested in the mode, median, and/or average sales volume among the respondents. Suppose we obtained the total sales value of eight seafood companies in a particular year as $90 million, $37 million, $24 million, $57 million, $68 million, $112 million, $78 million, and $68 million. The calculated mean is $67 million, the mode is $68 million, and the median is $68 million.

If the responses are categorical, say "yes" and "no" answers, the mean and the mode measures are irrelevant. However, the mean of each response is also the proportion of response. For example, if 950 people responded to a particular question requiring a "yes" and "no" answer and the "yes" responses are 456 and the "no" responses are 494, the mean of the "yes" responses will be 0.48 (or 48%) while the mean of the "no" responses will be 0.52 (or 52%), which are also the respective proportions of total responses.

Variability measures

Variability of responses to a particular question is another important measure. It provides information on how much difference and diversity exist among respondents on issues associated with particular questions. One can observe variability and diversity among responses by looking at the distribution of response frequencies and proportion of various responses to a question. If the distribution is over a wider range, it is an indication of differences and diversity. The common measure of variation or diversity in the responses is the standard deviation (commonly represented by ±), which measures how far responses are from the mean value. Figure 10.1 illustrates two normal distribution curves, showing the mean σ and the standard deviation μ. The values of the standard deviation are different; the curve with the larger standard

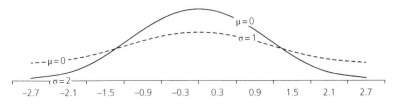

Fig. 10.1 Two normal curves, showing the mean, μ, and the standard deviation, σ.

deviation is more spread out. The value of standard deviation is always positive. It is zero if there is no difference or diversity in responses. Variance is another measure of differences and diversity.

Statistical inferences

It was pointed out earlier in the chapter that the basis of marketing research is to gain information about the entire population. However, because this is not often practical, information derived from a sample is extended to the population. Statistical procedures in which measures about the sample are used to make inferences about the population are known as statistical inference.

Statistical procedures involve estimation of parameters using the sample data, testing of hypotheses, and testing of the significance differences between estimated parameters. A parameter is a number that describes a population and a statistic is a number computed from the sample data. Sample statistics are therefore used to make assertions about unknown parameters. For example, if, in a survey of grocery outlets, 817 out of 950 respondents (sample) have fish counters, the sample proportion:

$$\hat{p} = \frac{817}{950} = 0.86$$

The sample statistic $\hat{p}$ is then used to estimate the unknown population parameter p. In theory, repeated random sampling or experimentation would result in a sampling distribution of the sample statistic, in this example, $\hat{p}$. With many samples drawn randomly from the population, the mean value of the sample statistic $\hat{p}$ will approach the true value of the population parameter p. In general, if the mean of the sampling distribution of a statistic is equal to the true value of the parameter being estimated, the statistic is said to be unbiased.

Just as variation or diversity measures such as the standard deviation measure how far responses are from their mean, there is also a measure of variability of a statistic describing the spread of its sampling distribution. This measure is called the standard error, s. Researchers often desire to have some level of statistical confidence in the estimated statistic; therefore a confidence interval is stipulated in the form of a percentage. Most researchers stipulate a confidence level of 95%, which corresponds to ±1.96 standard error. Others use 90% and 99%, which correspond to ±1.64 and ±2.58 standard errors, respectively. The percent levels of confidence are represented as z. Generally, a population estimate has a confidence interval of the form

$$\text{estimate} \pm z * s_{\text{estimate}}$$

and $z*s_{\text{estimate}}$ is known as the margin of error.

Relationships between variables or responses
Scatter-plots

The simplest way to examine relationships between two variables is a plot of the data. A scatter-plot reveals the relationship between two variables when one variable is plotted on each axis. Each individual data point appears as a point in the plot fixed by the values of the two variables for that individual. The scatter-plot can be examined for any direction, form, and strength in the relationship between the variables. The relationship may be positive (both variables increase or decrease together in the same direction), or negative (as one variable increases, the other decreases, or they move in opposite directions). A strong, moderate, or weak association may be observed between the variables or there may be no form of association between them. Where there is a strong association, it may be an indication that one variable depends on the other.

Scatter-plots are graphic depictions of relationships. Researchers may wish to obtain a numerical measure of the relationship instead of mere graphs. A common measure to examine the relationship is to calculate the *correlation, r*. Correlation measures the strength and direction of the linear association between two quantitative variables. A positive value indicates a positive association and a negative value indicates a negative association. The value of r lies between −1 and 1 and indicates the strength of the relationship by how close it is to −1 or 1.

Least-squares regression

A least-squares regression is a method of finding a straight line that summarizes the relationship between quantitative variables, where one variable is considered the dependent variable and the other is an explanatory variable. A least-squares regression line tries to fit a straight line as close as possible in a scatter-plot and the fitted line can be used to predict the value of the dependent variable for a given value of the explanatory variable. Mathematically, this can be expressed as

$$y = \alpha + \beta x + \varepsilon$$

where y is the dependent variable, x is the explanatory variable, β is an error term, ε is the intercept of the fitted line, and α is the slope of the fitted line. Not all relationships may be linear or appear to be in a straight line. The relationship

Table 10.1 Example of cross-tabulation relating to the statement "Customers prefer fresh fish to frozen fish."

	Strongly agree	Agree	Disagree	Strongly disagree	Total
National chain store	245	114	126	16	501
Regional chain store	115	19	135	18	287
Independent	66	118	16	12	212
Total	426	251	277	46	1000

may appear in the form of a curve, in which case a non-linear or curvilinear regression must be applied to fit the relationship. The simplest form of fitting a non-linear curve is to take the natural logarithm of the variables, that is,

$$\log(y) = \alpha + \beta \log(x) + \varepsilon$$

Cross-tabulations

A cross-tabulation table shows the relationship between two categorical variables with r rows and c columns. It is sometimes called an $r \times c$ table. Table 10.1 shows an example of responses from grocery supermarkets relating to the statement "Customers prefer fresh fish to frozen fish."

Suppose we want to test whether there are any differences in the responses given by the three groups of grocery outlets. We should first formulate a statistical null hypothesis that the responses from the three types of outlets are the same. Then we compare the observed counts of responses in the table with the *expected counts*. If the observed counts in the table are far from the expected counts, then there is evidence that the responses from the three types of outlets are different, that is, we do not accept the null hypothesis that the responses are the same. The expected count is calculated as:

$$\text{Expected Count} = \frac{\text{row total} \times \text{column total}}{\text{table total}}$$

The above equation generates the expected count for each cell.

The chi-square test

The chi-square test uses the observed counts and expected counts to determine if any differences are statistically significant. It is a measure of how far the observed counts in a cross-tabulation table are from the expected counts. The formula for chi-square, denoted as χ^2, is

$$\chi^2 = \sum \frac{(\text{observed count} - \text{expected count})}{\text{expected count}}$$

and the summation is over all $r \times c$ cells in the table. The chi-square analysis has been found to be a useful statistic for comparison when at least 20% of all expected counts are 5 or greater and there are no zero values of expected counts. Many statistical computer software applications will give a warning if fewer than 20% of calculated expected counts are less than 5.

Discrete choice analysis
Least-squares regression techniques are appropriate when the dependent variable is quantitative data. However, when the dependent variable is qualitative in nature, as is often obtained in survey data, the analysis requires different techniques. This is what is known variously as qualitative dependent variable analysis, limited dependent variable analysis, or discrete choice analysis. Good references for this type of analysis are Ben-Akiva and Lerman (1987), Maddala (1983), Greene (1997), and Louviere et al. (2000).

The technique is a linear probability model, in which the dependent variable is interpreted as the probability of occurrence. For example, suppose that we have gathered data from a survey that asked for responses on smoked tilapia. Suppose our dependent variable is a "yes" and "no" response to buying smoked tilapia and the dependent variables were related to pattern of fish purchase and demographics. Using a linear probability model to fit the data, we can predict the probability of buying smoked tilapia. Probabilistic models are based on the assumption that a choice among alternatives, such as "yes" and "no," is utility driven. In other words, individuals are assumed to choose an alternative that provides more utility than the other alternatives that were not chosen. Here are some examples of probabilistic models.

Logit model
In the example described above, the logit model specifies the probability that an individual will buy smoked tilapia as

$$prob(buy) = \frac{\exp(X\beta)}{1 + \exp(X\beta)} = \frac{1}{1 + \exp(-X\beta)}$$

where exp is exponent, X is a vector of explanatory variables, and β is a vector of estimated coefficients. The logit formulation is based on a logistic distribution of the error term. The probability of not buying will then be expressed as

$$prob(not\ buy) = 1 - \frac{\exp(X\beta)}{1 + \exp(X\beta)} = \frac{1}{1 + \exp(X\beta)}$$

The above formulation of the logit model can be expressed in a different way. The ratio of the prob(buy) to prob(not buy) is

$$\frac{prob(buy)}{prob(not\ buy)} = \exp(X\beta) \quad \text{or} \quad \ln\left[\frac{prob(buy)}{prob(not\ buy)}\right] = X\beta$$

In terms of predicting the probability of buying, the sign and magnitude of the coefficient indicate the direction and effect of the relevant explanatory variable on the probability of buying. Marginal effects are used to explain changes in probability given a change in the relevant independent variable. The elasticity measure gives the percentage change in the choice probability in response to a percentage change in the explanatory variable.

Probit model

The probit model is similar to the logit model except that it is based on the assumption of a normal distribution of the error term. In this formulation, the probability of buying is specified as

$$\text{prob}(\text{buy}) = \int_{-\infty}^{x\beta} \frac{\exp(-t^2/2)}{\sqrt{2\pi}} \, dt$$

Interpreting the coefficients and predicting the probability of buying are the same as for the logit formulation. Marginal effects and elasticity measures can also be used to interpret the effects of explanatory variables on the choice probability.

The logit and probit models are examples of binary choice (two choices) models. Other examples of binary choice models are the Gompertz or log log model, the Burr or Scobit model, and the complementary log log or extreme value model (see Greene, 1997, for differences in the models).

There are instances where the choices are greater than two. For example, suppose that we are interested in knowing which species of fish households will purchase among these alternatives: orange roughy, cod, buffalofish, tuna, salmon, and catfish. To analyze multi-choice response data, we would use extensions of the logit and probit models that are called multinomial logit and multinomial probit models.

Discrete choice models are commonly applied in marketing research to problems of how consumers choose among competing products. It helps to determine which attributes matter most to consumers when choosing among alternatives. Whether the project is qualitative research, exploratory research, or examining purchase motivation, product positioning, or market segmentation, discrete choice analysis will help to provide some answers to the question of "Why consumers buy what they buy."

The major advantage of discrete choice techniques is that they are based on the observation of consumer choices. These can either be real choices or simulated choices. In marketing, consumer choices are ultimately the important factors companies would want to know. The LIMDEP software by Econometric Software, Inc., of Plainview, New York, is commonly used to perform discrete choice analyses.

Conjoint analysis

An alternative methodology applied in marketing research to examine con-sumer choices and preferences is the conjoint technique. Conjoint is a generic term that refers to a number of paradigms in psychology, economics, and marketing that are concerned with the quantitative description of consumer preferences or value trade-offs. Conjoint analysis is sometimes referred to as "trade-off" analysis because individuals are forced to make trade-offs among different product attributes when completing conjoint questions. Through the trade-offs, inferences can be made as to how important or valuable different attributes are and how they influence individuals' decision-making processes. Good references for this type of analysis are Green (1974), Green and Srinivasan (1978), and Green and Wind (1975). There are several forms of the conjoint techniques.

The initial step in conjoint analysis is to identify the attributes that are critical to buyers when assessing the product or service. Focus groups and personal interviews of representative buyers can be utilized to determine the attributes. The choice of attributes to include in the experimental design should be distinct among products or services. The next step is to determine the number of attrib-utes and levels to include in the experimental design. For example, 4 attributes each with 4 levels would require 64 comparisons ($4 \times 4 \times 4$) so that care must be taken not to include too many attributes and levels. The combinations of the experimental design are included in a stimulus card and the number of stimulus cards presented to the respondents for evaluation. Each card represents a differ-ent combination of levels of attributes selected in the design. A sample of attrib-utes and the levels and a sample of a stimulus card representing one of the profile combinations are shown in Table 10.2 and Fig. 10.2, respectively. The SPSS software is commonly used to perform conjoint analyses.

Table 10.2 Example of levels of attributes for fish.

Fish attribute	Levels			
Price/lb	$1.50	$2.00	$2.50	$3.00
Color	Off-white	White	Pinkish white	Pinkish
Flavor	Mild	Fishy	Muddy	Musty
Texture	Oily	Moist	Moist & oily	Dry

Price:	**$2.00**
Color:	**Pinkish**
Flavor:	**Mild**
Texture:	**Moist & oily**

Fig. 10.2 An example of a stimulus card.

Traditional conjoint

In the traditional conjoint technique, respondents are shown different product/ service scenarios (stimulus cards) whose attributes vary according to an experimental design. Respondents are typically asked to rate or rank the product scenarios presented to them. Suppose respondents are presented with 18 such cards. They will be required to rank them from 1 to 18 in terms of preference. The sequential ranking procedure results in a ranked order for the number of cards from the least preferred combinations of attributes to the most preferred combinations. Once ranking data are collected, analysis involves quantifying the values assigned to each level. The method usually employed is the additive model that assumes that the utility of an alternative is formed by a linear combination of the utilities of its parts, that is, individual rankings (part-worths) for each attribute are added together to obtain a total worth value for each combination of the product or service.

$$
\begin{aligned}
\text{Total worth of a product/service} \quad = \quad & \text{Part-worth of level } i \text{ for attribute 1} \\
+ \quad & \text{Part-worth of level } i \text{ for attribute 2} \\
+ \quad & \text{Part-worth of level } i \text{ for attribute 3} \\
+ \quad & \text{Part-worth of level } i \text{ for attribute } n
\end{aligned}
$$

The specification of total worth of a product provides a means to estimate the importance and contribution of each attribute to the total utility of an alternative. This approach enables the assessment of the relative importance of various attribute levels in the context of preference and study of the effects of trade-offs among different attributes on consumer evaluations.

Discrete choice conjoint

In this technique, there is no ranking. Rather, respondents are provided with different pairs of product or service profiles and required to select the one they would most likely purchase. For example, respondents might be shown three different profiles of fish and asked to indicate the one they would purchase. Discrete choice conjoint is more commonly applied than the traditional conjoint because it is a more realistic exercise that mimics what actually takes place in the marketplace. The other advantage of discrete choice conjoint studies is that the alternatives often include the option to select "none" of the products, thus indicating that respondents do not like any of the products presented to them. Discrete choice conjoint also allows for much more complex statistical modeling to examine interactions among attribute levels, alternative-specific effects, and cross-effects (Fig. 10.3).

Best-worst conjoint

This is the least popular technique among the conjoint techniques. Respondents are typically shown the levels associated with attributes and are asked to select the one that they like best or the one that is most appealing, as well as the one

Please Choose ONLY ONE Alternative			
FISH ATTRIBUTE	ALTERNATIVE A	ALTERNATIVE B	ALTERNATIVE C
Price/lb	$1.50	$2.50	Neither A or B is preferred
Color	Off-white	White	
Texture	Oily	Dry	
Flavor	Mild	Fishy	
	⇩	⇩	⇩
I would choose	☐	☐	☐

Fig. 10.3 An example of a stated choice question.

they like least. The process is repeated several times with a different set of levels shown each time. After collection of the data, the utilities are calculated that indicate the relative value of the attributes and attribute levels. This conjoint technique is applicable to attributes that are abstract and cannot be easily quantified.

Traditional demand analysis

The neoclassical demand theory assumes that an individual consumer possesses a preference ordering for alternative bundles of commodities and that this ordering can be represented by an ordinal utility (U) function, $U = U(X)$ where X is a vector of bundles of commodities. It is required that this preference relationship satisfies some six axioms, which indicate rational consumer behavior and facilitate the maximization procedure:

1 Reflexivity: each bundle of commodities is at least as good as itself.
2 Completeness: the consumer has ability to rank all the bundles.
3 Transitivity: there is consistency in the consumer's ranking.
4 Continuity: the utility function is differentiable to the first and second order.
5 Non-satiation: more of the bundle of commodities is always preferred by the consumer.
6 Convexity: ensures diminishing marginal rate of substitution among bundles of commodities.

Details of demand theory and the basis for these assumptions can be found in any standard microeconomics or consumer theory textbook (e.g., Pindyck and Rubinfeld 2001). With the above assumptions satisfied, the individual consumer is assumed to face the choice of maximizing his or her utility function subject to a budget constraint. The problem of constrained utility maximization can be solved mathematically. The result is the derivation of demand relationships that give quantities as a function of prices and income or total expenditure.

An alternative approach to the consumer choice problem is one of selecting commodities to minimize the money outlay necessary to reach a predetermined utility level ($\bar{U}$). The solution to the minimization problem can also be solved mathematically to obtain compensated demand functions. These demand relationships provide a general characterization of the properties of demand functions, which includes adding up, homogeneity, symmetry, and negativity. In empirical analysis, these properties are usually imposed. This kind of analysis requires time series and cross-sectional quantitative data. There are several functional forms for demand model specifications that have been used to examine demand for food. The models include the linear expenditure system, the Rotterdam model, the direct translog, the indirect translog, the almost ideal demand system (AIDS), the quadratic AIDS, the inverse AIDS, the quadratic expenditure system, the general ordinary differential demand system, and Lewbel's demand system (see Theil 1975, 1976; Deaton and Muellbauer 1980; Lewbel 1990; Pollak and Wales 1992).

Aquaculture market synopsis: baitfish

Baitfish are small minnows that are sold to recreational fishermen who use them as bait to catch sport fish. Fish and crustaceans are raised and sold as bait all over the world. However, baitfish production in most countries is either small-scale, incidental, or simply the sale of fish and crustaceans that are too small to meet foodfish market requirements. However, the U.S. baitfish industry provides an example of baitfish production that has been developed into a large and important industry. Baitfish in the U.S. consists of crawfish for bait, fathead minnows, golden shiners, emerald or silver shiners, feeder and bait goldfish, and suckers.

The recreational fishing industry in the U.S. is a multi-billion dollar industry. In the 1950s and 1960s, when many new reservoirs were built, demand for live bait grew rapidly. The baitfish industry developed and grew to meet this demand. Nevertheless, competition continues from the sale of artificial lures for fishing and from wild-caught fish that are sold as bait in the U.S.

Baitfish farming is a unique type of aquaculture in many respects. The industry produces and sells vast numbers of various sizes of small fish. Arkansas alone produces over 6 billion baitfish annually (Stone et al. 1997a). Litvak and Mandrak (1993) estimated the retail value of baitfish sold in North America (including both farm-raised and wild-caught) to be $1 billion annually.

The majority of farm-raised baitfish sold in the U.S. is produced in ponds in Arkansas. Overall, $38 million of baitfish were sold in the U.S. in 2005 but this declined to about $33.1 million in 2013 according to the USDA Census of Aquaculture (USDA-NASS, 2014). In 2013, 63% were sold from Arkansas. Ninety-three percent of baitfish farms are small businesses. Most baitfish farms

Fig. 10.4 Grading and loading shed for baitfish. Courtesy of Dr. Nathan M. Stone.

are primarily family farms and partnerships. A few farms have diversified into distribution and wholesale functions and serve as market outlets for the smaller operations.

Baitfish are sold as a live product and differentiated by size. Different sizes of minnows are selected by recreational fishermen, depending on the type and size of sport fish they would like to catch.

Baitfish farmers have developed extensive marketing and distribution networks over time (Stone et al. 1997b). While some farms sell directly to fishermen, most sell through networks of wholesalers and distributors. Some large farms function as wholesale distributors for smaller farms (Fig. 10.4). Other distributors own their own holding facilities and have developed retail networks in a given sales area. Fish are then distributed from these warehouses to retail bait shops or other wholesale operators who then re-sell to bait shops (Figs 10.5 and 10.6). Baitfish are hauled by transport trucks long distances across the U.S. and handled several times en route. Thus, the fish must be vigorous and hardy enough to withstand the travel and handling and still be hardy and vigorous when bought by the consumers. The industry standard for customer service is to replace any fish losses incurred by the distributors and wholesalers regardless of the cause of the mortalities. A strong commitment to their customers is one of the characteristics of successful baitfish farms.

The greatest challenge to baitfish producers is that the demand for their product varies with the amount of recreational fishing and that is highly dependent on the weather. Moreover, the demand for different sizes of fish will depend on the weather conditions at different times of the year in different parts of the

Fig. 10.5 Retail bait shop, with hauling tank on back of pickup truck, Jimmy's Bait Shop, Pine Bluff, Arkansas. Courtesy of Dr. Nathan M. Stone, with permission of Jimmy's Bait Shop.

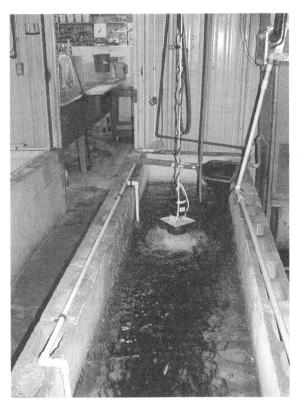

Fig. 10.6 Retail bait shop, River City Marine, Pine Bluff, Arkansas. Courtesy of Dr. Nathan M. Stone, with permission of River City Marine, Pine Bluff, Arkansas.

country. This requires baitfish farmers to maintain stocks of all sizes of fish at all times to have the supply ready to cover whatever the market will want that particular year.

Golden shiners (*Notemigonus crysoleucas*) are the major baitfish species raised. Nearly half of all baitfish raised in the U.S. are golden shiners. Feeder goldfish (*Carassius auratus*) are the second most commonly cultured baitfish. Goldfish are popular fish to keep as pets in either aquaria or pools in water gardens. Their value as ornamental fish is included in the synopsis in Chapter 7. However, goldfish are also raised on farms to sell as feeder fish (for customers to feed to pet carnivorous fish) or as trotline or other bait. Trotlines are a type of fishing tackle used in rivers and lakes and consist of a fishing line with a series of hooks that are baited with live fish overnight and checked in the morning. Fathead minnows (*Pimephales promelas*) are the other major species of baitfish and are sold most commonly as fishing bait, similar to sales of golden shiners.

Summary

Marketing research helps to find answers to questions such as: "What are the attitudes and desires of consumers?" "Is there a demand for our product?" "What is the volume of sales compared to competitors, that is, what is the share of the market for the product?" and "What products will consumers demand in future?" Answers to such questions allow a business to find out more about the current market situation relating to a product of interest as well as predict future market situations. The research can be designed in one of three forms: (1) exploratory research, (2) qualitative research, or (3) quantitative research. The type of research that is most appropriate depends upon the objectives. Whether the research effort is exploratory, qualitative, or quantitative, data need to be collected. Data to be collected should relate directly to the research objectives, research questions, and research hypotheses.

There are two basic types of data that can be gathered, primary and secondary data. Secondary data are information previously gathered by others while primary data are any systematic documentation of observations. The most common primary data collected in marketing research is the documentation of consumer attitudes and behavior using focus groups, interviews, or surveys. Surveys can be conducted in a variety of ways including telephone, mail, or face-to-face/in-person survey. Surveys can also be self-administered and may also combine several of these methods.

The basis of quantitative marketing research is to gain information about an entire group of people (i.e., population), such as consumers, households, and clients. It is usually necessary to draw a sample (portion of the population) in order to obtain information about the entire population using a sample design.

The sampling methods can be a simple random sampling, systematic random sampling, stratified random sampling, multi-stage or cluster sampling, or ad-hoc sampling.

Designing questionnaires should be done with care. Poorly worded questions, poorly structured questionnaires, and inappropriate questions can result in erroneous and misleading information. The choice of words, questions, formats, and hypothesis should be done with care. A questionnaire should begin with easy-to-answer questions. Subsequent questions should flow naturally and in a logical fashion. More sensitive questions on demographic information on age or income should be at the end. Questions can take the form of "Yes/No" questions, multiple-choice questions, Likert scale, rank order, rating scale, true/false, and/or semantic differential scale.

There are several different forms of collecting information related to consumer attitudes and preferences, and each type requires a different approach. People's choices manifest themselves in many ways, but are particularly expressed through active or passive responses of purchasing specific products or services. People's behavior can be examined using the neoclassical reference theory, revealed preference theory, and hedonic theory.

The development of a new product often depends on the type of product. Whether the product is evolving from a known or an existing product or if it is an entirely new product will affect its market development path. Repackaging a known food product or adding new flavors to an existing product is an evolutionary concept, while an entirely new product concept is a revolutionary concept. With a new product, testing is essential to identify its potential success and determine the probability that consumers will accept the product. It is always important for companies to monitor how their share of the market changes over time.

Data analysis can take the form of graphics that give a snapshot of all the data gathered; statistical summaries such as proportions; central measures such as mean, mode, and median; and variability measures such as variance and standard deviation. Data can also be subjected to scatter-plots. least-squares regression, and cross-tabulations. Chi-square tests can be used to determine if any differences between the observed counts and expected counts are statistically significant in a cross-tabulation. Other quantitative methods for data analysis include discrete choice analysis such as logit and probit, and conjoint analysis.

Study and discussion questions

1 Define marketing research. Outline three functions that marketing research can help to accomplish in developing a market for a new fish product.
2 A new seafood distribution company is about to be set. What form of market research would you recommend to the owners and why?

3 What are the advantages of secondary data over primary data? Suggest some ways of addressing some of the potential problems associated with using secondary data.

4 A focus group consists of 6–12 participants, a sample too small to be representative of the population. What is the usefulness of the information gathered from such a small sample of the population?

5 Response rate is a problem with mail surveys. What techniques can be applied to improve the response rate for mail surveys?

6 Mall-intercept interviews solicit the cooperation of shoppers to complete a survey. In today's society where there is always suspicion and mistrust, how can an interviewer gain the cooperation of busy shoppers?

7 Suppose the number of adult Hispanics living in Arkansas is 10,000 and we want to survey them to determine their attitudes toward catfish. Calculate the sample size to use for the survey assuming a 95% confidence level and a margin of error of 0.5.

8 What are the five types of sampling methods? Under what circumstances would each type be applicable?

9 What are the theoretical frameworks for examining consumer behavior? Which framework is more applicable for investigating the market for a new seafood product?

10 Why is it necessary for a company to monitor its market share regularly?

11 Describe the various measures of advertising effects. Which of these performs better in predicting the effectiveness of advertising?

12 Outline the steps a new business can follow to develop sales forecasts.

13 A sample of 15 clam consumers indicated the number of times that they have purchased clams from the grocery store within the past one year as: 27, 50, 33, 25, 86, 25, 85, 31, 37, 44, 20, 36, 59, 34, 28. Calculate the mean and mode of these observations. Calculate the variance and standard deviation (i.e., find the deviation of each observation from the mean, square the deviations, then obtain the variance and standard deviation).

14 Here are data from a survey of consumers of Asian origin in three cities in California on whether or not they buy fish from fish shops:

	Don't buy fish from fish shop	Buy fish from fish shop
City 1	400	1380
City 2	416	1823
City 3	188	1168

(a) Make a two-way table of city by whether or not they purchase from a fish shop.

(b) Calculate the proportion of Asians who buy fish from fish shops in each city.

(c) Find the expected counts, and check if the chi-square can be used. What null and alternative hypothesis does the chi-square test?

(d) What can you conclude from the data?

References

American Statistical Association. 1997. ASA Series: What is a Survey? Alexandria, Virginia.

Ardjosoediro, I. and F. Goetz. 2007. A value chain assessment of the aquaculture sector in Indonesia. US Agency of International Development (USAID) RAISE Plus IQC Task Order EDH-I-04-05-00004-00 (January 2007).

Ben-Akiva, M. and S.R. Lerman. 1987. *Discrete Choice Analysis: Theory and Application to Travel Demand*. MIT Press, Cambridge, Massachusetts.

Bennet, P. D. 1988. *Glossary of Marketing Terms*. Pp. 117–118. American Marketing Association, Chicago.

Bennett, E. 2001. Hooked on seafood. *Philadelphia Business Journal*, June 11.

Blankenship, A.B., G.E. Breen, and A. Dutka. (1998) *State of the Art Marketing Research*. 2nd edition. American Marketing Association, NTC Business Books, Chicago.

Christensen, V., J. Steenbeek, and P. Failler. 2011. A combined ecosystem and value chain modeling approach for evaluating societal cost and benefit of fishing. *Ecological Modeling* 222: 857–864.

Deaton, A. and J. Muellbauer, 1980. *Economics and Consumer Behavior*. Cambridge University Press, Cambridge, UK.

Green, P.E. 1974. On the design of choice experiments involving multifactor alternatives. *Journal of Consumer Research* 1:61–68.

Green, P.E. and V. Srinivasan. 1978. Conjoint analysis in consumer research: issues and outlook. *Journal of Consumer Research* 5:103–123.

Green, P.E. and Y. Wind. 1975. New ways to measure consumers' judgments. *Harvard Business Review*, July-August, 89–108.

Greene, W.H. 1997. *Econometric Analysis*. 3rd edition. Macmillan, New York.

Haley, R.I. and A.L. Baldinger. 1991. The ARF copy research validity project. *Journal of Advertising Research* 31:11–32.

Jacinto, E.R. and R. Pomeroy. 2011. Developing markets for small-scale fisheries: utilizing the value chain approach. In: Pomeroy, R.S. and N.L. Andrew, eds. *Small-scale Fisheries Management: Frameworks and Approaches for the Developing World*, pp. 160–177. CABI Publishing, Wallingford, UK.

Lancaster, K.J. 1971. *Consumer Demand*. Columbia University Press, New York.

Lewbel, A. 1990. Full rank demand systems. *International Economic Review* 31:289–300.

Litvak, M.K. and N.E. Mandrak. 1993. Ecology of freshwater baitfish use in Canada and the United States. *Fisheries* 18:6–13.

Louviere, J.J., D.A. Henser, and J.D. Swait. 2000. *Stated Choice Models: Analysis and Application*. Cambridge University Press, Cambridge, UK.

MacFadyen, G., A.M. Nasr-Alla, D. Al-Kenawy, M. Fathi, H. Hbicha, A.M. Diab, S.M. Hussein, R.M. Abou-Zeid, and G. El-Naggar. 2012. Value-chain analysis. 2012. An assessment methodology to estimate Egyptian aquaculture sector performance. *Aquaculture* s362–363:18–27.

Maddala, G.S. 1983. *Limited Dependent and Qualitative Variables in Econometrics*. Cambridge University Press, Cambridge, UK.

Pindyck, R.S. and D.L. Rubinfeld. 2001. *Microeconomics*. 5th edition. Pearson Prentice Hall, Englewood Cliffs, New Jersey.

Pollak, R.A. and T.J.Wales. 1992. *Demand System Specification and Estimation*. Oxford University Press, Oxford.

Porter, M.E. 1980. *Competitive Strategy: Techniques for Analyzing Industries and Competitors*. Free Press, New York.

Porter, M.E. 1985. *The Competitive Advantage: Creating and Sustaining Superior Performance*. Free Press, New York.

Sankaran, J.K. and M.V. Suchitra. 2006. Value-chain innovation in aquaculture: insights from a New Zealand case study. *R&D Management* 36:387–401.

Shank, J.K. and V. Govindarajan. 1993. *Strategic Cost Management: The New Tool for Competitive Advantage*. Free Press, New York.

Stone, N., E. Park, L. Dorman, and H. Thomforde. 1997a. Baitfish culture in Arkansas. *World Aquaculture* 28:5–13.

Stone, N., E. Park, L. Dorman, and H. Thomforde. 1997b. Baitfish culture in Arkansas: golden shiners, goldfish and fathead minnows. MP 386, Cooperative Extension Program, University of Arkansas at Pine Bluff, Pine Bluff, Arkansas.

Theil, H. 1975. *Theory and Measurement of Consumer Demand*, vol. I. North-Holland Publishing Company, Amsterdam.

Theil, H. 1976. *Theory and Measurement of Consumer Demand*, vol. II. North-Holland Publishing Company, Amsterdam.

Tran, N., C. Bailey, N. Wilson, and M. Phillips. 2013. Governance of global value chains in response to food safety and certification standards: the case of shrimp from Vietnam. *World Development* 45:325–336.

United States Department of Agriculture (USDA), National Agricultural Census. 2014. Census of aquaculture (2013). 2012 Census of Agriculture Volume 3, Special Studies, Part 2, AC-12-SS-2. United States Department of Agriculture, Washington, D.C.

Veliu, A., N. Gessese, C. Ragasa, and C. Okali. 2009. Gender analysis of aquaculture value chain in Northeast Vietnam and Nigeria. Agriculture and Rural Development Discussion Paper 44, The World Bank, Washington, D.C.

CHAPTER 11

Seafood demand analysis

Chapter 2 discussed the fundamental concept of demand and relevant elasticities, while Chapter 10 provided some overview on estimation of demand functions. The purpose of this chapter is to present a spectrum of creative, innovative, and interesting approaches used by researchers to analyze the demand for fish and other seafood products. A brief review of the theory of demand is presented to trace the evolution of different models used in food demand analysis. Empirical studies pertaining to seafood demand and preference articulation are selectively reviewed to illustrate concepts and highlight research applications.

Demand theory

The elements of traditional demand theory are first reviewed before the empirical literature on fish and seafood demand is surveyed. There are two main approaches, primal and dual, to develop theoretically consistent demand analyses. Demand functions can be derived indirectly either from utility maximization (known as the primal approach, which yields a Marshallian demand specification) or through expenditure minimization (the dual version of the former which yields a Hicksian compensated demand function). Figure 11.1 shows the linkage between the primal and dual approaches to demand function estimates for a two goods (x_1 and x_2) scenario, the details of which can be found in standard microeconomics textbooks (see, for example, Deaton and Muellbauer 1980a; Mas-Colell et al.1995).

In the primal approach, it is assumed that the consumers maximize utility by choosing quantities of n goods, $x_1, x_2, ...x_n$, subject to a linear budget constraint defined by total expenditure (m) and fixed market prices ($p_1, p_2, ...p_n$). The optimal solution of the primal problem is a system of Marshallian demand functions which show observable choices of a consumer ($x_1, x_2, ...x_n$) given market prices

Seafood and Aquaculture Marketing Handbook, Second Edition. Carole R. Engle,
Kwamena K. Quagrainie and Madan M. Dey.

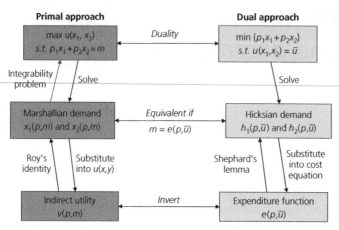

Fig. 11.1 Approaches to modeling seafood demand function.

and expenditures. Thus, the Marshallian demand equation for an individual consumer for commodity i (x_i) can be expressed as:

$$x_i = f(p_1, p_2, \ldots p_n, m).$$

The indirect utility function shows the maximum value of utility that a consumer attained given market prices $(p_1, p_2, \ldots p_n)$ and expenditures (m).

Under the dual approach, the consumers are assumed to choose the bundle of goods and services $(x_1, x_2, \ldots x_n)$ that will minimize expenditures required for a certain level of utility. The optimal solution of the dual problem is the Hicksian demand system, which considers that consumers are compensated for any price change. While Hicksian demand functions can be estimated empirically and are very useful for policy analysis, Hicksian demands are not directly observable in the marketplace.

Theoretical properties of demand

The Marshallian and Hicksian demand functions have four basic properties. The assumption that a consumer faces a linear budget constraint leads to the following two testable and desirable properties:

Property 1: Homogeneity The Marshallian demands are homogeneous of degree zero in prices and expenditures, and Hicksian demands are in prices. Homogeneity of degree zero in prices and expenditure requires that demand for all products be unchanged if the prices of the products and total expenditures all increase by the same percentage. This property is also called the "absence of money illusion."

Property 2: Adding up The total value of both Marshallian and Hicksian demands is total expenditure. The adding-up restriction satisfies Engel and Cournot aggregations. The Engel aggregation implies that income elasticities, weighted by the respective budget shares, sum to one. It ensures that when there is a shift in income, goods will be demanded in such a way that the whole income is absorbed. The Cournot aggregation condition ensures that budget constraints are held true when there is a change in price of a commodity, by adjusting the quantities demanded until the expenditure remains the same.

The other two properties, given below, are derived from the assumption that the expenditure function is concave in price (i.e., existence of consistent preferences).

Property 3: Symmetry The cross-price derivatives of the Hicksian demands (i.e., Slutsky substitution terms) are symmetric. This property stipulates that the compensated cross-price derivative of a product (say i) with respect to another product j equals the compensated cross-price derivative of product j with respect to product i.

Property 4: Negativity The matrix of own- and cross-price effects in Hicksian demands is negative semidefinite, implying that compensated own-price effects are negative.

Homogeneity, adding-up (Engel and Cournot aggregation), and symmetry restrictions are usually invoked or tested in empirical demand system models.

Approaches to modeling fish and seafood demand

Empirical investigations of seafood demand models have used single-equation or system approaches. Starting with the classic study by Bell (1968), earlier seafood demand models often used single-equation approaches. The most popular functional forms used in the single-equation approach to seafood demand analysis include linear, double-log, semi-log, and Box–Cox models. Some of these models are quantity-dependent, while others are price-dependent models (i.e., inverse demand functions). It is important to note that use of inverse demand models is quite common in fisheries where quantity is restricted by regulations (including quotas) and/or products are perishable in nature. However, in general, single-equation demand models are inconsistent with the standard utility maximization principle.

More recent empirical works have focused on the system approach, a technique pioneered by Stone in 1954. Although the demand system that Stone developed was consistent with the assumptions of neoclassical demand theory, the model restricts the nature of the relationship of the goods included in the system by assuming that the underlying preference ordering was additive. This implies that the marginal utility provided by the consumption of one good is independent of the consumption of the other goods; hence all goods are treated as substitutes, and inferior goods are excluded.

There are four main approaches to estimate seafood demand systems that are consistent with demand theory. These are: (1) use of Marshallian demand functions derived by maximizing the utility function subject to a budget constraint; (2) use of Marshallian demand functions derived from indirect utility function via Roy's identity; (3) use of Marshallian and Hicksian demand functions derived from a specified expenditure function; and (4) differential approximation of the Marshallian demand function. Popular functional forms used in estimating the demand system for fish and seafood are listed below.

1 Demand derived from specified utility function:
   (a) linear expenditure system (Stone 1954);
   (b) S-branch system (Brown and Heien 1972).
2 Demand derived from specified indirect utility function:
   (a) indirect addilog demand system (Houthakker 1960);
   (b) indirect translog demand system (Christensen et al. 1975).
3 Demand derived from specified expenditure function:
   (a) almost ideal demand system;
   (b) linear approximation to almost ideal demand system (Deaton and Muellbauer 1980b);
   (c) quadratic almost ideal demand system (Banks et al. 1997).
4 Demand derived from differential approximation:
   (a) Rotterdam model (Theil 1965; Barten 1966);
   (b) National Bureau of Research (NBR) demand system (Neves 1987);
   (c) Central Bureau of Statistic (CBS) demand system.

One of the most popular demand systems that uses a theoretically consistent demand model is the almost ideal demand system (AIDS). The demand system for the AIDS model was derived, by use of duality theory, from an optimal expenditure function defined as the minimum expenditure necessary to attain a specific level of utility at given prices.

This model was claimed to be more advantageous than its forerunners for the following reasons: (1) it gives an arbitrary first-order approximation to any demand system; (2) it satisfies the axioms of choice exactly; (3) it aggregates perfectly over consumers; (4) it has a functional form which is consistent with micro-level household budget data; (5) it is simple to estimate in its linear approximate form; and (6) it can be used to test for homogeneity and symmetry of demand parameters. In addition, although Deaton and Muellbauer (1980b) did not mention it, the AIDS is indirectly non-negative, allowing consumption of one good to affect the marginal utility of another good, whereas the linear expenditure system (LES) is directly additive, implying independent marginal utilities. Thus, the AIDS, in addition to the listed desirable properties, does not impose the severe substitution limitations implied by additive demand models such as the LES (Blanciforti and Green 1983).

Blanciforti and Green (1983) empirically compared the results generated by a simplified linear approximation of the AIDS and the LES model. One of their findings suggests that many commodities classified as luxury goods in the LES (income elasticities greater than 1) became necessities in the AIDS model (income elasticities less than 1). Specifically, the authors showed that AIDS possesses a property that income elasticities become more inelastic for necessities as their budget shares decrease (the reverse is true for LES). Thus, AIDS was concluded to be an attractive system for analyzing the demand for food commodities. However, the AIDS model requires a large number of parameters to be estimated which imposes constraints on the issue of sample size.

The AIDS model is specified as follows:

$$w_i = \alpha_i + \sum_j \gamma_{ij} \log p_j + \beta_i \log(m/P) \tag{11.1}$$

where:
w_i is the share in expenditure of the good i
p_j is the price of the good j
m is the income of the ith household
P is a price deflator of the income variable defined as follows:

$$\log P = \alpha_0 + \sum_k \alpha_k \log p_k + \frac{1}{2} \sum_j \sum_k \gamma_{kj} \log p_k \log p_j \tag{11.2}$$

Using the price index defined in equation (11.2) often raises empirical difficulties, especially when aggregate annual time-series data are used (Green and Alston 1990; Moschini 1995). One of the main reasons for the popularity of the AIDS model is that the price deflator P in equation (11.2) can be replaced by an index that will allow the estimation of a linear demand system. If prices are highly collinear (as they often are), then P may be approximately proportional to P*, i.e., P≈P*. Deaton and Muellbauer (1980b) suggest replacing P in the AIDS model by the Stone price index P* defined as:

$$\log P^* = \sum_k w_k \log p_k \tag{11.3}$$

The model that uses the Stone's price index is called the "linear approximate AIDS or LA/AIDS model." The LA/AIDS model has been used extensively in demand analysis, which includes the works of Blanciforti and Green (1983), Eales and Unnevehr (1988), and Moschini (1995).

Banks et al. (1997) suggested the quadratic extension of the Deaton and Muellbauer linear approximate AIDS model (1980b), hereinafter referred to as the QUAIDS model. The QUAIDS specification captures the non-linearity in consumption behavior of households for goods exhibiting threshold levels such as food commodities. At the same time, it relaxes the restriction imposed by linear demand functions regarding the allocation of marginal expenditures among commodities to be the same in rich and poor households (Beach and Holt 2001). Such assumptions limit the classification of goods into either necessities or luxuries and deny the possibility that some goods may be luxuries at low levels of income and necessities at higher levels of income. This type of consumption behavior may be observed in the case of high-value fish and other seafood products such as prawns, oysters, and crabs. Also, the computation of demand and income elasticities by income classes, i.e., low, medium and high income, is facilitated under the QUAIDS specification, since only one set of demand parameters needs to be estimated for the global sample. Subsequently, the demand elasticities by income group can be computed by simply varying the level of income in the elasticity formula.

Commodity grouping and separability

Generally, consumers make budget allocation decisions on large numbers of seafood products with different relative prices. Therefore, estimation of seafood demand systems can be difficult due to limited data and a relatively large number of parameters to estimate. There are two assumptions that can be made about how seafood products can be aggregated and separated into categories to make estimation possible:

1 Composite commodity theorem: One way to reduce the number of parameters to estimate in a seafood demand system is by combining various seafood items into a set of commodity aggregates. The composite commodity theory, proposed by Hicks (1936) and Leontief (1936), asserts that if all prices in a group move proportionately, then the corresponding group of commodities can be treated as a single good. The Hicks–Leontief composite group theorem requires that prices of all seafood categories within the same group have to be perfectly correlated, which may not hold true in most cases. Lewbel (1996) developed a generalized composite commodity theorem that relaxed the assumption of perfect collinearity of prices within a group and allowed a less perfect co-movement among intra-group prices.

2 Separability and multi-stage budgeting:An alternative approach to applying the composite group theorem is to assume that a group of closely related seafood products is separable from other foods. By assuming separability, commodities can be partitioned into groups so that preferences for products within a group can be described independently of the quantities in other groups. The separability assumption makes it possible to divide vast numbers of commodities into fewer workable groups. Several types of separability have been defined in the literature with varying assumptions on the nature of substitution between goods in different groups. The assumption of weak separability is widely followed in empirical seafood demand studies. Weak separability is based on the concept that the marginal rate of substitution of products belonging to the same group is independent of the consumption of goods within other groups. The weakly separable utility function can be represented by a "utility tree" with various subsets as its "branches" (Strotz 1957).

If commodities are assumed to be separable, multi-stage budgeting becomes possible. The multi-stage budgeting approach made use of the concept of Strotz (1957) who extended the idea of exhaustive expenditure to stages. In the first stage, the consumer is assumed to allocate expenditures to broad groups of commodities; then, in the second stage, the consumer is assumed to allocate expenditures within each of the broad groups to smaller groups. This process can continue, but most empirical analyses have been limited to two stages requiring the condition of weak separability, that is, the conditional ordering of goods on the independence of marginal utilities of goods within one group from consumption of goods in other groups.

This approach has been widely used to address a common problem in empirical demand system models, which requires a sizeable system of demand equations given the wide variety of consumption goods jointly purchased by households. The full demand system containing all these commodities warrants a large number of own- and cross-price parameters that are impractical to estimate given limited sample sizes. A solution forwarded in the literature is to estimate the model in stages, whereby expenditures on goods belonging to various food categories are estimated sequentially. Many studies have employed a multi-stage budgeting approach to estimate demand for seafood products. For example, Ioannidis and Whitmarsh (1987), Ioannidis and Matthews (1995), Jaffry et al. (1999), and Fousekis and Revell (2005) have used this approach to estimate fish demand in the United Kingdom. Dey (2000), Garcia et al. (2005), Kumar et al. (2005), Dey et al. (2008) and Dey et al. (2011) have used a three-stage budgeting approach for estimating fish demand in various Asian countries.

Other issues pertaining to estimating demand for seafood

The occurrence of zero observations is one of the most pressing issues in applied demand analysis and other microeconometric applications (Shonkwiler and Yen 1999). At the same time, the fact that the observed budget shares cannot take on negative values means that the dependent variable is censored. The problem of censored dependent variables was first recognized by Tobin (1958), who showed that the use of ordinary least squares (OLS) estimation for such models results in biased and inconsistent estimates. To address the problem, Tobin proposed a maximum likelihood (ML) estimation using the Tobit model. This technique is easy to carry out in the case of a single-equation demand estimation. However, the problem becomes more complex in the case of system demand models, which consist of a set of demand relations interrelated both through the error structure and cross-equation restrictions.

While theoretical literature exists for systems of equations with limited dependent variables (Amemiya 1974; Wales and Woodland 1983; Lee and Pit

1986, 1987), direct ML estimation of these models remains difficult when censoring occurs in multiple equations because of the need to evaluate multiple integrals in the likelihood function (Shonkwiler and Yen 1999). Heien and Wessells (1990) argued that it is possible to estimate models of this type by maximum likelihood, but such procedures generally are computationally prohibitive. Heien and Wessells (1990) provide a comprehensive survey of studies concerning the non-negativity constraint or the problem of censored dependent variables.

Heckman (1976, 1979) proposed a two-step estimation procedure for a system of equations with limited dependent variables, which was popularized by Heien and Wessells (1990) through the use of an inverse Mills ratio (IMR) in demand model estimation. The IMR is added in the model as a selectivity regressor (derived from a probit estimate in an earlier step) to remove the sample selection bias created by a significant number of zero consumption variables in the dataset. The demand system is then estimated using the seemingly unrelated regression (SUR) in the second step, hence the name Heckman two-step procedure. The first step involves a probit regression to compute for the probability that a given household will consume the good in question. The decision to consume is modeled as a dichotomous choice problem, i.e., $C_{ih} = f(P_{ih}, D_h)$ where C_{ih} is 1 if the hth household consumes that i^{th} food item and 0 otherwise; P_h is a vector of prices for the hth household and D_h is a vector of the demographic variables. This regression is then used to compute the inverse Mills ratio (IMR) for each consuming household. The IMR for the hth household who consumes and who do not consume the ith good are given by equations (11.4) and (11.5), respectively:

$$\text{for } C=1 : \alpha_j = \rho_{jo} + \sum_m \rho_{jm} d_m \tag{11.4}$$

$$\text{for } C=0 : \text{IMR} = \psi\left(P_{ih}, D_h\right) / \left[1 - \Psi\left(P_{ih}, D_h\right)\right] \tag{11.5}$$

where ψ and Ψ are the density and cumulative probability functions, respectively.

The IMR is used as an instrument that incorporates the censored latent variable in the second-stage estimation of the demand relations. Heien and Wessells (1990) compared the results generated by the censored model (with IMR) and the uncensored model. The authors concluded that the censored model provides substantially improved results in terms of goodness of fit and the conformity of price elasticities with prior expectation.

In spite of the popularity and extensive applications of the Heien and Wessells (HW) model (e.g., Heien and Durham 1991; Wellman 1992; Gao and Spreen 1994; Nayga 1995, 1996, 1998; Gao et al. 1996; Park et al. 1996; Wang et al. 1996; Salvanes and DeVoretz 1997; and Han and Wahl 1998), Shonkwiler and Yen (1999) criticized the model and claimed that "there is internal inconsistency in this model." In addition, the authors proposed an alternative consistent two-step estimation (CTS) procedure for systems of equations with limited dependent variables and conducted a Monte Carlo simulation to investigate and compare the performance of the CTS and the censored model proposed by Heien and Wessells (1990). Shonkwiler and Yen concluded that the CTS performs well compared to the HW procedure. The authors added that, although their CTS model only considered a three-equation linear system in the simulation, application of the methodology to the case of multiple and/or non-linear equations (e.g., "theoretically plausible" demand system) is equally straightforward. Another problem arising due to zero consumption is that of missing prices. In order to estimate a complete system, prices must be available for all items for all households.

However, for households not consuming a particular item, there will be no data on the price for that item. The usual procedure employed was to estimate the missing prices by performing a regression on the price of the item from those households who did consume it. Studies have used regional dummies, seasonal dummies, and income as regressors in such price models, and then used that model to estimate the missing price for those households who did not consume that particular item. The properties of estimates using price data obtained in this manner were discussed by Dagenais (1973) and Gourieroux and Monfort (1981). However, it should be pointed out that these properties hold only for non-censored variables.

Likewise, it had been recognized that food demand is influenced by the age structure of the population and various other demographic factors as cited by Heien and Wessells (1990). To incorporate demographic variables, the AIDS or LA/AIDS model can be modified by incorporating demographic variables in the budget share equations of the AIDS model as follows: $\alpha_j = \rho_{jo} + \sum_m \rho_{jm} d_m$, where d_m is the mth demographic variable. This method of incorporating demographic variables in the AIDS model is known as translation (Heien and Wessells 1990). The other widely used technique is demographic scaling. Translation preserves the linearity of the system, whereas scaling is a highly non-linear specification (Pollak and Wales 1981).

Data

Both cross-section and time series data have been used in estimating seafood demand systems. Earlier studies typically used time series data available from public institutes (see, for example, Bell 1968; Kabir and Ridler 1984; Hermann and Lin 1988; Barten and Bettendorf 1989; Bjørndal et al. 1994; Asche et al. 1997). Increasingly, survey data on cross-sections of households are being used to estimate seafood demand systems (see, for example, Dey, 2000; Garcia et al. 2005; Kumar et al. 2005; Dey et al. 2008, 2011). The recent availability of commercial scanner data allows significant advances in understanding demand for different seafood products (for different forms for the same species, for different species, and for different brands), and changing consumer buying patterns.

Scanner data

Scanner data collected on consumer purchases fall into two types: store (point-of- sale) scanner data, and household-based scanner data. Store scanner data are collected at cash registers and identify the products, quantities sold, and prices paid. Household-based scanner data come from a sample of households that scan all purchased products after each shopping trip. These household-based data provide information on household demographic characteristics (e.g., age, income, number of children, education level), the brand purchased, the package size, the price paid, and the store from which the product was purchased.

Nielsen Company (formally known as A.C. Nielsen, Inc.; commonly referred to as Nielsen) and Information Resources, Inc. (IRI) are the two major commercial suppliers of both store and household-based scanner data. The store scanner data service provided by IRI is called InfoScan; the one by Nielsen is called Scantrack. The in-home household scanner data collected by IRI is called the "Consumer Network", and the collection by Nielsen is called "Homescan." The Consumer Network and Homescan datasets are more complete than datasets of purchases of individual households collected through loyalty card use; the latter data collection does not include information on household demographics and is likely subject to more measurement errors because of infrequent use of loyalty cards or use of someone else's card for convenience. The entire Nielsen panel scans all products with a universal product code (UPC), and a subset of the panel also records purchases of random-weight or non-UPC products (e.g., fresh fruits and vegetables, bakery products produced and packaged in the store, and meat products cut and packaged in the store). The IRI in-home scanner data do not contain non-UPC random-weight perishable products.

Some of the important studies using scanner data for seafood demand analysis include Wessells and Wallstrom (1999), Chidmi et al. (2012), Singh et al. (2012), and Singh et al. (2014). Wessells and Wallstrom (1999) utilized panel data across 34 U.S. cities from 1988 through 1992 consisting of scanner data to test the stability of canned salmon demand. Chidmi et al. (2012) used national store-level scanner data from A.C. Nielsen and estimated substitution patterns across seafood categories at the U.S. retail market level. Using weekly national store-level scanner data acquired from A.C. Nielsen Inc., Singh et al. (2012) analyzed demand for 14 unbreaded frozen seafood products in the U.S. Singh et al. (2014) used market-level commercial scanner data obtained from A.C. Nielsen covering 52 U.S. cities to study the effects of season and space on the demand structure of unbreaded frozen finfish.

Elasticities and flexibilities of seafood demand

Chapter 2 discussed basic concepts of price and income elasticities of demand. Various studies have estimated the own-price elasticity (percentage change in quantity demanded for a product due to 1% change in price of the same product), cross-price elasticity (percentage change in quantity demanded for a product due to 1% change in price of another product), and expenditure elasticities (percentage change in quantity demanded due to 1% change in expenditure) of demand for diverse fish/seafood products. These concepts of elasticities are based on a quantity-dependent (ordinary) demand function.

Conceptually, there can be two types of price elasticities: "Marshallian" or "uncompensated" and "Hicksian" or "compensated" demand elasticity. Consumer demands can face two types of effects due to the price change: substitution effect and income effect. The substitution effect is the change in consumption in response to the price change holding real income (utility) constant. A change in price of a consumer good or service also has an income effect, that is, a reduction in price means a consumer has more income left than before if the same quantity is consumed. This change in real income due to the price change will change consumption (positively or negatively depending on the relationship between income and consumption). The compensated price elasticity measures demand when price changes are compensated by equivalent income changes such that the real income and utility remain unchanged. By contrast, the uncompensated price elasticity represents demand response when price changes are not compensated by income change, depicting the case where real income and total utility change while monetary income remains unchanged.

Flexibilities, which can be explained in a similar manner as elasticities, are concepts related to price-dependent (inverse) demand function. Price flexibility is the percentage change in the price of a product due to a 1% change in the quantity demanded of that product (own-price flexibility) or a related product (cross-price flexibility). Scale flexibility, a concept analogue to income elasticity in direct demand functions, is the percentage change in the normalized price of a product due to a 1% change in the scale of consumption bundle (or aggregate quantity index). A number of recent studies, including Eales et al. (1997), Lee and Kennedy (2008), Xie et al. (2009), Nguyen (2012), and Asche and Zhang (2013), have estimated flexibilities of seafood demand.

Mathematically, price elasticities are the inverse of price flexibilities. However, as pointed out by Houck (1965), the inverse of flexibility will be a consistent estimate of the respective elasticity only if the product in question has no substitute. Otherwise, the reciprocal of price flexibility will provide a lower limit of the elasticity in question. The benchmark value for price elasticity (flexibility) is -1. A seafood product with constant budget share and no substitutes will have price elasticity (flexibility) of demand of -1. If the price elasticity (flexibility) of a seafood product is between 0 and -1, demand is said to be inelastic. If, on the other hand, the price elasticity (flexibility) is less than −1 (greater than 1 in absolute value), demand is considered elastic.

As indicated in Chapter 2, an income elasticity of demand of 1 is the focal point. A product is considered a luxury good if its income elasticity of demand is higher than 1. For a similar reason, the benchmark value for scale flexibility is −1. If the scale flexibility of demand is less than −1 (greater than 1, in absolute value), the product in question is considered as a luxury good.

Estimates of elasticities and flexibilities of seafood demand

The number of studies on seafood demand has increased considerably in the last three decades or so, particularly from the 1980s. Gallet (2009) conducted a meta-analysis of demand for seafood products based on 168 previous studies, 160 of which were published during the period 1980–2007. Asche et al. (2007) provide a comprehensive review of this literature though 2005. Early studies are reviewed and summarized by Wessells and Anderson (1992) and Kinnucan and Wessells (1997). As Asche et al. (2007) noted, these reviews covered studies mostly from North America, the European Union, and Japan. Research on seafood demand is fairly new in developing countries (Dey et al. 2011).

Available recent estimates of income and own-price elasticities of fish demand are provided in Tables 11.1, 11.2, 11.3, 11.4 and 11.5, and are discussed in the following subsections. Estimated elasticities (flexibilities) vary substantially across studies due to differences in the source and type of data used (cross-section, time series or both), the model used, and the estimation procedure followed. The magnitudes of elasticities (flexibilities) also vary across different species and countries (even across regions within a specific country), indicating the relevance of estimation specific to species and geographic location.

Recent estimates of elasticities/flexibilities of seafood demand in developed countries

Recent studies on seafood demand in the U.S. and other developed countries, which were not covered in Asche et al. (2007), are reported in Table 11.1 and Table 11.2, respectively. Both direct demand and inverse demand models have been widely employed to analyze seafood markets in developed countries. Most of these studies used some variants of AIDS model as the analytical model.

Some of the important recent studies of seafood demand in developed countries are those of Dedah et al. (2011), Xie and Myrland (2011), Singh et al. (2011), Chidmi et al. (2012), Nguyen (2012), Singh et al. (2012), Asche and Zhang (2013), Nguyen and Jolly (2013), Singh et al. (2014), and Huang (2015). Several of these studies that were conducted in the U.S. used point-of-sale

Table 11.1 Price and income elasticities of seafood demand in the U.S.

Author	Product	Type of data	Method used	Elasticity/ flexibility	Species	Own-price elasticity (uncompensated)	Income elasticity/ scale flexibility
Huang (2015)	Blue crab	Time series data for the period 1994–2007 from MDNR	Nonlinear IAIDS model	Flexibility	#1 Male	−0.542 to −0.705	−0.967 to −1.141
					#2 Male	−0.535 to −0.187	−0.614 to −1.216
					Female	−0.646 to 0.343	−0.973 to −1.346
					SP (soft and peeled)	−0.969 to 2.468	−0.489 to 1.183
					Mixed	−0.505 to−0.372	−1.003 to −1.406
Singh et al. (2014)	13 Finfish species	Scanner data from A.C. Nielsen Inc. for the period June 2005 to June 2010	Dynamic AIDS model	Elasticity	Salmon	−0.83 to −2.17	0.71 to 1.05
					Tilapia	−1.09 to −2.31	0.84 to 1.47
					Whiting	−1.36 to −3.37	0.59 to 1.51
					Cod	−0.07 to −2.45	0.77 to 1.18
					Flounder	−1.12 to −2.30	0.22 to 1.56
					Pollock	−0.21 to −2.77	−1.99 to 1.23
					Catfish	−0.70 to −5.00	0.39 to 1.43
					Halibut	−0.73 to −1.23	−0.08 to 1.19
					Orange roughly	−0.88 to −1.06	0.18 to 1.26
					Mahi mahi	−0.58 to −1.20	0.71 to 1.13
					Tuna	−0.88 to −2.89	0.43 to 1.26
					Swordfish	−0.56 to −1.14	0.48 to 1.16
					Perch	−0.63 to −2.55	0.74 to 1.29
Asche and Zhang (2013)	White fish	Data from NMFS and U.S. International Trade Commission from 1994 to 2011	IAIDS	Flexibility	Cod	−0.21 to −0.51	−0.969 to −0.993
					Haddock	−0.11 to −0.17	−0.934 to −0.941
					Pollock	−0.16 to −0.21	−0.991 to −1.006
					Tilapia	−0.5 to −0.06	−1.028 to −1.230

(Continued)

Table 11.1 (*Continued*)

Author	Product	Type of data	Method used	Elasticity/ flexibility	Species	Own-price elasticity (uncompensated)	Income elasticity/ scale flexibility
Nguyen et al. (2013)	Crustaceans	Scanner data from June 2008 to June 2010	LA/AIDS	Elasticity	Shrimp	−1.585	0.823
					Crab	−0.766	1.191
					Crawfish	−0.583	0.721
					Lobster	−0.928	1.856
Chidmi et al. (2012)	Six seafood species	Scanner data from June 2008 to June 2010	A non-linear AIDS model	Elasticities	Catfish	−1.946	1.375
					Crawfish	−1.360	1.657
					Clams	−2.057	1.471
					Shrimp	−0.759	0.874
					Tilapia	0.773	1.618
					Salmon	−1.457	0.841
Singh et al. (2012)	14 Un-breaded frozen seafood products	Scanner data from A.C. Nielsen Inc. for the period June 2007 to June 2010	A log-linear version of Paasche's index used in AIDS model	Elasticities	Shrimp	−0.93	1.15
					Salmon	−0.28	0.44
					Crab	−0.19	0.96
					Catfish	−0.78	0.81
					Tilapia	−0.83	0.61
					Flounder	−0.28	0.32
					Cod	−1.1	0.40
					Whiting	−1.67	0.52
					Perch	−1.52	0.72
					Tuna	−0.98	0.33
					Pollock	−2.08	0.48
					Lobster	−0.91	1.77
					Scallop	−1.83	0.79
					Clam	−0.06	1.00

Study	Data category	Data source	Model	Measure	Item	Value	Value 2
Dedah et al. (2011)	Oyster from 4 regions	Quarterly time series data from 1985 to 2008 from NMFS	IAIDS	Elasticities	Gulf oyster	-0.683	1.17 to 1.13
					Chesapeake oyster	-0.639	0.06 to 0.31
					Pacific oyster	-0.273	0.64 to 0.72
					Imported oyster	-0.283	-0.97 to -0.32
Singh et al. (2011)	Four crustaceans	Time series data from NMFS and trade database from FAO, 1950 to 2007	An error correction AIDS model	Elasticities	Shrimp	-0.98 to -0.98	
					Crab	-0.32 to -0.50	
					Lobster	-0.29 to 0.45	
					Crayfish	-0.06 to -1.04	
Muhammad and Hanson (2009)	Fish forms	NASS data for January 1996 to January 2007	Rotterdam	Elasticities	Fresh catfish whole	-1.272	0.768
					Fresh catfish Fillet	-1.651	0.746
					Fresh catfish other form	-1.676	1.119
					Frozen catfish whole	-1.750	0.717
Lee and Kennedy (2008)	Fish species	Time series data from NMFS and other published sources for the period 1989 to 2005	Five different inverse demand models used but report based on DINBR model	Flexibility	Domestic crawfish	-0.809	
					Imported crawfish	-0.990	
					Catfish	-0.231	-1.761
					Shrimp	-0.085	-0.949
					Oyster	-0.872	-0.228

(Continued)

Table 11.1 (*Continued*)

Author	Product	Type of data	Method used	Elasticity/ flexibility	Species	Own-price elasticity (uncompensated)	Income elasticity/ scale flexibility
Quagrainie (2003)	Fish form	Time series data from January 1994 to 2001 from NASS	Dynamic AIDS	Elasticities	Catfish whole	−0.863	0.486
					Catfish fillet	−1.022	1.201
Hanson et al. (2001)	Fish form	Time series data from 1994 to 1999 from NASS	Trans-log	Elasticities	Catfish whole	−0.037	0.991
					Catfish fillet	−1.010	1.007
					Catfish steak	−1.147	0.976
					Catfish nugget	−0.874	1.021

AIDS, almost ideal demand system; DINBR, Differential Inverse National Bureau of Research; FAO, United Nations Food and Agriculture Organization; IAIDS, inverse almost ideal demand system; LA/AIDS, linear approximate AIDS model; MDNR, Maryland Department of Natural Resources; NASS, USDA National Agricultural Statistical Service; NMFS, U.S. National Marine Fisheries Service.

Table 11.2 Price and income elasticities of seafood demand in Europe, Canada, Japan, and the UK.

Author	Product	Country	Type of data	Method used	Elasticity/ flexibility	Species	Own-price elasticity		Income elasticity/scale flexibility
							Uncompensated	Compensated	
Nguyen (2012)	Mussel	Total 5 European countries	Monthly time series data from 2002 to 2008 were collected from different sources and websites	Non-linear IAIDS model	Flexibility	Denmark mussel	−0.93 to −1.02		−0.858 to −0.983
						Italy mussel	−0.624 to −0.62		−1.065 to −1.050
						Spain mussel	−0.711 to −0.719		−1.133 to −1.102
						France mussel	−0.529 to −0.538		−1.087 to −1.067
						Netherlands mussel	−0.859 to −0.608		−0.493 to −0.439
Xie and Myrland (2011)	Fish	France	French household data from Norwegian Seafood Export Council for 2005–2009	LA/AIDS	Elasticity	Fresh salmon	−1.156	−0.993	0.43
						Frozen salmon	−0.222	−0.163	0.339
						Smoked salmon	−1.32	−0.541	1.638
Jaffry and Brown (2008)	Fish forms	UK	Time series data from February 1995 to 1999 from IRI	AIDS	Elasticity	Tuna brine	−0.571		0.959
						Tuna sauce	−0.194		0.342
						Tuna oil	−0.796		1.346
Lambert et al. (2006)	Fish and other meat product	Canada	Canada's food expenditure survey for 1992 and 1996	QAIDS	Elasticity	Fish	−0.43 to −0.82	−0.404 to −0.714	0.45 to 0.68

(Continued)

Table 11.2 (*Continued*)

Author	Product	Country	Type of data	Method used	Elasticity/ flexibility	Species	Own-price elasticity		Income elasticity/scale flexibility
							Uncompensated	Compensated	
Klasra and Fidan (2005)	Fish, along with meat product	Turkey	Household survey	AIDS	Elasticity	Anchovy	−1.004		0.012
						Horse mackerel	−0.403		0.014
						Atlantic bonito	−0.513		0.013
						Blue fish	−0.360		0.001
						Trout	−0.345		0.013
						Sea bream	−0.180		0.0008
						Hake	−0.152		0.008
						Other fish	−0.110		
Salvanes and DeVoretz (1997)	Fish and meat product	Canada	1986 Food Expenditure Survey	LA/AIDS	Elasticity	Fish	−0.877	−0.885	1.0382
Hayes et al. (1990)	Fish and other 3 meat items	Japan	From different departments during 1947 to 1978	LA/AIDS	Elasticity	Fish	−0.70	−0.31	0.78

AIDS, almost ideal demand system; IAIDS, inverse almost ideal demand system; IRI, Information Resources, Inc.; LA/AIDS, linear approximate AIDS model; QAIDS, quadratic extension of the linear approximate AIDS model.

Table 11.3 Price and income elasticities of seafood demand in Bangladesh.

Author	Product	Country	Type of data	Method used	Elasticity/ flexibility	Species	Own-price elasticity		Income elasticity
							Uncompensated	Compensated	
Dey et al. (2011)	8 Different fish types	Bangladesh	Household survey data from July 1998 to August 1999	Three-stage budgeting framework and QAIDS model	Elasticity	Indian carp	−0.36	−0.10	1.09
						Exotic carp	−0.15	−0.10	0.28
						Hilsa	−0.62	−0.48	1.44
						Assorted small fish	−0.07	−0.06	0.07
						High-value fish	−1.96	−1.86	2.14
						Tilapia	−0.97	−0.94	0.46
						Shrimp and prawn	−1.14	−0.97	2.92
						Live fish	−0.98	−0.75	2.51
Dey et al. (2008)	Fish products	Bangladesh	Survey data for the year 1999	QAIDS	Elasticity	High-value freshwater fish	−1.08 to −2.02		0.94 to 2.63
						Low-value freshwater fish	−0.83 to −1.08		0.59 to 1.15
						High-value marine fish	−1.49 to −2.78		1 to 3.07
						Low-value marine fish	−0.8 to −1.04		0.85 to 1.25
						Shrimp	−0.98 to −1.04		0.47 to 0.8
Dey (2000)	Fish products	Bangladesh	Household survey data	Three-stage budget framework with AIDS model	Elasticity	Live fish	−1.27 to −1.56	−1.00 to −1.45	0.83 to 2.13
						Carp	−2.07 to −2.90	−2.02 to −2.87	1.77 to 0.74
						Assorted small fish	−0.99 to −1.00	−0.59 to −0.42	0.55 to 1.44
						Shrimp	−0.29 to −0.6	−0.28 to −0.58	0.33 to 1.26
						Dried fish	−1.32 to 1.81	−1.23 to −1.80	0.15 to 1.32
Talukder (1993)	Fish with other 5 food items	Bangladesh	Bangladesh household survey data, 1981–82	Log-polynomial	Elasticity	Fish			1.334 to 1.098
Pitt (1983)	Fish with other 8 food items	Bangladesh	Household expenditure survey of Bangladesh 1973–74	Limited dependent variable model	Elasticity	Fish	−0.97 to −0.66	−0.89 to −0.62	1.02 to 0.50

AIDS, almost ideal demand system; QAIDS, quadratic extension of the linear approximate AIDS model.

Table 11.4 Price and income elasticities of seafood demand in selected Asian developing countries.

Author	Product	Country	Type of data	Method used	Elasticity/ flexibility	Species	Own-price elasticity		Income elasticity
							Uncompensated	Compensated	
Hovhannisyan and Gould (2011)	Seafood (fish, shrimp and other), other food products	Urban China	Household expenditure survey data for the years 1995 to 2003	GQ-AIDS using maximum likelihood estimation	Elasticity	Seafood	-0.558 to -0.658	-0.477 to -0.492	0.789 to 1.501
Dey et al. (2008)	Fish products	China	Household survey data from NBS for the years 1997 and 2001	QAIDS	Elasticity	High-value freshwater fish	-0.29		0.99
						Low-value freshwater fish	-0.39		0.99
						High-value marine fish	-0.44		1.05
						Low-value marine fish	-0.95		0.95
						Shrimp	-0.4635		1.36
Dey et al. (2008)	Fish products	India	Survey data	QAIDS	Elasticity	High-value freshwater fish	-0.99		1.36 to 1.63
						Low-value freshwater fish	-0.99		1.36 to 1.64
						High-value marine fish	-0.62 to -0.97		1.14 to 1.37
						Low-value marine fish	-0.94 to -0.94		1.35 to 1.65
						Shrimp	-0.96 to -1.00		1.14 to 1.39
						Others	-0.99 to -1.0		1.12 to 3.75
Dey et al. (2008)	Fish products	Indonesia	Household survey data from CBS for the year 1999	QAIDS	Elasticity	Low-value freshwater fish	-0.89 to -0.94		0.53 to 3.05
						High-value marine fish	-1.35 to -1.45		0.53 to 3.05
						Low-value marine fish	-0.10 to -0.37		0.53 to 3.05
						Shrimp	-1.02 to 1.06		0.53 to 3.05
Dey et al. (2008)	Fish products	Malaysia	Household survey data from DSM for the year 2000	QAIDS	Elasticity	High-value freshwater fish	-0.97 to -1.46		0.54 to 1.12
						Low-value freshwater fish	-0.97 to -1.08		1.18 to 2.34
						High-value marine fish	-0.58 to -0.98		0.40 to 0.69
						Low-value marine fish	-0.22 to -1.00		0.64 to 1.05
						Shrimp	-0.89 to -1.24		–
						Others	-0.99 to -1.08		0.22 to 0.92

Study	Product	Country	Data source	Method		Item		
Dey et al. (2008)	Fish products	The Philippines	Household survey data from NSO for the year 2000	QAIDS	Elasticity	High-value freshwater fish	−1.46 to 3.61	0.14 to 0.59
						Low-value freshwater fish	−1.40 to −1.87	0.48 to 0.49
						High-value marine fish	−1.48 to −1.73	1.54 to 2.54
						Low-value marine fish	−1.32 to −1.60	0.34 to 0.94
						Shrimp	−0.92 to −1.00	0.89 to 2.66
						Others	−0.78 to −1.19	0.90 to 1.99
Dey et al. (2008)	Fish products	Sri Lanka	Household survey data from DCS for the year 1996	QAIDS	Elasticity	High-value freshwater fish	−1.06 to −1.15	0.72 to 1.05
						High-value marine fish	−0.96 to −0.98	1.00 to 1.19
						Low-value marine fish	−0.79 to −0.84	0.86 to 1.01
						Shrimp	−0.83 to −0.86	1.00 to 1.03
Dey et al. (2008)	Fish products	Thailand	Survey data for the year 1999 and 2002	QAIDS	Elasticity	High-value freshwater fish	−0.46 to −0.65	0.04 to 0.52
						Low-value freshwater fish	−0.59 to −0.61	0.001 to 0.3
						High-value marine fish	−0.74 to −0.76	0.36 to 0.91
						Low-value marine fish	−1.20 to −1.32	0.35 to 0.77
						Shrimp	−0.66 to −0.74	0.35 to 0.99
						Others	−0.62 to −0.71	0.73 to 0.88
Dey et al. (2008)	Fish products	Vietnam	Survey data for the year 2002	QAIDS	Elasticity	High-value freshwater fish	−0.88 to −0.9	0.99
						Low-value freshwater fish	−0.92 to −1.74	0.66 to 0.98
						High-value marine fish	−0.94 to −1.09	1.04 to 1.14
						Shrimp	−2.21 to −3.06	0.96 to 0.98
Gale and Huang (2007)	Fish products	China	Household survey data for 2002–03 from Chinese NBS	Engel curve regression estimations using log-log inverse	Elasticity	Fish		0.35 to 0.57
						Shrimp		0.69 to 1.53
						Other aquatic products		0.66 to 1.36

(Continued)

Table 11.4 (*Continued*)

Author	Product	Country	Type of data	Method used	Elasticity/ flexibility	Species	Own-price elasticity		Income elasticity
							Uncompensated	Compensated	
Garcia et al. (2005)	Fish products	The Philippines	Country-wide Family Income and Expenditure Survey 2000	QAIDS	Elasticity	Anchovy	−1.34 to −1.78	−1.31 to −1.77	0.34 to 1.035
						Milkfish	−1.46 to −3.61	−1.26 to −3.61	0.145 to 0.745
						Round scad	−1.27 to −1.42	−1.15 to −1.37	0.332 to 0.838
						Tilapia	−1.40 to −1.87	−1.28 to −1.84	0.481 to 0.613
						Shrimp	−0.92 to −1.00	−0.88 to −0.91	0.886 to 2.659
						Squid	−1.17 to −1.24	−1.11 to −1.44	0.922 to 2.407
						Crabs	−0.39 to −0.47	−0.37 to −0.45	0.870 to 1.576
						Other fresh fish	−1.48 to −1.73	−0.99 to −1.37	1.536 to 2.139
						Canned fish	−1.02 to −1.06	−0.95 to 1.03	0.436 to 1.062
						Dried fish	−1.33 to −1.78	−1.20 to −1.74	0.391 to 1.076
						Salted fish	−1.22 to −1.70	−1.18 to −1.70	0.017 to 0.944
Kumar et al. (2005)	Fish	India	Household survey data from dietary-pattern survey in 6 states of the country	Three-stage budgeting framework with QAIDS model	Elasticity	IMC	−0.99	−0.36 to −0.60	1.36 to 1.79
						Other freshwater fish	−0.99	−0.83 to −0.89	1.36 to 1.80
						Prawn/shrimp	−0.96 to −1.00	−0.83 to −0.95	1.14 to 1.72
						Pelagic high-value fish	−0.78 to −0.99	−0.78 to −0.91	0.72 to 1.76
						Pelagic low-value fish	−0.78 to −0.99	−0.90 to −0.97	1.34 to 1.81
						Demersal high value	−0.46 to −0.96	−0.46 to 0.93	1.36 to 1.79
						Demersal low value	−0.82 to −0.93	−0.81 to 0.9	1.36 to 1.8
						Mollusks	−0.99 to −1.01	−0.96 to −0.99	1.12 to 3.75
Liu and Sun (2005)	Fish and other 3 meat products	Taiwan	Cross-sectional national household survey data for 1981 and 1991	AIDS	Elasticity	Fish	−1.599 to −1.652		0.578 to 0.633

Study	Commodities	Country	Data	Model		Commodity			
Jung and Koo (2002)	Fish products along with other commodity	South Korea	Time series data for 1980–98	LA/AIDS with 3SLS estimation	Elasticity	Fish Crustaceans Mollusks	−0.847 −0.695 −0.862	0.196 −0.097 −0.874	0.179 1.655 −0.207
Meenakshi and Ray (1999)	Meat, eggs, and fish as a group and other 4 food items	India	NSSO data from 1972–73 to 1987–88	AIDS	Elasticity	Meat, eggs, and fish	−0.913 to −1.965		1.074 to 1.227
Gao et al. (1996)	Fish and other 8 food commodities	China	Rural household survey data from China's State Statistical Bureau for 1990	A demand system for food commodities with upper level AIDS model and lower level GLES	Elasticity	Fish	−0.807		0.892

AIDS, almost ideal demand system; CBS, Central Bureau of Statistics, Indonesia; DCM, Department of Statistics, Malaysia; DCS, Department of Census and Statistics, Sri Lanka; GLES, generalized linear expenditure systems; GQ-AIDS, generalized quadratic almost ideal demand system; LA/AIDS, linear approximate AIDS model; NBS, National Bureau of Statistics, People's Republic of China; NSO, National Statistics Office, the Philippines; NSSO, National Sample Survey Office, India; QAIDS, quadratic extension of the linear approximate AIDS model; 3SLS, three stage least squares.

Table 11.5 Price and income elasticities of seafood demand in selected non-Asian developing countries.

Author	Product	Country	Type of data	Method used	Elasticity/ flexibility	Species	Own-price elasticity		Income elasticity
							Uncompensated	Compensated	
Nguyen and Jolly (2013)	Fish products	Caribbean region	Time series data from different sources from 1976 to 2006	Co-integration and error correction model	Elasticity	Fish	−0.203		
Regoršek and Erjavec (2007)	Fish, with meat as a unit, and 6 other food commodities	Slovenia	Household budget survey year 2001	LA/AIDS	Elasticity	Fish with meat	−0.206 to −0.397	−0.128 to 0.038	0.924 to 0.968
Dhehibi et al. (2005)	Fish	Tunisia	Time series data covering 1975–2000	Double-log demand model	Elasticity	Fish	−0.367		0.273
Agbola et al. (2003)	Fish with meat, fish products along with other food items	South Africa	1993 South Africa Integrated Household Survey	LA/AIDS	Elasticity	Meat and fish / Fresh fish / Tinned fish	−0.894 to −1.309 / −0.03 to −1.15 / 0.01 to 2.61	−0.53 to −0.949 / −1.14 to −0.02 / −0.02 to 2.63	1.413 to 1.389 / −0.12 to 1.44 / −2.97 to 0.90
Ackah et al. (2003)	Fish with other food commodities	Ghana	Household survey data 1991–92, 1998–99	LA/AIDS	Elasticity	Fish	−0.874 to −0.988		0.699 to 0.781
Gould et al. (2002)	Fish with other 4 food commodities	Mexico	Nationwide household survey using weekly diary	Trans-log indirect utility function	Elasticity	Fish		−0.472 to −0.492	
Fayyad et al. (1995)	Fish with other food items	Egypt	Time series data	LA/AIDS	Elasticity	Fish	−0.173		0.936

LA/AIDS, linear approximate AIDS model.

(store-level) scanner data on seafood, and estimated elasticities for disaggregated products (Table 11.1).

Chidmi et al. (2012) found own-price elastic demand for catfish and salmon, and own-price inelastic demand for tilapia in the U.S. Singh et al. (2012) estimated own-price elasticities of demand (in absolute value) for salmon, catfish, tilapia, flounder, and tuna lower than 1, and for cod, whiting, perch, and pollock greater than 1. Singh et al. (2014) showed that the demand elasticities varied across species, seasons, and geography in the U.S., and that the majority of finfish products were either relatively own-price elastic or unitary elastic in most of the seasons and divisions in the country. Asche and Zhang (2015) and Huang (2015), based on price-dependent models, found most of the seafood products in the U.S. to be price inelastic. Recent studies conducted in Europe and other developed countries also show that the majority of seafood products are relatively price inelastic (Table 11.2). Xie et al. (2009) estimated the world demand curves faced by major exporters of fresh farmed salmon and found that the demand for Norwegian fresh farmed salmon and frozen fresh farmed salmon were own-price inelastic. Their results also suggested that the demand for farmed salmon has become less price elastic over time.

Recent estimates of income (expenditure) elasticity and scale flexibility show that the seafood demand in developed countries, in general, is income/scale inelastic (Tables 11.1 and 11.2). These results reveal that most of the seafood products are not considered as luxury goods in developed countries. Chidmi et al. (2012) estimated expenditure elasticities of demand for salmon, catfish, and tilapia at the levels of 0.84, 1.37, and 1.61, respectively. Singh et al. (2012) found that demand for finfish (i.e., salmon, catfish, tilapia, flounder, cod, whiting, perch, tuna, and pollock) was expenditure inelastic in the U.S.; however, expenditure elasticity of demand for salmon was lower than for catfish and tilapia. Singh et al. (2014) reported that the expenditure elasticity of demand was greater than 1 for tilapia, flounder, and catfish, and less than 1 for salmon, whiting, cod, pollock, halibut, orange roughy, tuna, swordfish, and perch in the U.S. Xie et al. (2009) found that the estimated expenditure elasticities were less than 1 for fresh farmed salmon from the UK (0.85), Chile (0.92), "rest of the world" (0.69), and frozen farmed salmon (0.83); whereas the expenditure elasticity was greater than 1 for Norwegian fresh farmed salmon (1.25).

Singh et al. (2014) revealed that the responsiveness of seafood demand to changes in its substitute product prices varies over seasons and U.S. census divisions. Their results, for example, show that tilapia is a substitute for catfish in the East North Central, New England, and Pacific divisions, but it has a complementarity for catfish in the Mountain, West North Central, and East South Central divisions. The analyses highlighted the importance of studying consumer demand behavior at the species level, across seasons and geography. Xie et al. (2009) found that the cross-price elasticities of demand of various fish are rather high in developed countries.

Recent estimates of elasticities/flexibilities of seafood demand in developing countries

Tables 11.3, 11.4, and 11.5 report estimates of seafood demand elasticities in developing countries, based on 20 studies conducted over the last two decades or so covering 18 countries/territories. In comparison to the developed countries, it is quite evident that not many studies have been implemented to estimate the demand for seafood products in developing countries.

As reported in the previous section, recent seafood demand studies in developed countries have covered most of the important species such as salmon, shrimp, groundfish, lobster, and catfish, to name a few. However, there are no such "flagship" species in developing countries. This may be due to the fact that most of the developing countries are located in the tropical region where the species diversity is higher and no single species dominates the others consistently in terms of harvest or value. Also, there are very few studies that analyze the demand for different processed forms of fish products.

As in developed countries, the AIDS model or its variant is the most widely used analytical tool in estimating seafood demand systems in developing countries. Most of these models are direct demand (quantity dependent) models. Own-price elasticities show the usual negative sign in all the cases reviewed. The uncompensated elasticities are larger than compensated elasticities, which is theoretically consistent. Similar to the situation in developed countries, estimated elasticities vary across geographic location, species studied, data used, and methodology followed. This is partly due to the fact that fish is a heterogeneous product.

On average, it can be said that the demand for fish products is mostly price inelastic in key fish-consuming Asian countries (Tables 11.3 and 11.4). However, there is wide variation in elasticities across species and countries. Studies on Chinese demand indicate an inelastic demand for different fish products with respect to their own prices.

Dey et al. (2008), in their cross-country comparative study, found that, in general, the elasticity of fish demand tends to be lower among households with higher incomes in Asia. These suggest that the poorer households exhibit more demand responsiveness given changes in fish prices than the more affluent households. Among the low-income households in Asia, only the low-value marine fish and dried fish showed inelastic demand. This suggests that the poorer households respond more to price changes of the more expensive fish types than of the low-value species. Among the more affluent sector, only the high-value fish types, that is, high-value freshwater and marine species and shrimp, showed elastic demand. Demand for the rest of the fish types was found to be price inelastic.

The income elasticities of various fish types in different countries yielded all positive values, except for mollusks in Korea and tinned and freshwater fish in South Africa. This implies that fish in general (whether fresh or dried) is considered a normal good in these developing countries. The average income elasticities were found to be mostly elastic with values greater than 1 in various countries (e.g., Bangladesh, China, India, Indonesia, and the Philippines), suggesting that fish is considered a luxury item in these countries.

Dey et al. (2008) found that income elasticities for all fish types registered elastic values among low-income households. Conversely, high-income households yielded inelastic values for all fish types. These suggest that fish consumption among the poorer households responds more to income than that of richer households. This further suggests that increases in per capita income of the poorer households will boost demand for fish in Asia.

It is important to note that, with the same set of data, different functional forms could result in widely different elasticity estimates; a long-run elasticity is always higher than a short-run elasticity because of the greater time available to the consumers to adjust to price change; and, with disaggregated markets, the range of elasticity estimates tends to widen because individual estimates will reflect quite unique market conditions, as aggregation averages out some variabilities of price sensitivities in the market scenario.

Aquaculture market synopsis: crawfish

Freshwater crawfish are important segments of the aquaculture industry in the U.S., Australia, Europe, and China. Crawfish have been consumed for centuries in Europe, and in North America by Native Americans (Lutz et al. 2003).

World supply of crawfish has increased dramatically since 2002 (Fig. 11.2). Wild capture of crawfish has remained stable for many years, and the substantial increase in world supply is from the reported aquaculture production of crawfish. Farmed production of crawfish in 2012 was more than seven times greater than volumes captured from the wild. Only six countries (China, U.S., Mexico, Indonesia, Italy, and the Ukraine) had records of crawfish production in 2012,

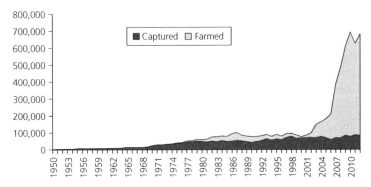

Fig. 11.2 Global supply of crawfish, 1950–2012. Source: FAO (2014).

Fig. 11.3 Female red swamp crawfish with eggs. Courtesy of Dr. Greg Lutz.

with China producing 93% of total world production of farmed crawfish. However, the FAO (2014) reports that the dramatic increase in supply of crawfish from China is likely from capture fisheries and incidental catch from harvest seining of foodfish ponds, although it is reported as aquaculture production.

Data from the Food and Agriculture Organization of the United Nations show five different species of crawfish raised around the world, including red swamp crawfish (*Procambarus clarkii*), noble crawfish (*Astacus astacus*), signal crawfish (*Pacifastacus leniusculus*), red claw crawfish (*Cherax quadricarinatus*), and yabby crawfish (*Cherax destructor*). Of these, 99.9% of all farmed crawfish are red swamp crawfish. The red swamp crawfish (Fig. 11.3) is native to the U.S. and is cultured in nine states (USDA 2014). However, 96% of total production in the U.S. is located in Louisiana where crawfish constitute an important culinary tradition for the Cajun culture in that part of the U.S. It is prepared in locally popular dishes such as the traditional crawfish boil and in étouffé.

Crawfish in Louisiana were sold commercially beginning in the late 1800s from wild-caught supplies (Lutz et al. 2003). Over time, the market shifted from local and household consumption to sales in urban areas of Baton Rouge and New Orleans. Growers began to re-flood rice fields, woodlands, and marshland to produce crawfish in the 1950s. By the mid-1960s, a crawfish peeling industry had developed with continued increases in acreage.

Wild crawfish are caught from the Atchafalaya Basin between Louisiana and Alabama. The wild catch exhibits dramatic fluctuations from year to year that are dependent on weather conditions. Wild-caught crawfish move through the same market channels as do farm-raised crawfish. Thus, the price of farm-raised crawfish has been affected strongly by the fluctuations in the wild-caught supply.

Farm-raised crawfish are typically raised with some other forage crop for feeding (Fig. 11.4). The forage crop can be a flooded area with natural

Fig. 11.4 Crawfish ponds with forage crops used for feeding crawfish. Courtesy of Dr. Greg Lutz.

vegetation, a rice crop that has been harvested with the stubble remaining, or a grain crop that is planted especially to serve as forage for the crawfish crop. Crawfish emerge from burrows under the pond when the pond is flooded and begin to forage. When the pond is drained, the crawfish return to the burrows to wait for the next period of flooding.

Crawfish marketing channels in the U.S. include live sales, sales to processing plants, and exports of whole boiled crawfish to Scandinavian countries (Lutz et al. 2003). Crawfish are harvested using traps (Fig. 11.5) and specially designed boats (Fig. 11.6), graded at the pond bank, and then packed live into onion sacks. Most crawfish farmers sell most of their product to buyers that specialize in the distribution of crawfish, although they typically will also sell a portion of their crop directly to the public, to restaurants, or to small seafood buyers.

The crawfish industry has faced several challenges over its history. During 1999–2001, drought conditions in Louisiana resulted in yields less than half of typical yields in previous years. Moreover, the crawfish peeling, or processing, sector has shrunk from 90–100 processors in Louisiana in 1996 to about 15 in 2003 (Lutz et al. 2003).

The difficulties of the peeling sector are related to the lower-priced imports of crawfish from China. In 2009, approximately $71.9 million of frozen and peeled crawfish meat were imported into the U.S. from China (Tordsen 2013). Crawfish in Louisiana historically were peeled and sold as fresh or frozen tail meat when crawfish harvests exceeded the demand for live, whole product. Thus, peeled crawfish tail meat served to moderate the seasonality of prices for crawfish. Peeled crawfish tail meat from China arrived initially during the period when

Fig. 11.5 Trap used to harvest crawfish. Courtesy of Dr. Greg Lutz.

Fig. 11.6 Crawfish harvesting boat. Courtesy of Dr. Greg Lutz.

yields had decreased and processors were having difficulty finding raw material to process. Tariffs were levied on imported Chinese tail meat in the late 1990s, and the tariffs have been subsequently renewed. However, Chinese exporters have shifted to quick-frozen whole boiled crawfish for sale in both traditional and non-traditional markets throughout the U.S. By 2003, the U.S. crawfish industry had largely recovered from the drought years and acreage and production levels had increased substantially (Lutz and Romaire 2003), but Hurricanes Katrina and Rita caused further damage to crawfish production infrastructure in 2005. However, the Louisiana crawfish industry has recovered substantially from those damages as well. While consumption of imported crawfish from China has continued to increase, demand for whole crawfish as a specialty seafood product has continued to be strong in traditional crawfish market areas.

While much attention is paid to the U.S. crawfish industry, crawfish of the genus *Cherax* have been cultured in Australia in a manner similar to that of the U.S. industry for a number of years. The Australian crawfish reach a much larger size and thus occupy somewhat different market niches. Consumers in several European countries, including Austria and Sweden, have a long tradition of catching and eating crawfish (*Astacus* spp.). Crawfish growers in a number of countries have targeted European as well as U.S. markets.

Summary

Three important results emerge from the review of demand elasticities reported in this chapter. First, fish is clearly a heterogeneous product, as shown by the wide disparity in the estimated income and price elasticities for the different fish types. Second, consumer behavior on seafood consumption varies across regions and countries. Third, the estimated price and income elasticities vary across income groups, particularly in Asia. Specifically, both price and income elasticities for all fish types tend to be higher among the poorer sector of the economy compared to the more affluent members of society in Asia. This implies that the poor often consider fish as a luxury commodity while the rich consider it as an ordinary food item.

The aquaculture/seafood industry needs to develop market specific strategies in order to gain further market share. Estimated elasticities show that the responsiveness of seafood demand to changes in its own and substitute product prices vary over seasons and regions. Understanding consumer demand behavior across seasons and over space is essential as (1) seafood demand varies over species, season, and space, and (2) not only does the degree of competition among seafood products vary considerably over space, but substituting products themselves change.

A simple, "back-of-the-envelope" analysis suggests that, as per capita incomes and populations grow in most Asian countries, there will be tremendous increases in fish demand. If there will be no commensurate increase in the supply of fish, then the price of fish in the market is expected to go up, which will hurt consumers, with worrisome consequences for the protein intake of the poor. However, suppose fish supply also increases dramatically, probably from aquaculture sources, then prices would be expected to fall, other factors being constant, which may be disadvantageous to fish farmers. For seafood products with elastic own-price demand, a price decline shall be followed by rising gross incomes of fish suppliers. There is a need for detailed market-specific and disaggregated analysis of seafood supply and demand to provide necessary guidance to the seafood industry.

Study and discussion questions

1 What is the most commonly used demand model for seafood? Why is this the case?
2 Describe and contrast the two main approaches (primal and dual) to developing theoretically consistent demand analyses.
3 List and describe the four basic theoretical properties of demand functions.
4 Contrast single-equation and system approaches to estimation of seafood demand.
5 Describe the two assumptions that can be made about how seafood products can be aggregated and separated into categories to make estimation possible.
6 Explain the appropriate techniques to use when zero observations are found in a dataset.
7 Explain and contrast seafood demand models estimated with time series data with those estimated with cross-sectional data, such as scanner data.
8 Describe and contrast Marshallian and Hicksian demand elasticities.
9 How do flexibilities differ from elasticities? Use seafood examples in your answer.
10 What differences have been found in seafood demand analyses for developed versus developing countries?

References

Ackah, C. and S. Appleton. 2003. Food price changes and consumer welfare in Ghana in the 1990s. CREDIT Research Paper, No. 07/03. Centre for Research in Economic Development and International Trade, University of Nottingham, Nottingham, UK.

Agbola, F.W., M. Pushkar, and K. McLaren. 2003. On the estimation of demand systems with large number of goods: An application to South Africa household food demand. Proceedings of the XXXXIst Annual Conference of the Agricultural Economic Association of South Africa (AEASA), October 2–3, Pretoria, South Africa.

Amemiya T. 1974. Multivariate regression and simultaeous eequation models when the dependent variables are truncated normal. *Econometrica* 42:999–1012.

Asche, F. and D. Zhang. 2013. Testing structural changes in the US whitefish import market: An inverse demand system approach. *Agricultural and Resource Economics Review* 42:453–470.

Asche, F., K.G. Salvanes, and F. Steen. 1997. Market delineation and demand structure. *American Journal of Agricultural Economics* 79:139–150.

Asche, F., T. Bjorndal. and D.V. Gordon. 2007. Studies in the demand structure for fish and seafood products. In: Weintraub, A., C. Romero, T. Bjorndal, and R. Epstein, eds. *Handbook of Operation Research in Natural Resources*, pp. 295–314. Springer, New York.

Banks, J., R. Blundell, and A. Lewbel. 1997. Quadratic Engel Curves and Consumer Demand. *The Review of Economics and Statistics* 79:527–539.

Barten, A.P. 1966. Theorie en empirie van een volledig stelsel van vraagvergelijkiingrn. Doctoral dissertation, University of Rotterdam.

Barten, A.P. and L.J. Bettendorf. 1989. Price formation of fish: An application of an inverse demand system. *European Economic Review* 33:1509–1525.

Beach, R. and M. Holt. 2001. Incorporating quadratic scale curves in inverse demand systems. *American Journal of Agricultural Economics* 83:230–245.

Bell, F. 1968. The pope and the price of fish. *American Economic Review* 58:1346–1350.

Bjørndal, T., D.V. Gordon, and K.G. Salvanes. 1994. Elasticity estimates of farmed salmon demand in Spain and Italy. *Empirical Economics* 4:419–428.

Blanciforti, L. and R. Green. 1983. The almost ideal demand system: A comparison and application to food groups. *Agricultural Economics Research* 35:1–10.

Brown, M., and D. Heien. 1972. The S-branch utility tree: a generalization of the linear expenditure system. *Econometrica* 40:737–747.

Chidmi, B., T. Hanson, and G. Nguyen. 2012. Substitution between fish and seafood products at the US national retail level. *Marine Resource Economics* 27:359–370.

Christensen, L.R., D.W. Jorgenson, and L.J. Lau. 1975. Transcendental logarithmic utility functions. *American Economic Review* 65:367–383.

Dagenais, M.G. 1973. The use of incomplete observations in multiple regression analysis. *Journal of Econometrics* 1:317–328.

Deaton, A. and J. Muellbauer. 1980a. *Economics and Consumer Behavior*. Cambridge University Press, New York.

Deaton, A. and J. Muellbauer. 1980b. An almost ideal demand system. *The American Economic Review* 70:312–326.

Dedah, C., W.R. Keithly, Jr., and R.F. Kazmierczak. 2011. An analysis of US oyster demand and the influence of labeling requirements. *Marine Resource Economics* 26:17–33.

Dey, M.M. 2000. Analysis of demand for fish in Bangladesh. *Aquaculture Economics & Management* 4: 63–81.

Dey, M.M., Y.T. Garcia, K. Praduman, S. Pyumsombun, M.S. Haque, L. Li, A. Radam, A. Senaratne, N.T. Khiem, and S. Koeshendrajana. 2008. Demand for fish in Asia: a cross-country analysis. *The Australian Journal of Resource Economics* 52:321–338.

Dey, M.M., M.F. Alam, and F.J. Paraguas. 2011. A multistage budgeting approach to the analysis of demand for fish: An application to inland areas of Bangladesh. *Marine Resource Economics* 26:35–58.

Dhehibi, B., L. Lachaal, and A. Chebil. 2005. Demand analysis for fish in Tunisia. An empirical approach. Paper prepared for presentation at the XIth Congress of the EAAE (European Association of Agricultural Economists). The Future of Rural Europe in the Global Agri-Food System, Copenhagen, Denmark, August 24–27, 2005.

Eales, J.S. and L.J. Unnevehr. 1988. Demand for beef and chicken products: separability and structural change. *American Journal of Agricultural Economics* 70:521–532.

Eales, J., C. Durham, and C. R. Wessells. 1997. Generalized models of Japanese demand for fish. *American Journal of Agricultural Economics* 79:1153–1163.

Food and Agricultural Organization of the United Nations (FAO). 2014. Cultured Aquatic Species Information Programme. *Procambarus clarkia* (Girard, 1852). FAO, Rome.

Fayyad, B.S., S.R. Johnson, and M. El-Khishin. 1995. Consumer demand for major foods in Egypt. CARD Working Paper 95-WP 138. Center for Agricultural and Rural Development, Iowa State University, Ames, Iowa.

Fousekis, P. and B.J. Revell. 2005. Retail fish demand in Great Britain and its fisheries management implications. *Marine Resource Economics* 19:495–510.

Gale, H.F. and K. Huang. 2007. Demand for food quantity and quality in China. Economic Research Report No. 32. US Department of Agriculture, Washington, D.C.

Gallet, C.A. 2009. The demand for fish: a meta-analysis of the own-price elasticity. *Aquaculture Economics & Management* 13:235–245.

Gao, X.M. and T. Spreen. 1994. A microeconometric analysis of the U.S meat demand. *Canadian Journal of Agricultural Economics* 42:397–412.

Gao, X.M., E.J. Wailes, and G.L. Cramer. 1996. A two-stage rural household demand analysis: Microdata evidence from Jiangsu Province, China. *American Journal of Agricultural Economics* 78:604–613.

Garcia, Y.T., M.M. Dey, and S.M.M. Navarez. 2005. Demand for fish in the Philippines: a disaggregated analysis. *Aquaculture Economics & Management* 9:141–168.

Gould, B.W. and H.J. Villarreal. 2002. Adult equivalence scales and food expenditures: an application to Mexican beef and pork purchases. *Applied Economics* 34:1075–1088.

Gourieroux, C. and A. Monfort. 1981. On the problem of missing data in linear models. *Review of Economics and Statistics* 48:579–586.

Green, R. and J.M. Alston. 1990. Elasticities in AIDS models. *American Journal of Agricultural Economics* 72:442–445.

Han, T. and T. Wahl. 1998. China's rural household demand for fruit and vegetables. *Journal of Econometrics* 72:85–134.

Hanson, T., D. Hite, and B. Bosworth. 2001. A translog demand model for inherited traits in aquacultured catfish. *Aquaculture Economics & Management* 5:3–13.

Hayes, D.J., T.I. Wahl, and G.W. Williams. 1990. Testing restrictions on a model of Japanese meat market. *American Journal of Agricultural Economics* 72:556–566.

Heckman, J.J. 1976. The common structure of statistical models of truncation, sample selection and limited dependent variables and a simple estimator for such models. *Annals of Economic and Social Measurement* 5:475–492.

Heckman, J.J. 1979. Sample selection bias as a specification error. *Econometrica* 47:153–161.

Heien, D. and C. Durham. 1991. A test of the habit formation hypothesis using household data. *Review of Economics and Statistics* 8:189–199.

Heien, D. and C.R. Wessells. 1990. Demand system estimation with microdata: A censored regression approach. *Journal of Business & Economic Statistics* 8:365–371.

Hermann, M. and B. Lin. 1988. The demand and supply of Norwegian Atlantic salmon in the United States and the European Community. *Canadian Journal of Agricultural Economics* 36:459–471.

Hicks, J.R. 1936. *Value and Capital*. Oxford University Press, Oxford.

Houck, J. P. 1965. The relationship of direct price flexibilities to direct price elasticities. *Journal of Farm Economics* 47:789–792.

Houthakker, H.S. 1960. Additive preferences. *The Econometric Society* 28:244–257.

Hovhannisyan, V. and B.W. Gould. 2011. Quantifying the structure of food demand in China: An econometric approach. *Agricultural Economics* 42:1–18.

Huang, P. 2015. An inverse demand system for the differentiated blue crab market in Chesapeake Bay. *Marine Resource Economics* 30:139–156.

Ioannidis, C. and K. Matthews. 1995. Determination of fresh cod and haddock margins in the UK 1988–1992. Cardiff Economic Research Associates, for Sea Fish Industry. Edinburgh, Scotland, UK.

Ioannidis, C. and D. Whitmarsh. 1987. Price formation in fisheries. *Marine Policy* 11:143–145.

Jaffry, S. and J. Brown. 2008. A demand analysis of the UK canned tuna market. *Marine Resource Economics* 23:215–227.

Jaffry, S., S. Pascoe, and C. Robinson. 1999. Long-run price flexibilities for high value UK fish species: a cointegration systems approach. *Applied Economics* 31:473–481.

Jung, J. and W.W. Koo. 2002. Demand for meat and fish products in Korea. In the Annual Meeting of American Agricultural Economics Association, Long Beach, California, July 28–31, 2002.

Kabir, M. and N.B. Ridler. 1984. The demand for Atlantic salmon in Canada. *Canadian Journal of Agricultural Economics* 32:560–568.

Kinnucan, H. and C.R. Wessells. 1997. Marketing research paradigms for aquaculture. *Aquaculture Economics & Management* 1:73–86.

Klasra, M.A. and H. Fidan. 2005. Competitiveness of major exporting countries and Turkey in the world fishery market: A constant market share analysis. *Aquaculture Economics & Management* 9:317–330.

Kumar, P., M.M. Dey, and F.J. Paraguas. 2005. Demand for fish by species in India: Three-stage budgeting framework. *Agricultural Economics Research* 18:167–186.

Lambert, R., B. Larue, C. Yélou, and C. Criner. 2006. Fish and meat demand in Canada: Regional differences and weak separability. *Agribusiness* 22:175–199.

Lee, L.F. and M.M. Pitt. 1986. Microeconometric demand systems with binding nonnegativity constraints: The dual approach. *Econometrica* 54:1237–1242.

Lee, L.-F. and M.M. Pitt. 1987. Microeconometric models of rationing, imperfect markets, and non-negativity constraints. *Journal of Econometrics* 36:89–110.

Lee, Y., and P.L. Kennedy. 2008. An examination of inverse demand models: An application to the US crawfish industry. *Agricultural and Resource Economics Review* 37:243–256.

Leontief, W. 1936. Composite commodities and the problem of index numbers. *Econometrica* 4:39–49.

Lewbel, A. 1996. Nesting the AIDS and translog demand systems. *International Economic Review* 30:349–356.

Liu, K.E. and C.H. Sun. 2005. A globally flexible, quadratic almost ideal demand system: An application for meat demand in Taiwan. *American Agricultural Economics Association* 35:1025–1036.

Lutz, C.G. and R. Romaire. 2003. Louisiana aquaculture as of 2003. Agricultural Experiment Station, Louisiana State University, Baton Rouge, Louisiana.

Lutz, C.G., P. Sambidi, and R.W. Harrison. 2003. Crawfish industry profile. Agricultural Marketing Resource Center, Iowa State University, Ames, Iowa.

Mas-Colell, A., M.D. Whinston, and J.R. Green. 1995. *Microeconomic Theory*. Oxford University Press, New York.

Meenakshi, J.V. and R. Ray. 1999. Regional differences in India's food expenditure pattern: a completed demand systems approach. *Journal of International Development* 11:47–74.

Moschini, G. 1995. Units of measurement and the stone index in demand system estimation. *American Journal of Agricultural Economics* 77:63–68.

Muhammad, A. and T.R. Hanson. 2009. The importance of product cut and form when estimating fish demand: the case of U.S. catfish. *Agribusiness* 25:480–499.

Nayga, R. 1995. Microdata expenditure analysis of disaggregate meat products. *Review of Agricultural Economics* 17:275–285.

Nayga, R. 1996. Wife's labor force participation and family expenditures for prepared food, food prepared at home, and food away from home. *Agricultural and Resource Economic Review* 25:179–186.

Nayga, R. 1998. A sample selection model for prepared food expenditures. *Applied Economics* 29:345–352.

Neves, P. 1987. Analysis of consumer demand in Portugal, 1958–1981. *Memorie de Maitrise en Sciences Economiques*. Université Catholique de Louvain, Louvain-la-Neuve, France.

Nguyen, T. 2012. An inverse almost ideal demand system for mussels in Europe. *Marine Resource Economics* 27:149–164.

Nguyen, G.V. and C.M. Jolly. 2013. A co-integration analysis of seafood import demand in Caribbean countries. *Applied Economics* 45:803–815.

Nguyen, G.V., T.R. Hanson, and C.M. Jolly. 2013. A demand analysis for crustaceans at the US retail store level. *Aquaculture Economics & Management* 17:212–227.

Park, J.L., R.B. Holcomb, K.C. Raper, and O. Capps, Jr. 1996. A demand systems analysis of food commodities by U.S. households segmented by income. *American Journal Agricultural Economics* 78:290–300.

Pitt, M.M. 1983. Food preferences and nutrition in rural Bangladesh. *The Review of Economics and Statistics* 65:105–114.

Pollak, R.A. and T.J. Wales. 1981. *Demand System Specification and Estimation*. Oxford University Press, Oxford.

Quagrainie, K.K. 2003. A dynamic almost ideal demand model for US catfish. *Aquaculture Economics & Management* 7:263–271.

Regoršek, D. and E. Erjavec. 2007. Food demand in Slovenia. Paper prepared for presentation at the Mediterranean Conference of Agro-Food Social Scientists. 103rd EAAE Seminar "Adding Value to the Agro-Food Supply Chain in the Future Euromediterranean Space." Barcelona, Spain, April 23– 25, 2007.

Salvanes, K.G. and D.J. DeVoretz. 1997. Household demand for fish and meat products: separability and demographic effects. *Marine Resources Economics* 12:37–56.

Shonkwiler, J.S. and S. Yen. 1999. Two-step estimation of a censored system of equations. *American Journal of Agricultural Economics* 81:972–982.

Singh, K., M.M. Dey, and G. Thapa. 2011. An error corrected almost ideal demand system for crustaceans in the United States. *Journal of International Food and Agribusiness Marketing* 23:271–284.

Singh, K., M.M. Dey, and P. Surathkal. 2012. Analysis of a demand system for unbreaded frozen seafood in the United States using store level scanner data. *Marine Resource Economics* 27:371–387.

Singh, K., M.M. Dey, and P. Surathkal. 2014. Seasonal and spatial variations in demand for and elasticities of fish products in the United States: An analysis based on market level scanner data. *Canadian Journal of Agricultural Economics* 62:343–363.

Stone, R. 1954. Linear expenditure systems and demand analysis: an application to the pattern of British demand. *Economic Journal* 64:511–527.

Strotz, R.H. 1957. The empirical implications of a utility tree. *Econometrica* 25:269–280.

Talukder, R.K. 1993. Patterns of income induced consumption of selected food items in Bangladesh. *The Bangladesh Development Studies* 21:41–53.

Theil, H. 1965. The information approach to demand analysis. *Econometrica* 33:67–87.

Tobin, J. 1958. Estimation of relationships for limited dependent variables. *Econometrica* 26:24–36.

Tordsen, C. 2013. Crawfish profile. Louisiana State University Agricultural Center (can be obtained from glutz@agctr.lsu.edu, psambil@lsu.edu, and rharrison@agctr.lsu.edu). Accessed 12/21/2014.

United States Department of Agriculture (USDA). 2014. Census of Aquaculture (2013), National Agricultural Statistics Service, USDA, Washington, D.C., www.agcensus.usda.gov/ Publications/2012/Online_Resources/Aquaculture/aquacen.pdf.

Wales, T.J. and E.T. Woodland. 1983. Estimation of consumer demand systems with binding non-negativity constraints. *Journal of Econometrics* 21:263–286.

Wang, J., X.M. Gao, E.J. Wailes, and G.L. Cramer. 1996. Consumer demand for alcoholic beverages: cross-section estimation of demographic and economic effects. *Review of Agricultural Economics* 18:477–488.

Wellman, K.F. 1992. The U.S. retailed demand for fish products: an application of the Almost Ideal Demand System. *Applied Economics* 24:445–457.

Wessells, C.R. and J.L. Anderson. 1992. Innovations and progress in seafood demand and market analysis. *Marine Resource Economics* 7:209–228.

Wessells, C.R. and P. Wallstrom. 1999. Modeling demand structure using scanner data: implications for salmon enhancement policies. *Agribusiness* 15:449–461.

Xie, J., and O. Myrland. 2011. Consistent aggregation in fish demand: A study of French salmon demand. *Marine Resource Economics* 26:267–280.

Xie, J., H.W. Kinnucan, and O. Myrland. 2009. Demand elasticities for farmed salmon in world trade. *European Review of Agricultural Economics* 36:425–445.

CHAPTER 12

Policies and regulations governing seafood and aquaculture marketing

The regulatory environment for seafood and aquaculture has become more complex and more stringent in recent years. The increasing globalization of the seafood trade has heightened discussions related to policies and regulations in both exporting and importing nations. Regulatory conflicts among countries have increased due to disparities among the many exporting nations in the developing world and the major importing countries that tend to be more developed countries. Other conflicts have arisen among interest groups such as consumer groups concerned over food safety and environmental groups concerned with environmental impacts.

This chapter first discusses and contrasts several regulatory frameworks and associated permitting systems and compliance costs. Food safety concerns are summarized along with the roles of the World Trade Organization, the U.S. Food and Drug Administration, the Directorate-General for Health and Consumer Protection of the European Commission, the Ministry of Health and Welfare in Japan, and the use of hazard analysis of critical control points (HACCP) programs. Policies related to organic standards and green labeling programs for seafood, marketing and transportation of live aquatic animals, and aquatic animal health and biosecurity are then discussed. The chapter concludes with a synopsis of the growth in mariculture of grouper, snapper, tuna, and cobia.

Regulatory frameworks for seafood and aquaculture

The regulatory framework and its effects on development of aquaculture have long been a concern (Bowden 1981). There is ample literature that describes the effects of increasing numbers of regulations on businesses in the U.S. (Christainsen and Haveman 1981; Gray 1987; Antle 2000), the EU (Directorate-General for Internal Policies 2009), Australia (Harris 1998), and New Zealand (Stewart 2012). However, in spite of a comprehensive and stringent regulatory environment in

Seafood and Aquaculture Marketing Handbook, Second Edition. Carole R. Engle,
Kwamena K. Quagrainie and Madan M. Dey.
© 2017 John Wiley & Sons, Ltd. Published 2017 by John Wiley & Sons, Ltd.

the U.S., EU, and other developed countries, aquaculture frequently is subjected to a more confusing array of regulations because it is less well understood than other businesses. For example, Bowden (1981) points out that cattle ranching is viewed strictly as a farming enterprise and regulated as an agricultural activity while many forms of fish farming, including fish ranching, are regulated by natural resource or fish and game agencies, not by agricultural agencies.

In the U.S., the Joint Subcommittee on Aquaculture identified 17 different federal agencies with regulations for aquaculture but also recognized that state-level regulations are more numerous (JSA 1993). De Voe (1997) estimated that there were more than 1200 laws in the U.S. that affect aquaculture.

Regulatory business permits vary from country to country and within countries. Permits or licenses can be required for possession, processing, or depuration, as well as other activities. Licenses may be required, fees charged, or taxes levied depending upon the specific regulation. Such costs are considered direct costs of regulations. Engle and Stone (2013) identified various categories of regulatory permits that included environmental, food safety, legal and labor standards, interstate transport of aquatic products, fish health, and culture of commercially harvested species.

The total cost of compliance with regulations, however, extends far beyond the direct costs of the permits, fees, and licenses. Some regulations may require additional capital investment for effluent treatment infrastructure while others may result in a less efficient scale of production or management. Managers must allocate time to comply with regulations, and workers must spend additional time on record-keeping (Coppock 1996). Time spent on compliance activities represents a non-cash opportunity cost (Hurley and Noel 2006) or an additional cash expense if new personnel are hired for such functions. Most important may be the cumulative effect of the total "suite" of regulations with which individual farms must comply (Hurley and Noel 2006).

Food safety

A 2008 Wall Street Journal–Harris Interactive Poll found that "65% of American consumers doubted the safety of imported food from developing countries". Consumer concerns over food safety have increased greatly in recent decades. Recent consumer food scares in the UK alone were related to: salmonella; bovine spongiform encephalitis (BSE or "mad cow disease"); hormone implants; genetically modified organisms (GMOs); antibiotic residues; used cooking oil; sewage waste; polychlorinated biphenyl (PCBs); dioxin; foot (hoof) and mouth disease; chloramphenicol; nitrofurans; mycotoxin mycophenolic acid (MPA); and nitrofen. In the U.S., some of the more prominent scares have been related to: *Escherichia coli* in bagged salads in 2012; *Cyclospora* in salad mix in 2013; salmonella in peanut butter, cucumber, and chicken in

2013; hepatitis A in frozen berries and pomegranate seed mix; and *Listeria* in cheeses in 2013 and 2014. China has also experienced regular food safety scares, even in U.S.-based major chains such as Kentucky Fried Chicken, McDonald's, and Walmart. Each of these resulted in dramatic decreases in sales of product that resulted in financial losses to companies producing and marketing the products affected.

Some food safety problems are caused by natural phenomena. Harmful algae blooms in natural waters can result in decreased supplies of shellfish as beds are closed and delays are incurred in re-seeding the stock (Conte 1984; Kahn and Rockel 1988; Tester and Fowler 1990). Additional losses are incurred when demand for products decreases when public announcements and public warnings appear (Brown 1969; Hamilton 1972; Sherrel et al. 1985). Public announcements that shellfish from some areas are toxic may cause consumers to fear and avoid related products (Swartz and Strand 1981). Wessells et al. (1995) distinguished between "acute" hazards that pose an immediate health hazard and those that result from a slow accumulation over a period of time. For acute hazards, it was shown that consumers based decisions on immediate, not past, news. However, in the case of a persistent accumulation of toxins, the demand impact of total cumulative information may be greater than in cases with acute effects. For example, direct losses from one farm were 8% of total average annual sales during an acute hazard event in Montreal. An additional 6.5% of total average annual sales were lost over the succeeding 3 months from decreased demand for the product.

Concerns over additives or residues in seafood products can prompt governments to ban their use or presence in both domestic and imported product. For example, chloramphenicol in shrimp imported into the EU from China resulted in an EU ban on imports from the entire country, not just from the one company where the problem was first identified. After imposition of the restrictions shrimp exports from Indonesia shrank by 64%, 21% from Thailand, 39% from Malaysia, and 14% from Vietnam (Asia Pulse 2003). In reaction to the ban, shrimp producers in Southeast Asia and China threatened to boycott shrimp exports to the EU. They claimed that levels of chloramphenicol in meat, milk, and flour exported from the EU were of similar levels to those found in imported shrimp. The EU removed its policy requiring shrimp from Indonesia to be free of chloramphenicol in September 2003 due to a determination that no country could comply fully with the conditions.

The European Commission, in 2006, expanded its ban on livestock use of antibiotics for human medicine to include a ban on antibiotic use to promote growth in animals. This was the last step to reduce development of resistance of human microbes to antibiotics and now bans all use of antibiotics for non-medicinal purposes. Regulation EC/178/2002 established rules for hygiene of food in the EU from production through processing, and distribution with compulsory traceability (Europa 2015).

In the U.S., the U.S. Food and Drug Administration (FDA) has authority over the safety of seafood, whether farmed or wild-caught. However, the FDA has limited ability to inspect foreign exporters to the U.S. A Government Accounting Office report (2011) showed that, while the FDA inspects 20% of domestic seafood processing establishments a year (or once every 5 years), it inspected only 0.5%/year (once every 200 years) of foreign exporters.

Exports of farmed *Pangasius* to major world seafood markets, including the EU, Japan, Russia, and the U.S., have continued to grow in spite of ongoing documentation of the use of antibiotics banned for livestock feeds due to risk of increased microbe resistance. In a 2013 study, Rico et al. (2013) found that 100% of farms surveyed used antibiotics. The survey documented use of 17 different antibiotics that belonged to 10 different classes of antibiotic. The antibiotics found included those important in human medicine such as penicillins, aminoglycosides, cephalosporins, quinolones, tetracyclines, amphenicols, polymyxin, diaminopyrimidines, rifamycins, and sulfonamides. The ongoing use of banned substances in *Pangasius* from Vietnam and other species traded globally, particularly from Vietnam and China, has triggered numerous import alerts and bans on imports by the EU, Japan, Russia, and the U.S.

Some specific regulations have been enacted to ensure the safety of shellfish products. In the U.S., for example, four states have regulations or permits for purging (depuration), transplant, and safe food handling of shellfish. In Connecticut, shellfish depuration and transplant licenses are required to operate a depuration plant and to sell processed shellfish. Transplant licenses are required to relay oysters from prohibited areas into private shellfish beds in approved areas. Florida requires a special activity license for depuration of oysters and clams in controlled purification facilities. The state of California also has shellfish safety regulations that require safe handling of shellfish while the state of Virginia has food quality sanitation regulations that govern the inspection of food manufacturers, warehouses, and retail food stores, food product sampling, and food product label review.

Industry-initiated programs

Certification programs have developed over time to both reassure buyers of the safety of seafood and also act as a means to encourage suppliers to practice sustainable ways to capture or grow seafood products. Certification programs evolved from earlier quality assurance (QA) programs. Most QA programs included systems of internal and external audits that were used to inspect products to ensure safety and quality. While on a global basis, the percentage of seafood that is certified is still small (4.6% of global aquaculture production; 7% of global wild-caught landings) (Bush et al. 2013), it is growing. In the U.S., 60% of fishery landings are certified.

The Catfish Farmers of America and the U.S. Trout Farmers Association developed catfish and trout quality assurance programs in the 1990s (Brunson 1993).

The Catfish Quality Assurance program was developed in 1993 as an educational program designed to maintain consumer confidence with farm-raised catfish. The program was intended for all catfish producers to ensure the safety and quality of farm-raised catfish. The Trout Quality Assurance program was organized somewhat differently and was based on the hazard analysis of critical control points (HACCP) concept that is discussed in greater detail later in this chapter.

The Interstate Shellfish Sanitation Conference (ISSC) is an organization of representatives from the shellfish industry, state and federal agencies, and universities to foster and promote shellfish sanitation. It is a voluntary cooperative effort to establish uniform standards and procedures for handling shellfish. The emphasis of the ISSC is on sanitary controls on shellfish harvesting, processing, and distribution. The states take the primary role for enforcement by monitoring waters for contamination and pollution, inspecting processing facilities, and preventing poaching.

The U.S. Department of Commerce (USDOC) offers an optional fee-for-service quality assurance inspection. USDOC inspectors will, upon request, inspect processing plants and facilities, and grade aquaculture products for quality assurance (50 CFR Part 260).

Supermarkets in the EU have established processes to ensure the safety of food products. The Euro Retailer Produce Working Group (EUREP) developed a mechanism for developing production standards for commodities entering the retail trade through their outlets. GLOBALG.A.P. (formerly EUREPGAP Good Agricultural Practices) operates with HACCP guidelines from the FAO with governance under the ISO Guide 65 for certification. The original EUREPGAP program was extended to aquaculture products in 2001 with a focus on quality, labeling, traceability, and food safety with third-party verification required. Production units are assessed by independent third-party licensed certification companies.

In France, some shellfish wholesalers have created trademarks, labels, and signs that purport to establish and certify the quality and safety of cultured products (Girard and Mariojouls 2003). France has official procedures for certifications such as that established on a local scale for mussels from the Mont St. Michel Bay region of France. The national shellfish farmers association of France, Comité National de la Conchyliculture (CNC), established a certification list for "bouchot" mussels (Girard and Mariojouls 2003). One of the most recognized quality certification programs in France is the Label Rouge program that was created in 1965 by the Centre de Développement des Certifications des Qualités Agricoles et Alimentaires (CERQUA) (Label Rouge 2004). To be approved for the French Label Rouge, the product must be demonstrated to be of superior quality as determined by appropriate taste tests.

The Global Aquaculture Alliance (GAA) established a Best Aquacultural Practices Certification Program (GAA 2014) that includes standards that address food safety, environmental impacts, social welfare, and animal health and welfare. While the original emphasis was on shrimp, the GAA BAP has since expanded to also address salmon, tilapia, and *Pangasius*.

Most certification programs focus on environmental sustainability, food safety, social responsibility, and traceability. While the FAO has developed guidelines for aquaculture certification and ecolabeling, existing certification programs have widely differing standards. One dilemma with third-party certification is that products with the same general label are implied to be equally safe and sustainable. In reality, labeling standards can differ by species and location even within the same program. Much of the impetus for certification programs has come from importers and retailers that sell seafood from developing nations with weak environmental, health and safety regulatory frameworks. Such retailers often seek third-party certification for reasons of corporate responsibility, to reduce risk, liability, and pressure from environmental non-governmental organizations (NGOs), and to help consumers to identify products. Thus, certification programs "provide buyers some insurance against food scares and a due diligence defense" (FAO 2011). Some environmental groups have developed certification programs to influence the way food products are grown.

The growth in certification programs has resulted in requirements for some buyers of U.S. farmed products to also pay costs of certification. However, U.S. government regulations already cover the key certification areas of environmental sustainability, food safety, social responsibility, and traceability. Moreover, government regulations are enforced through civil and/or criminal penalties for violations. Government programs often involve testing that is not frequently the case with certification programs. Nevertheless, if a grower's buyer requires a specific type of certification, suppliers will need to comply.

The future course of certification programs in the overall seafood market is unclear. Observers of aquaculture certification programs have stated that: "It is doubtful that aquaculture certification will become fully viable unless one or both of the following occur: 1) clear evidence is developed revealing that the better practices and certification enhance efficiency enough to offset the added costs of participation in these programs, and/or 2) more consumers become willing to pay a premium for "environmentally friendly" products, and a fair portion of the higher price filters down to farmers" (Boyd and McNevin 2012).

Regulation of food safety
The United Nations
The Codex Alimentarius Commission (CAC) of the United Nations Food and Agriculture Organization (FAO) and the World Health Organization (WHO) has been responsible for implementing the Joint FAO/WHO Food Standards Programme. The Codex Alimentarius Commission is divided into two types of committees: (1) nine general subject matter committees that deal with general principles, hygiene, veterinary drugs, pesticides, food additives, labeling, methods

of analysis, nutrition, import/export inspection and certification systems; and (2) commodity committees that deal with a specific food class or group.

The World Trade Organization (WTO)

The WTO agreement, in the Final Act of the Uruguay Round, developed an agreement on the Application of Sanitary and Phytosanitary Measures (SPS). The SPS agreement confirms the right of WTO member countries to apply measures necessary to protect human, animal, and plant life and health.

The International Organization for Standardization (ISO)

The ISO is a network of national standards institutes from 148 countries that works in partnership with international organizations, governments, industry, business, and consumer representatives (ISO 2015). The ISO 9000 series is an accreditation program for the food industry.

The United States

In the U.S., state health departments develop guidelines related to materials and conditions of buildings, equipment, and temperatures in processing and transportation of processed products. Local county sanitarians enforce these guidelines and have jurisdiction over sanitary conditions in processing plants. As consumer awareness and concern over food safety have grown, additional regulations by national authorities have been put in place.

The U.S. Food and Drug Administration (FDA) was created from the 1906 Food and Drugs Act (FDA 2015). It regulates the production and marketing of most food products, including fish. It is responsible for protecting the public health by assuring the safety, efficacy, and security of human and veterinary drugs, biological products, medical devices, the nation's food supply, cosmetics, and products that emit radiation. It is also responsible for advancing public health by helping to speed innovations that make medicines and foods more effective, safe, and affordable. The FDA provides accurate, science-based information to the public as needed to issue medicines and foods to improve their health. The FDA has developed regulations that deal with food production and marketing, food name and ingredients, food quality, manufacturing practices, packaging, and labeling.

Moreover, FDA specifies product labeling requirements, including the content of the product label information, the label's layout, and its size. A fundamental requirement of labeling is that the information be displayed in a prominent and visible manner.

Hazard Analysis of Critical Control Points (HACCP)

HACCP programs were developed in the EU in 1996 and in the U.S. in 1997. In the U.S., processing plants are required by the FDA to have an HACCP plan in place. The plan must identify areas with potential for product contamination or

safety problems. The U.S. seafood HACCP rule covers all processors and importers, but fishing vessels, common carriers, and retailers are not required to have
HACCP plans. For FDA purposes, processors are defined as seafood-related entities classified as establishments in the FDA inventory and foreign processors that
export to the U.S.

The HACCP rule requires every processor to conduct an analysis of potential
hazards to determine whether food safety problems might occur. If it is deemed
that no food safety hazards are likely, the processor does not need a HACCP plan,
but the burden of proof is on the processor. If the hazard analysis reveals a need,
the processor must have a written HACCP plan that is specific to the plant's location and the types of products prepared. Food safety hazards that are reasonably
likely to occur may include toxins, microbes, chemicals, pesticides, drug residues, physical hazards, or decomposition. Critical control points can occur both
inside and outside the processing plant and must be identified. Critical limits, or
safe operating parameters, must be defined for each critical control point, monitoring procedures established, and corrective action plans developed. Verification
procedures must be put in place and carried out at least annually to ensure that
the HACCP plan is up-to-date and that ongoing implementation is adequate.
Verification procedures may include reviewing consumer complaints, calibrating
monitoring devices, and end-product testing.

A record-keeping system must be developed to document monitoring, corrective actions, and verification procedures. Records must state the name and
location of the processor and the date and signature of the person making the
record. Plans, HACCP records, and sanitation records must be available to FDA
inspectors for review and copying. Plans and records in the possession of the
FDA are not available for public disclosure due to the Freedom of Information
Act. Some of the HACCP functions (plan development, plan reassessment and
modification, and reviewing HACCP records) must be performed by an individual who has been trained in HACCP through either course materials or job experience equivalents.

Importers must verify that their overseas suppliers follow HACCP rules by
obtaining the product from a country with which the U.S. has an HACCP-based
agreement regarding inspection programs, developing product specifications for
safety, and taking steps that might include: (1) obtaining the processor's HACCP
and sanitation records; (2) third-party certification; (3) sending inspectors overseas to ensure that the product meets requirements; or (4) end-product testing.

Molluscan shellfish have special requirements within the FDA HACCP rule.
Shellfish must be harvested from waters approved by a "shellfish control authority." Shellfish must be purchased from harvesters in compliance with local
licensing requirements, or they can be tagged.

In 2005 the U.S. implemented country-of-origin labeling (COOL) for fish and
shellfish that requires retailers such as supermarkets to notify customers of the
origin of seafood and whether it is wild or farm-raised (USDA-AMS 2014).

Administration and enforcement of COOL is by the Agricultural Marketing Service of USDA (USDA-AMS). While COOL was opposed by food wholesalers and retail organizations, it gives seafood suppliers an opportunity to include product information for retailers on the backs of point-of-sale (POS) tags.

Fish and seafood used as an intermediate product (i.e., as an ingredient in other processed foods) are excluded if they have undergone a change such as cooking, curing, or smoking or if they have been combined with other commodities such as with a breading or a tomato sauce. For example, shrimp that are dusted lightly with flour are excluded from COOL requirements. Restaurants, cafeterias, and other food service establishments also are exempt from mandatory COOL requirements. However, there are states that have passed labeling laws that require restaurants to inform customers of the origin of certain species.

The National Marine Fisheries Service (NMFS) administers grade and quality standards for fish. The NMFS also conducts inspection and certification services. These are voluntary and are funded by fees charged to industry. For example, NMFS establishes minimum flesh content requirements for breaded and battered products.

Organic standards

The International Federation of Organic Aquaculture Movements (IFOAM), founded in France in 1972, operates in 108 countries through 750 member organizations (FAO 2014). Through its Organic Guarantee System, it accredits third-party certifiers. IFOAM added a chapter on aquaculture to its Basic Standards in 2005. Several groups certified by IFOAM cover aquaculture commodities, including: Agrior, operating in Israel (tilapia, carp, red drum, sea bass, sea bream, sea lettuce); Debio, operating in Norway (salmon, trout, cod); and Organic Agriculture Certification Thailand (Nile tilapia and butter fish). KRAV (Sweden) (KRAV 2015) certifies aquaculture, but its aquaculture standards are not accredited by IFOAM. The KRAV program includes salmonids, perch, and blue mussels.

Organic farming has developed into one of the fastest-growing segments of agriculture in the EU (European Union 2004; European Food Safety Authority 2004). The European Commission introduced an organic logo in 2000 to inform consumers that the product meets its conditions established for organic farming. Producers use the logo voluntarily but must pass inspections to ensure that: (1) at least 95% of the product's ingredients were produced organically; (2) the product complies with official inspections; (3) the product is delivered from the producer in a sealed package; and (4) the product bears the producer's name and inspection code. In Europe, supervision is the responsibility of each country.

There are several associations in the EU and New Zealand that certify aquaculture products as organic. France has adopted organic aquaculture standards that have been applied to rainbow trout for domestic sales and export. Naturland (the Association for Organic Agriculture) is an international association of farmers that promotes organic agriculture. Founded in Germany in 1982, it grew to 40,700 farmers cultivating more than 137,000 ha globally in 2013 (Naturland 2014). Naturland farms raise organic trout (Germany, France, Italy, Spain), organic salmon (Ireland, Scotland), organic shrimp (Ecuador, Peru, Brazil, Vietnam, India, Indonesia), organic tilapia (Israel, Ecuador), organic *Pangasius* (Vietnam), and organic sea bass and gilthead sea bream (Greece, Croatia). In the UK, The Soil Association (The Soil Association 2015) includes aquaculture organic standards for Atlantic salmon, trout, and arctic char, shrimp, bivalves, and carp. Bio Suisse (Switzerland) (Bio Suisse 2015) adopted organic aquaculture standards in 2000 for trout and salmon in Europe and *Pangasius* in Vietnam.

BioGro (BioGro 2015) in New Zealand has organic standards for finfish, shellfish, and crustacean farms, but the aquaculture standards are not part of BioGro's IFOAM accreditation.

Organic markets for all types of products are growing rapidly in the U.S. More than $35 billion of organic products were sold in the U.S. in 2013 (Organic Trade Association 2014), an increase of 25% from the $28 billion of sales reported in 2012 (Greene 2013). The U.S. National Organic Standards Board (NOSB) formed an aquaculture advisory group in 2000 and a National Organic Aquaculture Work Group was formed to work toward developing national standards for organic aquaculture (Brister 2004a, b). However, despite efforts for more than a decade, the U.S. has still not adopted organic standards for aquaculture.

Green labeling and standards

Various groups have issued guides as to the "sustainability" of various types of seafood. Chapter 4 includes additional details on sustainability certification of seafood. The Monterey Bay Aquarium, as part of its Seafood Watch program, features 94 of the most popular seafood species in the U.S. on its web site (Seafood Watch 2015). The site includes reports on each species and a series of pocket guides as to sustainability of each species. Pocket guides have been developed on a national as well as regional (West Coast U.S., Central U.S., Southeast, Hawaii, Northeast, Southwest, and one for sushi) basis. Sustainability of wild-caught species is evaluated based on impacts on the species itself, impacts on other species, management effectiveness, and impacts on the habitat and ecosystem. For farmed species, the assessment criteria include the quality of data, effluents, habitat, use of chemicals, feed use, escapes, diseases, and sources of stocks used. U.S. farm-raised catfish is recommended due to the diet fed and its control over water quality in the ponds where raised. Examples of species included on the list of "best" choices are: farm-raised catfish (U.S.), striped bass (U.S.), rainbow trout (U.S.), tilapia (Ecuador, U.S.), and shrimp (U.S.), wild-caught Alaska salmon, spiny lobsters (Mexico), and yellowfin tuna caught by pole and by troll (U.S.). Species classified as to be avoided, based on the aquarium's definition of sustainability, include: Pacific cod, orange roughy, swordfish imported into the U.S., shrimp imported into the U.S., shark, and spiny lobster from Belize, Brazil, Honduras, and Nicaragua. However, while wild-caught salmon is recommended, farm-raised salmon is not. The main objections to farm-raised salmon include criticisms over the use of fishmeal in the diet and the use of Atlantic salmon in net pens on the West Coast. The Blue Ocean Institute also publishes a "Guide to Ocean Friendly Seafood" (Blue Ocean Institute 2004). Farmed clams, mussels, oysters, wild-caught Alaskan salmon, mackerels, striped bass, mahi mahi, and tuna (yellowfin and bigeye) are listed in the top category. The problematic list includes, as examples: sharks, shrimp imported into the U.S., farmed salmon, orange roughy, and Atlantic bluefin tuna.

The Marine Aquarium Council (MAC) has developed a certification program for ornamental fish. The MAC certification includes certification of industry operators throughout the supply chain, including collectors, exporters, and importers (Marine Aquarium Council 2004). MAC product certification requires that marine ornamentals be harvested from a certified collection area and sold to MAC-certified buyers at the next level of the marketing chain. Key emphases are on ensuring health of the ecosystem in the collection area, and handling procedures that ensure the health of the fish being sold. In response to growing concerns over the capture fishery for marine ornamental fish, standards were developed for certification for supplying marine ornamental fish for the aquarium trade.

Marketing and transportation of live aquatic animals

In the U.S., marketing of live aquatic animals has come under increased scrutiny. Concerns relate to the spread of non-native, possibly invasive species, over-harvesting of species with already diminished stocks, and the spread of pathogens. In the U.S., the major statutes used include:

1 The Endangered Species Act of 1973. This statute deals with any activity that might affect endangered or threatened species or their habitat.
2 Lacey Act Amendments of 1981. Under this law, it is unlawful to import, export, sell, acquire, or purchase fish, wildlife, or plants taken, possessed, transported, or sold (1) in violation of U.S. or Indian law or (2) in interstate or foreign commerce involving any fish, wildlife, or plants taken, possessed, or sold in violation of state or foreign law. The Lacey Act is enforced through both civil and criminal penalties depending on the knowledge of the defendant, the type of violation, and the value of the fish involved. The Lacey Act has been invoked in situations involving shipments of fish through or into states

that prohibit their entry. While the Lacey Act was developed to protect wildlife, it is applied to farm-raised fish that are shipped across state lines. Lacey Act penalties can include mandatory incarceration of up to five years and fines up to $500,000 for a business (Rumley 2012).

3 The Migratory Bird Treaty Act. The Migratory Bird Treaty Act regulates the use of lethal control methods on migratory birds, including those that cause aquaculture crop losses. The U.S. Fish & Wildlife Service (USFWS) issues permits for the control of these migratory birds.

4 The USFWS also maintains Migratory Bird Treaties with Japan, Canada, and the Soviet Union.

In the U.S., state agencies also issue permits and regulations dealing with nonnative species that include stocking licenses, general importation permits, and restrictions on possession, sale, importation, transportation, and release. Some states have special importation permits regarding specific species of aquatic animals such as grass carp (or white amur) (*Ctenopharyngodon idella*), crawfish (*Procambarus* spp.), piranha (*Pygocentrus nattereri*), and rudd (*Scardinius erythrophthalmus*).

Bighead carp (*Hypophthalmichthys nobilis*), silver carp (*Hypophthalmichthys molitrix*), and black carp (*Mylopharyngodon piceus*) were listed as injurious species under The Lacey Act (bighead carp in 2010, and silver and black carp in 2007). Thus, it is not legal to transport them across state lines. No other fish species have been added to the injurious species list since 2010. Some states prohibit exotic species while other states have developed "clean" lists of specific species that are allowed with and without permits.

In the U.S., there are separate jurisdictions for game and sportfish that can create regulatory problems for farmers who wish to culture those species. In some cases, farms have been required to include gill tags that clearly label each fish as farmed, while in other cases, sale of that species is completely banned from those states.

The EU has established regulations that protect aquatic animals during transport (European Commission 2011). The UK requires a fish health certificate and that the shipment be authorized by the Fish Health Inspectorate (FHI). Additional rules have been developed for non-native species, and biosecurity plans are required for koi (*Cyprinus carpio*). To export to the EU, the national authority of the exporting country sends a request to the Directorate-General for Health and Consumer Protection of the European Commission for approval to export fish and seafood. The Convention on International Trade in Endangered Species (CITES) requires a permit to transport endangered species internationally.

Aquatic animal health and biosecurity

Concerns have increased in recent years over the potential spread of aquatic pathogens through transport of live aquatic animals. Scrutiny and testing of aquatic animals has increased as a result.

The principal regulatory standard for aquatic animal health worldwide is the World Organisation for Animal Health (OIE). OIE standards for fish health, inspection, and certification are considered to be the most rigorous and to constitute the primary standards worldwide. As a consequence, many countries and individual states within countries require inspection of animals prior to issuance of licenses and permits to import live animals. Testing programs can focus either on each lot of fish or on tests at the farm level. Testing of each individual sales lot of fish can be onerous and impractical, given the requirement for cell culture testing for viruses.

An example of a proactive certification program for aquatic animal health is the Arkansas Baitfish Certification Program. This farm-level testing program requires a two-year history of the farm being free of specified pathogens. Each farm in the program is tested twice a year and must maintain its negative status for the pathogens of concern to remain in the program. The program also inspects farms for the presence of specified aquatic nuisance species. The certificate is issued by a third party, the Arkansas Department of Agriculture, based on laboratory testing of samples collected by a private veterinarian who delivers fish to approved and certified testing laboratories.

Aquaculture market synopsis: mariculture of grouper, snapper, tuna, and cobia

Mariculture of species such as grouper, tuna, cobia, snapper, and grouper has grown rapidly since about 2002 (Fig. 12.1). The most dramatic growth has been that of various species of groupers, but ongoing research is expected to result in continued expansion of production of tuna, cobia, and snapper.

Grouper are raised in marine net pens and in ponds, but net pen production has been the more common culture system. Grouper production increased by an average rate of 30% per year from 1970 to 2012, but with a single-year increase of 120% from 2002 to 2003. Farmed grouper production contributed 31% of the

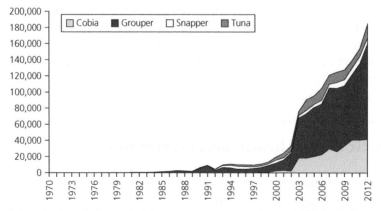

Fig. 12.1 Growth of mariculture of cobia, grouper, snapper, and tuna, 1970–2012. Source: FAO (2014).

total global supply of grouper in 2012. The FishStatJ database (FAO 2015) lists 13 countries with production of grouper in 2012. Of these, China produces 62% of all farmed grouper, followed by Taiwan with 19%, Indonesia with 10%, Malaysia with 5%, and Thailand with 1%. Production of six different species of grouper was reported in FishStatJ, although data for production of the Hong Kong grouper (*Epinephelus akaara*) appeared only from 1970 to 1995. More than 94% of all grouper raised is listed only as "grouper," with some production reported of orange-spotted (*Epinephelus coioides*), greasy (*Epinephelus tauvina*), humpback (*Cromileptes altivelis*), and spotted coral (*Plectropomus maculatus*) grouper.

Farmed snapper production increased to 7284 metric tons in 2012. While composing a small percentage of total global supply, farmed production of snappers has nearly doubled since 2005. Malaysia produced 91% of all snapper produced, with 4% from Taiwan and smaller percentages from six other countries. FishStatJ reports production of four species of snapper in addition to a general snapper category. Of these, mangrove red snapper (*Lutjanus argentimaculatus*) composed 60% of total farmed supply in 2012, followed by John's snapper (*Lutjanus johnii*) with 32% of production, the general snapper category with 7%, and very small percentages of Russell's snapper (*Lutjanus russellii*) and spotted rose snapper (*Lutjanus guttatus*).

Tuna production reached 16,887 metric tons in 2012. Japan produced 57% of all farmed tuna in 2012, followed by Australia (15%), Mexico (11%), and Croatia (7%). Of the four species raised, Pacific bluefin tuna (*Thunnus orientalis*) composed 68% of all farmed production in 2012, Atlantic bluefin tuna (*Thunnus thynnus*) 17%, and southern bluefin tuna (*Thunnus maccoyii*) 15%.

Cobia (*Rachycentron canadum*) production was 41,774 metric tons in 2012, the highest year of production. Growth has been slow, but steady, since 1995. More than 90% of farmed cobia production in 2012 was in China, 5% in Vietnam, and 3% in Taiwan, with additional production reported from Panama, Colombia, and Singapore.

Salmon continues to be the marine finfish species with the greatest amount of production overall, but new production technologies for a variety of commercially important species have led to growth of farmed production of several marine finfish species. Among these, some of the greatest commercial growth has been in marine net pen production of grouper, snapper, tuna, and cobia.

Summary

As aquaculture industries have grown and developed, the number and type of regulations that affect the marketing of aquaculture products have grown over time. Those related to food safety have been the most comprehensive, but issues of transport and sale of live aquatic animals have attracted increased regulatory attention. Several industry segments have developed industry-enforced quality assurance programs and codes of practice. Certification programs have developed to reassure buyers of the safety and sustainability of seafood species, but there is little standardization among them. Nevertheless, use of antibiotics banned in livestock feeds in several major seafood exporting countries continues to be widespread. National and local regulatory agencies have created a variety of permitting, licensing, and bonding requirements for all phases of the aquaculture marketing chain. Compliance costs, particularly the indirect farm adjustments that must be made, can be considerable deterrents from expanding aquaculture production. Organic farming of aquaculture products has grown, but has been hampered in the U.S. by lack of progress on the development of national standards for organic aquaculture.

Study and discussion questions

1 What are the major areas of aquaculture marketing that are regulated?
2 What are quality assurance programs? Who initiates them, and what is their purpose?
3 What is the major regulatory agency in the U.S. for food and public health concerns?
4 What does HACCP stand for, and what are the major components of an HACCP plan?
5 What international agencies are involved with aquaculture marketing standards or regulations?
6 What is The Lacey Act in the U.S.? List the key provisions and penalties for violations.
7 What agency is considered the definitive authority for fish health testing and inspection in the U.S. and in the EU?
8 Name two major organic certification organizations and list which aquaculture species are certified as organic under their programs. Discuss whether it is possible to sell seafood in the U.S. as "certified organic" and what conditions would be necessary to do so.
9 Why have aquaculture certification programs developed, what did they develop from, and why are they being used?
10 What are the some of the greatest costs associated with regulations and policies? Include examples of both direct and indirect compliance costs.

References

Antle, J.M. 2000. No such thing as a free lunch: the cost of food safety regulation in the meat industry. *American Journal of Agricultural Economics* 82:310–322.
Asia Pulse Pte. Limited. 2003. European Union relaxes stance on Indonesian shrimp. Asia Pulse Pte. Limited.
Bio-Suisse. 2015. Available at www.bio-suisse.ch/en/home.php.
BioGro. 2015. Available at www.biogro.co.nz/organic-standards.

Blue Ocean Institute. 2004. Guide to ocean friendly seafood. Currently the Sustainable Seafood Program Seafood Choices. Available at safinacenter.org/programs/sustainable-seafood-program/seafood-choices/

Bowden, G. 1981. *Coastal Law and Policy: A Case Study of California*. Westview Press, Boulder, Colorado.

Boyd, C.E. and A.A. McNevin. 2012. An early assessment of the effectiveness of aquaculture certification and standards. In: Berry, M., B. Cashore, J. Clay, M. Fernandez, L. Label, T. Lyon, P. Mallett, K. Matus, P. Melchett, M. Vandenbergh, J.K. Vis, and T. Whelan. *Toward Sustainability: The Roles and Limitations of Certification*, pp. A-35–A-69. Resolve, Inc., Washington, D.C.

Brister, D.J. 2004a. The importance of organic aquaculture. *Ecology and Farming* 35:16–18.

Brister, D.J. 2004b. Two organic aquaculture groups convene meetings. *The Organic Standard* 36:7–9.

Brown, J.D. 1969. Effect of a health hazard scare on consumer demand. *American Journal of Agricultural Economics* 51:676–678.

Brunson, M. 1993. Catfish quality assurance. Catfish Farmers of America and National Fisheries Institute, Publication 1873, Mississippi State University, Mississippi.

Bush, S.R., B. Belton, D. Hall, P. Vandergeest, and F.J. Murray. 2013. Global food supply. Certify sustainable aquaculture? *Science* 341:1067–1068.

Christainsen, G.B. and R.H. Haveman. 1981. Public regulations and the slowdown in productivity growth. *The American Economic Review* 71:320–325.

Conte, F.S. 1984. Economic impact of paralytic shellfish poison on the oyster industry in the Pacific United States. *Aquaculture* 39:331–343.

Coppock, R. 1996. Research Update: Farmers say regulations complicate farming. *California Agriculture* 50:5.

DeVoe, M.R. 1997. Marine aquaculture regulation in the United States: Environmental Policy and Management Issues. UJNR Technical Report No. 24. Pp. 1–15 in *Interaction in the Environment*. 1995. Proceedings of the Twenty-fourth U.S.-Japan Aquaculture Panel Symposium, Corpus Christi, Texas.

Directorate-General for Internal Policies. 2009. Regulatory and legal constraints for European aquaculture. Policy Department, European Union Parliament.

Engle, C.R. and N.M. Stone. 2013. Competitiveness of U.S. aquaculture within the current U.S. regulatory framework. *Aquaculture Economics & Management* 17:251–280.

Europa. 2015. General principles of food law. Available at www.fsai.ie/legislation/food_legislation/general_principles_of_food_law.html.

European Commission. 2011. Report from the Commission to the European Parliament and the Council on the impact of Council Regulation (EC) No 1/2005 on the protection of animals during transport. Available at eur-lex.europa.eu/legal-content/EN/TXT/?uri=CELEX%3A52011DC0700. Accessed March 28, 2016.

European Food Safety Authority. 2004. Available at www.efsa.europa.eu.

European Union. 2004. White paper on food safety. Available at ec.europa.eu/food/food/intro/white_paper_en.htm.

Food and Agriculture Organization of the United Nations (FAO). 2011. Technical guidelines on aquaculture certification. Available at www.fao.org/docrep/015/i2296t/i2296t00.htm.

Food and Agriculture Organization of the United Nations (FAO). 2014. International Federation of Organic Agriculture Movements. Available at www.ifoam.bio/. Accessed March 28, 2016.

Food and Drug Administration (FDA). 2015. Available at www.fda.gov.

Global Aquaculture Alliance (GAA). 2014. www.gaalliance.org. Accessed March 28, 2016.

Girard, S. and C. Mariojouls. 2003. French consumption of oysters and mussels analyzed within the European market. *Aquaculture Economics & Management* 7:319–333.

Government Accounting Office. 2011. Seafood Safety: FDA needs to improve oversight of imported seafood and better leverage limited resources. Report to Congressional Requesters No. GAO-11-286, Government Accounting Office, Washington, D.C.

Gray, W.B. 1987. The cost of regulation: OSHA, EPA and the productivity slowdown. *The American Economic Review* 77:998–1006.

Greene, C. 2013. Growth patterns in the U.S. organic industry. Amber waves. October 24, 2013. Economic Research Service, United States Department of Agriculture. www.ers.usda.gov/amber-waves/2013-october/growth-patterns-in-the-us-organic-industry#.VvGSQ3rSQYw. Accessed January 6, 2015.

Hamilton, J.L. 1972. The demand for cigarettes: advertising, the health scare, and the cigarette advertising ban. *Review of Economics and Statistics* 54:401–411.

Harris, D. 1998. Aquaculture regulatory reform task force discussion paper. Aquaculture Regulatory Reform Task Force, Victoria, Australia.

Hurley, S.P. and J. Noel. 2006. An estimation of the regulatory cost on California agricultural producers. Annual Meeting of the American Agricultural Economics Association, Long Beach, California.

International Organization for Standardization (ISO). 2015. Available at www.iso.org.

Joint Subcommittee on Aquaculture (JSA). 1993. Aquaculture in the United States: status, opportunities and recommendations. Report to Federal Coordinating Council on Science, Engineering, and Technology, Washington, D.C.

Kahn, J.R. and M. Rockel. 1988. Measuring the economic effects of brown tides. *Journal of Shellfish Research* 7:677–682.

KRAV. 2015. Available at www.krav.se.

Label Rouge. 2004. Qui somme-nous. Label Rouge. Available at agriculture.gouv.fr/signes-de-qualite-le-label-rouge.

Marine Aquarium Council. 2004. MAC certification. Marine Aquarium Council. Available at www.marineaquariumcouncil.org.

Naturland. 2014. Available at www.naturland.de.

Organic Trade Association. 2014. 2014 Press Releases: American appetite for organic products breaks through $35 billion mark. Available at www.ota.com/news/press-releases/17165. Accessed March 28, 2016.

Rico, A., T.M. Phu, K. Satapornvanit, J. Min, A.M. Shahabuddin, P.J.G. Henriksson, F.J. Murray, D.C. Little, A. Dalsgaard, P.J. Van den Brink. 2013. Use of veterinary medicines, feed additives and probiotics in four major internationally traded aquaculture species in Asia. *Aquaculture* 412–413:231–243.

Rumley, E.R. 2012. Aquaculture and the Lacey Act. SRAC Publication No. 5005, Southern Regional Aquaculture Center, Mississippi State University, Mississippi.

Seafood Watch. 2015. Available at www.montereybayaquarium.org.

Sherrel, D., R.E. Reidenbach, E. Moore, J. Wagle, and T. Spratlin. 1985. Exploring consumer response to negative publicity. *Public Relations Review* 11:13–28.

Stewart, J. 2012. Not made in the United States of America. *Fish Farming International* 2:22–27.

Swartz, D.G. and I.E. Strand. 1981. Avoidance costs associated with imperfect information: the case of Kepone. *Land Economics* 57:139–150.

Tester, P. and P.K. Fowler. 1990. Effects of the toxic dinoflagellate *Ptychodiscus brevis* on the contamination, toxicity and depuration of *Crassostrea virginica* and *Mercenaria mercenaria*. In: Graneli, E., D.M. Anderson, L. Edler, and B.G. Sundstrom, eds. *Toxic Marine Phytoplankton*. Elsevier, New York.

The Soil Association. 2015. Available at www.soilassociation.org.

United States Department of Agriculture, Agricultural Marketing Service (USDA-AMS). 2014. Country of origin labeling (COOL). Available at www.ams.usda.gov/cool.

Wessells, C.R., C.J. Miller, and P.M. Brooks. 1995. Toxic algae contamination and demand for shellfish: a case study of demand for mussels in Montreal. *Marine Resource Economics* 10:143–159.

Glossary

Absolute quotas: Regulations that limit the quantity of an imported good to a certain time period and volume.

Acute hazard: Exposure to a substance or condition that may result in injury.

Adding up property of demand: A property of demand that stipulates that the sum of all total expenditure elasticities, when weighted by the corresponding budget share, must add up to unity.

Administered pricing: System in which prices are announced as non-negotiable selling or buying prices.

Ad valorem tariff: Tax levied on value of a commodity, expressed as a percentage.

Advertising: Organized programs and presentations designed to communicate product attributes to consumers to encourage sales.

Agent: Individual or firm that represents either buyers or sellers in the marketplace; agents do not take title to goods.

Agricultural cooperative: A user-owned and user-controlled business from which benefits are derived and distributed equitably on the basis of use by the owners.

Agricultural Marketing Service (AMS): Division of the U.S. Department of Agriculture responsible for grading and testing of agricultural products.

Almost ideal demand system (AIDS): A frequently used consumer demand model.

Antidumping duties: Levies on products that are deemed to be imported at less than fair market value.

Arkansas Baitfish Certification Program: Program of the Arkansas Department of Agriculture that includes third-party testing and inspection for specified aquatic pathogens and aquatic nuisance species.

Asymmetric information: Condition in which one participant in the market has greater knowledge of prices and quantities than do other market participants.

ATA: American Tilapia Association.

Autarky: Condition of such restrictive trade policies and restrictions that no trade occurs; the country's economy exists in isolation from the rest of the world.

Seafood and Aquaculture Marketing Handbook, Second Edition. Carole R. Engle, Kwamena K. Quagrainie and Madan M. Dey.
© 2017 John Wiley & Sons, Ltd. Published 2017 by John Wiley & Sons, Ltd.

Away-from-home consumption: Food dispensed for immediate consumption outside of the consumer's home. Includes all food consumed in food service facilities, such as restaurants, hotels, cruise ships, and schools.

Biosecurity: Processes that protect against introduction of animal diseases.

Bio Suisse: Association of organic farmers in Switzerland.

Birds Directive of European Commission (Council Directive 2009/147/EC): European Union policy on conservation of wild birds.

Bouchot mussels: Mussels cultured in France on fixed wooden poles. Seed mussels that have been collected on ropes are wrapped around wooden poles that have been driven into the ocean bottom. They are transferred to plastic net tubes that are wrapped again around the poles.

Bovine spongiform encephalitis (BSE): Also known as mad cow disease. Fatal neurodegenerative disease.

Brackish water: Water with salinity between 0.5 and 35 ppt (parts per thousand).

Bretton Woods Agreement: Agreement made in Bretton Woods, USA, in 1944 which established a post-war fixed currency rate between countries and the International Monetary Fund.

Broad-line distributors: Merchant wholesale operators that handle a broad line of groceries, health and beauty aids, and household products. Also referred to as general-line and full-line distributors.

Brokers and agents: Wholesale operators who buy or sell as representatives of others for a commission and typically do not physically handle the products or take title to the goods.

Business-to-business (B2B): Refers to direct market transactions between two independent businesses.

Captive supplies: Livestock acquired by meat packers through forward basis contracts.

Cardinal utility: Cardinal utility enables a consumer to specify the actual numeric level of utility or "satisfaction" obtainable.

Carryover stocks: Stocks left from one marketing year and held for sale in the next.

Catfish Quality Assurance: Food safety assurance plan specific to catfish production.

CERQUA: Centre de Développement des Certifications des Qualités Agricoles et Alimentaires, France.

Certification programs: Set of processes that provide assurance that a product meets specific requirements.

Ceteris paribus: Latin expression meaning "holding all other factors constant," or "all else being equal."

CFA: Catfish Farmers of America.

Chainstore: A company with more than 11 stores under one ownership and name.

Checkoff program: Program that adds a fee to either feed sales or product sales for use in advertising or research related to that particular commodity.

Chloramphenicol: Antibiotic used for major bacterial infections, including typhoid fever.

CITES: Convention on International Trade in Endangered Species of Wild Fauna and Flora.

Code of Practice: Set of written specification of management practices.

Codex Alimentarius Commission: Commission established by the United Nations to develop international food standards.

Collaborative planning, forecasting, and replenishment (CPFR): Supply chain technology that involves sharing sales forecasts of the manufacturer with the retailer, and tailoring orders and deliveries accordingly.

Collective action: Action taken jointly by a group of people to achieve goals that the group has in common.

Commission merchant: Middleman who takes a load of a commodity to market, sells it for the best price, deducts a commission, and sends the balance back to the growers.

Commodity: Economic good that can be legally produced and sold by a large number of individuals as opposed to differentiated products that belong to a specific seller.

Commodity market: Market in which primary products are traded as opposed to manufactured products.

Comparative advantage: An economic principle that states that a country should specialize in producing and exporting those goods which it can produce at relatively lower cost and should import those goods for which it has a relatively high cost of production.

Competition-oriented pricing: Prices set based on prices for similar and competing goods.

Competitive market: Market in which numerous firms supply a product that is homogeneous or standardized.

Complementary product: Products that consumers tend to consume at the same time.

Computer-aided telephone interviewing (CATI): Interviewing system in which responses are entered directly into a computer database while the interview is taking place.

Concentration: The degree to which a decreased number of firms in the industry control a high portion of the sales.

Conjoint analysis: Sometimes referred to as "trade-off" analysis because respondents are forced to make trade-offs among different product attributes; inferences are made from the quantified trade-offs as to how important or valuable different attributes are and how they influence respondents' decision-making processes.

Consolidation: Reduction in the number of firms in an industry as a result of mergers.

Convenience store: Small, self-service store located near a residential area that offers a limited line of goods.

Conventional distribution channel: Channel consisting of one or more independent producers, wholesalers, and retailers, each a separate business seeking to maximize its own profits even at the expense of profits for the system as a whole.

COOL: Country-of-origin labeling, USDA-AMS.

Cooperative: Business that is owned and controlled by those working in it and whose benefits are allocated equally among the owners/members; an organization that is owned and controlled by the people who use its products, supplies, or services.

Cost-plus pricing: Pricing system in which a set margin is added to costs of production to determine selling price.

Countervailing: An action designed to offset (countervail) the effect of another action.

Countervailing duties: Duties levied on imported products that receive an unfair subsidy from a foreign government.

Cournot aggregation: Restriction on the derivative of a linear budget constraint of a household demand system with respect to prices because total expenditure cannot change in response to a change in prices.

Cross-price elasticity: Responsiveness of quantity demanded in one good to changes in price of a related good.

Cyclospora: Sporozoan that causes diarrhea.

Delivery rights: A tradable share that requires delivery of a certain quantity and quality of a product for a specified period at some negotiated price. Some contracts for delivery rights specify production standards.

Demand: Various quantities of a good or service that consumers are willing and able to take off the market (purchase) at varying prices.

Demand-oriented pricing: Accompanies market segmentation in which higher prices are charged for those products considered to be of higher quality and lower-cost products are sold to market segments that seek out lower prices.

Department store: Large retail outlet with entire departments of different categories of consumer goods.

Depuration: Process to purify a product.

Determinants of demand: Factors that determine the specific relationship between price and quantity demanded.

Differentiated product: Economic good that belongs to a single seller and that has unique characteristics.

Dioxin: Persistent toxic heterocyclic hydrocarbon.

Directorate-General for Health and Consumer Protection of the European Commission: European Union agency responsible for food safety.

Direct sales: Sales of product from farm to end buyer without intermediate buyers.

Discount coupons: Coupons, often provided by manufacturers to promote their products, that can be used to purchase goods in supermarkets at a lower price.

Discount pricing: Price reductions offered from advertised prices.

Discount store: Store that offers lower-priced merchandise.

Disintermediation: Bypassing intermediaries to sell directly to final buyers.

Distribution channel: Various market levels through which products move from farm or boat to end consumer.

Dockage rates: Percentage reduction in price of fish, often at a processing plant, for fish that do not meet purchase specifications, i.e., too small, too large, etc.

Double-barreled question: A question that asks the respondent to address more than one issue at a time.

Dressed fish: Fish that has been deheaded, eviscerated, and skinned.

Dual approach: Use of cost functions instead of production functions to analyze production relationships.

Dumping: Selling products at prices below the cost of production and below normal domestic prices.

Ecolabel: Product label that indicates that it meets certain environmental standards.

Economics: Allocation of scarce resources to meet the unlimited needs and wants of human beings.

Economies of scale: Condition in which average per-unit costs decrease as the size of a business increases; decreasing average costs with increasing output levels.

Economy of size: Larger companies can operate at relatively lower costs by having cost advantages.

Efficient consumer response (ECR): A collaborative relationship in which any combination of retailer, wholesaler, broker, and manufacturer works together to seek out more ways to distribute manufactured food products. The purpose of ECR is to drive the order cycle and all the other business processes with point-of-sale data and other consumer-oriented data, giving an accurate read on consumer demand.

Efficient food service response (EFR): Technology system in the food service supply chain that links food manufacturers to distribution warehouses, and to restaurant outlets.

Elasticity: Measure of degree of change in one variable as a related variable changes.

Elasticity of demand: Degree of responsiveness of quantity demanded to a given change in price.

Elasticity of supply: Degree of responsiveness of the quantity supplied to changes in the price of the good.

Electronic data interchange (EDI): A technological system that allows businesses to order merchandise, streamline delivery, and reduce overall costs. The system requires that suppliers and retailers use compatible computer systems.

Endangered Species Act: U.S. law passed to prevent extinction of plants and animals.

Engel aggregation: The weighted sum of income elasticities of an item in a consumer's basket is equal to unity.

Entrepreneurship: Assuming control over the decision-making, organization, and operation of a business including the associated risks and benefits.

Equilibrium: Point of intersection of demand and supply curves.

Equilibrium price: Price at which buyers and sellers agree on the quantity to be offered and that desired; all product clears the market at the equilibrium price.

Escherichia coli: A bacterium that lives in human intestines. Some strains, such as *E. coli* 0157-H7, cause serious food-borne illness.

EUREPGAP: Set of farm management practices developed by European supermarket chains.

European Commission: Executive body of the European Union.

European Food Safety Authority: European Union agency responsible for food safety.

Eviscerate: Remove internal organs and other internal body contents.

Evolutionary concept: Product concept involving an existing product that can be modified to suit particular needs or improve on particular experiences of customers in the use of the product.

Exclusive dealing: When a processor or supplier forbids an intermediary to carry products of competing suppliers or processors.

Exclusive economic zone (EEZ): Imposition by a country of a 200-mile fishing zone along their coast line that is reserved for fishermen from their own country; fishing exploitation rights reside exclusively with that country.

Existing demand: Quantities that would be purchased of a particular product for a range of specific prices.

Exploratory research: Informal research that has no structure to the process of gathering data and information, e.g., observation, reading periodicals, and surfing the Internet.

Export subsidies: Payments by a government to a business that exports certain products.

External opportunities: Means to advance the business's goals that come from outside the business.

External shocks: Occurrences outside the farm business or economy that cause economic effects on the farm business or economy.

External threats: Events with potential to negatively affect the farm business.

Farm-retail price spread: Difference between the retail price of food products and the farm value of an equivalent quantity of food sold by farmers.

Farm-value share: Amount of food agriculture products multiplied by the unit prices of those goods divided by the retail price of food.

Fillet: Piece of fish cut along one side of the fish along the backbone.

Focus groups: Informal techniques to assess consumer preferences and needs, new product concepts, and purchase behavior for a good or service.

Food marketing bill: Difference between total consumer expenditures for all domestically produced food products and what farmers receive for equivalent farm products.

Form utility: Value added to products as they are transformed into products for final sale.

Free trade: Voluntary exchange of goods between and among different countries that occurs in the absence of regulations that either promote or constrain the exchange of goods.

Fresh water: Water with salinity less than 0.5 ppt (parts per thousand).

Futures contracts: Standardized, legally binding agreements to either deliver or receive a certain quantity and grade of a specific commodity during a designated delivery period.

Futures market: A contractual agreement made between two parties through a regulated futures exchange where the parties agree to buy or sell an asset at a certain time in the future at a mutually agreed upon price.

General Agreement on Tariffs and Trade (GATT): International agreement negotiated originally in Geneva, Switzerland, with the intent to increase international trade by reducing barriers to trade.

General equilibrium analysis: Analysis in which a number of variables are allowed to vary, and changes in price may affect other prices.

General tariff: Duty levied on imported products that applies to countries not eligible for preferential or most-favored-nation status.

Generalized system of preferences: Framework under which developed countries give preferential treatment to manufactured goods imported from certain developing countries.

General-line food service wholesaler: Business that provides products to restaurants, hospitals, schools, hotels, and other food service establishments.

General-line grocery wholesaler: Business that purchases both food and non-food products for sale to retailers that do not have warehouses.

Generally Recognized as Safe (GRAS): FDA category for food additives that have been shown through scientific studies or experience in common use to be safe for human use in food.

Generic advertising: Promotion of a general type of commodity without specification of particular brand or processor.

Genetically modified organism (GMO): A product that has been subjected to genetic engineering methods.

Giffen good: Product for which the quantity demanded goes down (up) as prices go down (up).

Global Aquaculture Alliance: International association to advance environmentally and socially responsible aquaculture.

GLOBALG.A.P.: Formerly EUREPGAP; internationally recognized criteria for safe and sustainable food products.

Green labeling: Product labeling that claims environmentally sound production practices.

Hazard analysis of critical control points (HACCP): Food safety system designed to prevent contaminants and other health hazards in food products.

Hedonic theory: Consumer theory that assumes that the qualities of a product are the ultimate source of utility for consumers and that a product is described solely by its characteristics.

Heterogeneous products: Products with attributes that are different from each other; products that do not substitute for each other.

H & G: Headed and gutted.

HGS: Headed, gutted, and skinned.

Hicksian demand: Demand of a consumer over a bundle of goods that minimizes their expenditure while delivering a fixed level of utility.

Homogeneity: Property of a dataset such that the statistical properties of any one part of an overall dataset are the same as any other part.

Homogeneous products: Products with nearly identical characteristics.

Horizontal marketing system: A distribution channel in which two or more companies at the same level of the marketing chain (with similar marketing functions) join together to pursue a new marketing opportunity.

Hypermarket: Largest of the supermarket-type grocery stores with up to 200,000 sq. ft. of selling space in groceries, sporting goods, auto supplies, etc., selling up to 40% of sales of general merchandise.

IFOAM: International Federation of Organic Aquaculture Movements; an umbrella organization for the organic agriculture movement.

Income elasticity: Measure of the response of the quantity demanded to changes in income.

Incomprehensible question: A question that respondents cannot understand, probably because of the concept or wording.

Industry concentration: Percentage of business (share of total value of shipments) accounted for by a number of businesses in the industry.

Inelastic: Demand or supply condition in which the quantity is not sensitive to changes in price or income.

Inelastic demand: Demand condition in which the quantity demanded is not sensitive to changes in price.

Inferior good: Product for which demand decreases (increases) as incomes increase (decrease).

Inferior good pricing strategy: Pricing strategy for a good for which demand decreases as income increases.

Informative advertising: Ad that has the appearance of a newspaper article.

Injurious species: Animals that cause harm to human beings.

Integrated multi-trophic aquaculture (IMTA): Aquaculture co-production of animal and plant species for which one serves as fertilizer or food for the other crop.

Intermediary: Middleman in the marketing chain who adds value to the product by either assembling units into large volumes, processing or transporting products, or identifying and servicing customers at the next level of the marketing chain.

Internal strengths: Strengths of a farm business that are internal to the business.

Internal weaknesses: Weaknesses of a farm business that are internal to the business.

International Organization for Standardization (ISO): International body that sets commercial standards.

Inverse Mills ratio: The ratio of the probability density function to the cumulative distribution function of a distribution.

Joint Subcommittee on Aquaculture: Coordinating group in U.S. of federal agencies created by the National Aquaculture Act of 1980 (P.L. 96-362, Sec. 6; 16 USC 2805).

KRAV: Swedish organization that certifies aquaculture.

Label Rouge: Label of a quality assurance program in France.

Lacey Act: U.S. law that bans sale of wildlife obtained illegally.

Laissez-faire policy: No regulations that would either restrict or encourage exchange of goods between and among different countries.

Law of demand: Economic principle that the quantity demanded of a product will decrease as the price increases and increase as the price decreases.

Leading (or loaded) question: A question that forces or directs a respondent to a response that he or she may not normally give.

Likert scale: A technique that presents a set of statements to respondents to which a respondent expresses agreement or disagreement using a scale, usually a 5-point scale. The most common scale is where 1 = strongly disagree, 2 = disagree, 3 = not sure, 4 = agree, and 5 = strongly agree.

Linear expenditure system (LES): A convenient, linear model for representing consumer response to price and income.

Listeria: Bacterium that can cause human health problems by contaminating food products.

Livehauler: Business that buys live fish from producers, transports live fish, and sells to fee-fishing businesses, grocery stores, or other outlets.

Loss-leader: Product priced below cost to draw customers into the store to have opportunities to sell other, more profitable goods.

Luxury good: Opposite of inferior good. As income increases, demand increases more rapidly.

Magnuson Fishery Conservation and Management Act: Public Law 94-265, defining U.S. rights and authority regarding fish and fishery resources, including agreements regarding foreign fishing and international fisheries, and the national fishery management program.

Marginal utility: The "satisfaction" gained from the consumption of one extra unit of a good.

Market: Location where goods are exchanged; where goods and services are bought and sold.

Market channel: Path through which a product moves from farm to end consumers.

Market equilibrium: The price and quantity at which all product is removed from the market.

Market failure: An occurrence when the market is characterized by destructive competition; structural imperfections such as monopoly and monopsony; externalities relating to commodity promotion, grades, and standards; and uncertainty relating to information needs, e.g., asymmetric information.

Market intelligence: The art of obtaining updates about relevant developments in the market.

Market penetration price: Pricing strategy in which a low price is offered initially with the goal of rapidly increasing sales.

Market performance: Measure of allocation and production efficiency and technological advancement.

Market plan: Blueprint for the target market, projected sales volume, geographic market, and advertising plan.

Market power: Ability to influence the price received or the price paid; the opposite of a price taker that has no influence over price.

Market segmentation: Strategy by which the market for a product is divided into separate sub-markets.

Market skimming strategy: Pricing strategy based on charging the greatest price the customer will pay.

Market structures: Organizational characteristics of a market.

Market-penetration pricing: Pricing strategy to quickly create high sales volume by setting a low initial product price.

Marketing: Performing all functions related to assembling, processing, transporting, and advertising goods from the point of production through to consumption by the end user.

Marketing bill: A USDA measure of the amount of total consumer dollar expenditures incurred by marketing functions as compared to that received by farmers.

Marketing channels: Routes of product flows and customer value delivery systems in which each channel member adds value for the customer; a combination of interrelated intermediaries (individuals and organizations) who direct the physical flow of products from producers to the ultimate consumers.

Marketing function: Role within a company related to strategic market planning, product development, promotion, and distribution.

Marketing margin: Costs (including profit) incurred from services and value added as products move through the marketing chain.

Marketing order: Marketing orders and agreements are legal instruments issued by the United States Department of Agriculture (USDA) Secretary that are designed to stabilize market conditions for certain agricultural commodities by regulating the handling of those commodities in interstate or foreign commerce. Marketing orders and agreements are administered by the Agricultural Marketing Service (AMS), an agency within the USDA, and are authorized by the Agricultural Marketing Agreement Act of 1937, as amended, 7 U.S.C. §§ 601-14; 671–74.

Marketing plan: Document that describes the business's marketing activities.

Marketing strategy: Plan of action to increase sales and competitiveness of products.

Marshallian demand: Mathematical equation that specifies what the consumer would buy in each price and income or wealth situation, assuming that utility is maximized. Also called Walrasian demand or uncompensated demand function.

Merchant wholesalers: Operators of firms primarily engaged in buying groceries and grocery products, and reselling to retailers, institutions, and other businesses.

Migratory Bird Treaty Act: U.S. law designed to protect migrating birds.

Ministry of Health and Welfare, Japan: Cabinet level ministry of Japanese government.

Miscellaneous wholesaler: Establishment specializing in the wholesale distribution of a narrow range of dry groceries such as canned foods, coffee, tea, or spices. Also referred to as a systems distributor.

Monopolistic competitive market: Market in which a relatively large number of firms operate competitively by supplying differentiated products.

Monopolistic market: Market with only one firm as supplier of a unique product for which there is no close substitute.

Monopsony: One buyer control.

Multi-stage budgeting: Econometric approach to analysis that assumes that consumers first allocate their budget among needs and then allocate that budget among subgroups of each need.

Multi-stage or cluster sampling: A random selection process in which the sample is chosen in stages.

Mycotoxin mycophenolic acid (MPA): Antibiotic produced by penicillin.

National Marine Fisheries Service: U.S. federal agency responsible for stewardship and management of the nation's living marine resources and habitat.

Naturland: Germany-based international association of organic farmers.

Neoclassical preference theory: Theory assuming that the decision-making process involves a comparison of two alternatives, *a* and *b* in a choice set *C* using a preference ordering.

NFI: The National Fisheries Institute.

Niche market: A portion of the market that focuses on a specific product.

Non-tariff phytosanitary issues: Includes product contaminants and adulterants.

Non-governmental organization (NGO): Non-profit voluntary citizens group.

Non-native species: Species of plant or animal living outside its native range.

Normal good: Product for which the quantity demanded goes up as the price goes down.

Nugget: In some fish markets, a processing cut that includes the belly flaps.

Office International des Epizooties (OIE): Intergovernmental organization of 152 member countries to reduce spread of animal diseases.

Oligopolistic market: Market in which few firms operate and products may be differentiated.

Opportunity cost: The value foregone from spending one's resources on a particular project.

Ordinal utility: Ordinal utility enables a consumer to order commodity combinations by level of utility or "satisfaction" obtainable (first, second, third).

Organic standards: Rules established for production of organic products.

Partial equilibrium analysis: Analysis in which most of the key parameters are held constant in order to understand the relationships of other variables one at a time.

PBO: Pinbone-out products.

Perceived-value pricing: Pricing the product based on non-price factors such as quality, healthfulness, environmental sustainability, or prestige.

Perfectly elastic demand: Demand relationship in which consumers' willingness to purchase a product disappears if price rises.

Perfectly inelastic demand: Demand relationship in which the product price remains the same regardless of quantity.

Persuasive advertising: Type of promotion designed to convince a consumer to purchase it.

Pinbone-out fillet: Fish fillet with pinbones removed.

Place utility: Increasing attractiveness of product by making it available in a location frequented by that group of consumers.

Poikilothermic: Lacking the ability to control body temperature.

Point of purchase: Place where sales occur.

Point of sale: Retail location where goods are sold.

Polychlorinated biphenyl (PCB): Synthetic organic chemical compound banned in the U.S. as a persistent organic pollutant.

Population: Entire group of individuals from whom information is required.

Post-larvae (PL): Term used to describe the size and stage of shrimp stocked into growout ponds. This stage in the shrimp's life cycle is the first one in which the shrimp has transformed from a floating, planktonic stage to a bottom dweller with walking legs.

Potential demand: Quantities that consumers might purchase at specific prices if the product were available.

Premium price: A high price set to match favorable perceptions of buyers.

Price checkoff: Mandatory or voluntary program that requires the affected individual or business to pay a flat fee per unit of sale or some specified percentage of sale value of the products sold by the individual or business.

Price determination: Interaction of demand and supply in market.

Price elasticity of demand: Same as elasticity of demand.

Price leader: Company whose price is adopted by other companies selling the same product.

Price penetration: Pricing strategy in which a low price is charged to gain increased market share.

Price taker: Firm that is unable to affect market prices and must accept prevailing prices.

Price-cost margin: Difference between price and cost of production.

Price-quality matrix: Table that correlates pricing strategies to various levels of quality that can help to develop an effective pricing strategy.

Primal approach: Direct derivation of demand function using utility maximization.

Product differentiation: Products are distinguishable through physical attributes, functional features, material make-up, packaging, advertising, and branding.

Product life cycle: Period of time for a product to be introduced to market, grow sales, and eventually be removed from market.

Product lines: Group of related products manufactured by a single company.

Product positioning: Identifying the most successful target markets and segments for a specific product.

Product-space map: Illustration of related products in a market.

Production capacity utilization rate: The ratio of total capacity utilized relative to the total processing capacity available.

Promotion: Advertising.

Protectionist policy: Restricting imports into a country to support local business. Often includes tariffs, quotas, subsidies, or tax cuts.

Psychological pricing: Establishing prices that either look better or convey a certain message to the buyer.

Qualitative research: Formal and structured research that deals with words, images, and subjective assessment. It is concerned with obtaining explanations to certain issues of subjects of interest.

Quality assurance: Formal system to achieve desired quality at each stage of production.

Quantitative research: Formal and structured research that deals in numbers, theory, and objective measures to provide statistical predictability of results to the target population.

Questionnaire (survey instrument): Formalized framework consisting of a set of questions and scales designed to generate primary raw data.

Quick response (QR) system: Supply chain system for the grocery retail industry used to shorten the retail order cycle; i.e., the total time from the point merchandise is recognized as needed to the time it arrives at the store.

Quota: Limit to the total quantity that can be imported of a particular good for a given period of time.

Rank order question: Question requiring the respondent to rank a set of factors in a certain order, e.g., low to high, usually using numbers. These types of questions allow certain product attributes or brands to be ranked based upon specific characteristics. Example: "Rank the following shrimp attributes in terms of importance to your purchase decisions, where 1 is the most important and 4 is the least important: quality ___, freshness ___, price ___, and size ___."

Rating scale question: Question requiring a respondent to rate a product or brand along a well-defined and evenly spaced continuum. Rating scales are often used to measure the direction and intensity of attitudes. Example: "Which of the following categories best describes the taste of lime-flavored marinated tilapia fillet? very tasty __; somewhat tasty __; neither tasty or sour __; somewhat sour __; very sour __."

Rational individual: Individuals are assumed to have preference orderings that satisfy six axioms: reflexivity, completeness, transitivity, continuity, non-satiation, and convexity.

Reminder advertising: Marketing strategy that uses short messages designed to reinforce key product attributes.

Retailing: Selling product to the end consumers.

Revealed preference theory: Preference theory that utilizes actual behavior of consumers to "reveal" the preference of consumers.

Revolutionary concept: Product concept that involves discovery of consumer needs that have not been met by existing products.

Safeguard remedies: Actions taken when increasing volumes of imports threaten to injure a U.S. industry or the creation of a U.S. industry.

Salmonella: Bacterium that causes food poisoning.

Saltwater: Water with salinity levels of 35 ppt.

Sample: Part of the population that is studied in order to gather information about the entire population.

Sample design: Method used to choose the sample from the population.

Sample size: The number of samples to use that is assumed to be representative of the population.

Scan-based trading (SBT): Electronic-based sales-sharing system that tailors orders and deliveries using retailer checkout counter scan systems. A technological system that provides food manufacturers instant information on their inventory in retailer outlets when the goods are scanned and sold. Inventory is therefore on a consignment basis from vendors. The system allows food manufacturers to monitor inventory levels for replenishment and bill retailers for their inventory only after the goods are scanned. Also known as pay-on-scan (POS).

Self-distribution retailer: Large independent retailer or small independent retailers that band together in the form of a cooperative to provide its own wholesaling. Self-distributing retailers own distribution centers and buy directly from food manufacturers and producers.

Selling agent: Individual who sells a product on the basis of a commission.

Semantic differential scale question: Question that asks respondents to rate a product, brand, or attribute based upon some point scale that has two extreme adjectives at each end. Example:

Which of the following categories best describes the taste of lime-flavored marinated tilapia fillet? (Check only one)				
Very tasty				Very sour
(_5_)	(_4_)	(_3_)	(_2_)	(_1_)

Shephard's lemma: Economic concept in which a consumer will buy a unique ideal amount of each item to minimize the price for obtaining a certain level of utility given the price of goods in the market.

Shrimp futures: Financial exchange for trading contracts to buy and sell shrimp.

Simple random sampling: The sample of individuals is chosen from the population in such a way that each person in the population has an equal chance of selection.

Skimming: Introducing the product at a relatively higher price for more affluent, quality-conscious consumers, and then lowering the price as the market becomes saturated.

Slotting fees/allowances: Slotting allowances and slotting fees describe a family of marketing practices that involve payments by manufacturers to persuade downstream channel members to stock, display, and support new products.

SMI: Salmon Marketing Institute.

Specialty wholesaler: Establishment primarily engaged in the wholesale distribution of items such as frozen foods, bakery, dairy products, poultry products, fish, meat and meat products, or fresh fruits and vegetables.

Specific tariff: Fixed charge per unit of imported good, regardless of its value.

Speculative stocks: Inventory held in anticipation of higher prices.

Steaks: Processing cut of fish that consists of a cross-section slice that includes the backbone.

Stock-keeping unit (SKU): Identification system, usually alphanumeric, of a particular product that allows it to be tracked for inventory purposes.

Strata: Subdivisions of similar individuals of a population. Each subdivision is known as a stratum.

Structured question: Closed-ended question that requires the respondent to choose from a predetermined set of responses or scale points.

Subsidies: Payments by a government to a business that produces a particular good.

Substitute good: Product that shares sufficient attributes with another product such that consumers readily choose one or the other depending upon price.

Substitute product: Competing product.

Superior good: Product for which demand increases (decreases) as income levels increase (decrease).

Superstore: A large supermarket that seeks to supply all the products, food and non-food, that consumers want.

Supply: Quantity of goods and services that producers are willing and able to offer in the marketplace at specific prices.

Supply chain management: Managing the flow of resources, final products, and information among input suppliers, producers, re-sellers, and final consumers.

Surimi: Minced, washed fish product formed into various seafood analog products with flavorings.

Surrogate: Market economy country at a level of economic development comparable to that of the non-market economy and a significant producer of comparable merchandise.

Survey: Method used to gather systematic information from a sample of a target population.

Tariff: Tax levied on imports, often passed to reduce quantities of imports.

Tariff-rate quota: Regulation that allows a certain volume to be imported at a reduced tariff rate.

TCI: The Catfish Marketing Institute.

Terms of trade: Terms of trade measure the rate of exchange of one good or service for another when two countries trade with each other.

TMI: The Tilapia Marketing Institute.

Tobin's q: Value of the market value of a firm to its replacement cost.

Traceability: Process to verify chain of custody of a fish product.

Trade barrier: Policy, regulation, program, or law that makes it more difficult for imports to enter a country.

Trade liberalization: Reduction or elimination of policies that restrict, encourage, or otherwise change what the trade would be without government intervention.

Trout Quality Assurance: Set of practices for trout production to ensure environmental sustainability.

Tying agreement: Agreement in which a supplier supplies a product to a channel member with the stipulation that the channel member must purchase other products as well.

UCCnets: Registry and synchronization service of UCC that helps to improve the accuracy of members' supply chain product and location information. Suppliers provide product, location, and trading partner information to the UCCnet Registry service and the system then validates the data with demand side partners, ensuring that all trading partners are using identical UCC standards.

Unanswerable question: Question that requires some specific information to respond but the respondent does not have access to the information.

Uniform Code Council (UCC): A not-for-profit standards organization that administers the Universal Product Code (U.P.C.) and provides a full range of integrated standards and business solutions for over 250,000 member companies doing business in 25 major industries.

Unstructured question: Open-ended question formatted to allow respondents to respond in their own words. There is no predetermined list of responses available to aid or restrict the respondents' answers.

Uruguay Round: Agreement that created the World Trade Organization (WTO) after negotiations among 100 nations from 1986 to 1993.

U.S. Food and Drug Administration (USFDA): U.S. federal agency responsible for food safety in the U.S.

USTFA: United States Trout Farmers Association.

Utility: Refers to the level of "satisfaction" obtainable from "consuming" a bundle of goods or products.

Value-for-money pricing strategy: Setting price based on the product's benefits to consumers.

Vertical coordination: Method by which goods and services may be exchanged between different stages of production. Units at different stages of production owned by the same firm and product flows coordinated through administrative means.

Vertical distribution channel: Distribution channel structure in which producers, wholesalers, and retailers act as a unified system. One channel member owns the others, has contracts with them, or has so much power that they all cooperate.

Vertical integration: When a firm operates at more than one level of a series of levels leading from raw materials to the final consumer in the business chain.

Volume discount: Reduction in price based on the purchase of a large quantity.

Voluntary export restraint (VER): Regulation established by a government to limit the volume of a good that can be exported.

Warehouse club: Hybrid wholesaler and retailer that sells food, appliances, hardware, office supplies, and similar products to members (both individuals and small businesses) at prices slightly above wholesale.

Warehouse food store: Discount supermarket that sells at lower prices than traditional supermarkets but with fewer services offered to customers.

Whole-dressed fish: Processing form in which head, scales, and guts are removed.

Wholesale club: Retailer selling annual membership fees and a variety of grocery and non-food items at deep discounts.

Wholesaler: Intermediate level of the market supply chain that includes collecting product from multiple producers to sell to larger retail buyers.

Wholesaling: Assembling smaller units of product into larger volumes to facilitate larger sales to larger companies.

World Health Organization (WHO): Agency of the United Nations charged with preventing international spread of disease.

World Trade Organization (WTO): Replaced the GATT institutions in 1995; created by the Uruguay Round Agreement; administers the provisions of the GATT.

Annotated bibliography of aquaculture marketing information sources

Agricultural marketing

Books

Abbott, J.C. 2009. *Agricultural Marketing Enterprises for the Developing World*. Cambridge University Press, Cambridge, UK.

> This book focuses on marketing enterprises from tropical areas. It covers a wide range of marketing systems from subsistence production for home consumption up to transnational joint venture.

Asche, F. and T. Bjorndal. 2011. *The Economics of Salmon Aquaculture*. 2nd edition. John Wiley and Sons, Oxford, UK.

> *The Economics of Salmon Aquaculture* was updated with this second edition. It traces the production process and productivity changes through to a detailed discussion of markets and competitiveness. The book draws heavily from both the authors' years of experience studying the growth of the salmon industry as well as from the scientific literature. While focused on details of the global salmon industry, there are many lessons for other segments of aquaculture around the world.

Kohls, R.L. and J.N. Uhl. 1985. *Marketing of Agricultural Products*. Macmillan Publishing Company, New York.

> This book has been a classic agricultural marketing textbook for a number of years with various editions. It is a good source for studying the fundamental principles of agricultural marketing.

Rhodes, V.J. 2007. *The Agricultural Marketing System*. 6th edition. Holcomb Hathaway, Inc., Scottsdale, Arizona.

> This book presents a good overview of the agricultural marketing system in the United States. It is written in a concise and reduced form that allows the reader to concentrate on the most critical information. The book does expect the reader to have an understanding of fundamental economic principles. It covers factors of each stage of the agricultural supply chain.

Vercammen, J. 2010. *Agricultural Marketing: Structural Models for Price Analysis*. Routledge, London.

> This book explores relationships among prices of agricultural commodities as well as other commodities such as oil and metals. This book requires understanding of basic economic theory.

Periodicals

Meat and Seafood Merchandising, P.O. Box 2074, Skokie, Illinois 60076.

> This is a magazine issued by Vance Publishing. It covers the latest trends in the retail grocery sector related to displaying and advertising all types of meat and seafood items.

Seafood and Aquaculture Marketing Handbook, Second Edition. Carole R. Engle,
Kwamena K. Quagrainie and Madan M. Dey.
© 2017 John Wiley & Sons, Ltd. Published 2017 by John Wiley & Sons, Ltd.

Aquaculture marketing

Books

Asche, F. and T. Bjorndal. 2011. *The Economics of Salmon Aquaculture*. 2nd edition. John Wiley and Sons, Oxford, UK.

Comprehensive book on the salmon industry with detailed information on marketing relationships and trends of the global salmon industry.

Shaw, S.A. 1986. Marketing the products of aquaculture. FAO Fisheries Technical Paper 276, Food and Agriculture Organization of the United Nations, Rome.

This manual provides practical advice on choosing products and markets, product forms, and retail issues such as displaying fish for customers.

Journals that publish scientific articles related to marketing aquaculture products

Aquaculture Economics & Management

This journal is the only one devoted exclusively to issues related to the economics of aquaculture. This includes marketing issues. This journal has devoted several special issues to aquaculture marketing.

The literature on aquaculture marketing is widely dispersed among the various aquaculture and agricultural economics journals. To conduct a thorough literature analysis, the following journals should be searched:

Journal of the World Aquaculture Society

Journal of Applied Aquaculture

Aquaculture

North American Journal of Aquaculture

Aquaculture Research

Reviews in Aquaculture

American Journal of Agricultural Economics

Journal of Applied and Resource Economics

Agribusiness

Conjoint analysis

Books

Louviere, J.J., D.A. Henser, and J.D. Swait. 2000. *Stated Choice Methods: Analysis and Application*. Cambridge University Press, Cambridge, UK.

Articles

Green, P.E. 1974. On the design of choice experiments involving multifactor alternatives. *Journal of Consumer Research* 1:61–68.

Green, P.E. and V. Srinivasan. 1978. Conjoint analysis in consumer research: issues and outlook. *Journal of Consumer Research* 5:103–123.

Green, P.E. and Y. Wind. 1975. New way to measure consumers' judgments. *Harvard Business Review* July-August: 89 –108.

Data sources for aquaculture products and markets

Food and Agricultural Organization of the United Nations (FAO)
The FAO puts out the most comprehensive statistical reports on world aquaculture and world fisheries. These are revised and reprinted every six years with interim updates. The last printed version was in 2012. The FAO reports fisheries and aquaculture production by species, by country, by region, and by type of environment (freshwater, marine, etc.). It is available in printed form, but complete information is also available on the web. For details, see the Annotated Webliography in this book.

United States Department of Agriculture (USDA)
USDA publishes data on several of the leading aquaculture industry segments on a regular basis. For example, catfish data are provided on supply, sales, prices, inventory (broodfish, fingerlings/fry, stockers, small food-size, medium food-size and large food-size), processor sales (by product form), acres, numbers of farms, and inventory estimates. Similarly, data are compiled on trout sales, weight, and the value of foodfish, stockers, fingerlings, and eggs. Quantities and value of ornamental fish, trout, salmon, shrimp, oysters, mussels, clams, and tilapia imported into the U.S. for the past several years are reported. Imports of tilapia, salmon, and shrimp by country are also reported. U.S. export quantities and value are reported for the past several years on oysters, mussels, clams, ornamental fish, trout, salmon, and shrimp.

USDA. 1998. Census of aquaculture (1998). 1997 Census of Agriculture Volume 3, Special Studies, Part 3, United States Department of Agriculture, Washington, D.C.
   The USDA conducted its first ever census of aquaculture in 1997. The aquaculture census data were published in 1998 and are available in hard copy. Subsequent censuses of aquaculture were conducted in 2005 and 2013. Data collected include acreage, total production and value of production of many aquaculture species by state and by region.

Demand analysis

Books
Deaton, A. and J. Muellbauer. 1980. *Economics and Consumer Behavior*. Cambridge University Press, Cambridge, UK.
Pollak, R.A. and T.J. Wales. 1992. *Demand System Specification and Estimation*. Oxford University Press, Oxford, UK.
Theil, H. 1975. *Theory and Measurement of Consumer Demand*, vol. I. North-Holland Publishing Company, Amsterdam.
Theil, H. 1976. *Theory and Measurement of Consumer Demand*, vol. II. North-Holland Publishing Company, Amsterdam.

Articles

Lewbel, A. 1990. Full rank demand systems. *International Economic Review* 31:289–300.

Discrete choice analysis

Books

Ben-Akiva, M. and S.R. Lerman. 1987. *Discrete Choice Analysis: Theory and Application to Travel Demand.* The MIT Press, Cambridge, Massachusetts.

Greene, W.H. 1997. *Econometric Analysis.* 3rd edition, Macmillan Publishing Company, New York.

Maddala, G.S. 1983. *Limited Dependent and Qualitative Variables in Econometrics.* Cambridge University Press, Cambridge, UK.

Extension materials on holding, transportation of fish for market

There are many extension materials available both in written form and downloadable from the Internet on marketing alternatives for fish farmers. One of the most accessible sets is through the Regional Aquaculture Center (RAC) networks. The RAC networks of scientists and extension specialists have developed a series of fact sheets that are available free of charge. The following is a sampling of fact sheets that provide detailed information and recommendations on specific components of marketing channels for aquaculture. These are available through extension aquaculture specialists in each state, through the Cooperative Extension Service, and can be downloaded from the Internet at http://srac.msstate.edu/publications.html.

Cichra, C.E., M.P. Masser, and R.J. Gilbert. 1994. Fee-fishing: an introduction. Southern Regional Aquaculture Center Publication No. 479, Stoneville, Mississippi.

Cichra, C.E., M.P. Masser, and R.J. Gilbert. 1994. Fee fishing: location, site development and other considerations. Southern Regional Aquaculture Center Publication No. 482, Stoneville, Mississippi.

Cole, B., C.S. Tamaru, R. Bailey, C. Brown, and H. Ako. 1999. Shipping practices in the ornamental fish industry. Center for Tropical and Subtropical Aquaculture Publication No. 131, University of Hawaii, Hilo, Hawaii.

Engle, C.R. and N.M. Stone. 1997. Developing business proposals for aquaculture loans. Southern Regional Aquaculture Center Publication No. 381, Stoneville, Mississippi.

Gilbert, R.J. 1989. Small-scale marketing of aquaculture products. Southern Regional Aquaculture Center Publication No. 350, Stoneville, Mississippi.

Higginbotham, B.J. and G.M. Clary. 1992. Development and management of fishing leases. Southern Regional Aquaculture Center Publication No. 481, Stoneville, Mississippi.

Jensen, G.L. 1990. Sorting and grading warmwater fish. Southern Regional Aquaculture Center Publication No. 391, Stoneville, Mississippi.

Jensen, G.L. 1990. Transportation of warmwater fish. Southern Regional Aquaculture Center Publication No. 390, Stoneville, Mississippi.

Jensen, G.L. 1990. Transportation of warmwater fish: procedures and loading rates. Southern Regional Aquaculture Center Publication No. 392, Stoneville, Mississippi.

Masser, M.P., C.E. Cichra, and R.J. Gilbert. Fee-fishing ponds: management of food fish and water quality. Southern Regional Aquaculture Center Publication No. 480, Stoneville, Mississippi.

Regenstein, J.M. 1992. Processing and marketing aquacultured fish. Northeastern Regional Aquaculture Center Fact Sheet No. 140-1992, University of Massachusetts, Amherst, Massachusetts.

Riepe, J.R. 1999. Marketing seafood to restaurants in the North Central region. North Central Regional Aquaculture Center Fact Sheet Series No.110, Iowa State University, Ames, Iowa.

Riepe, J.R. 1999. Supermarkets and seafood in the North Central Region. North Central Regional Aquaculture Center Fact Sheet Series No.112, Iowa State University, Ames, Iowa.

Strombom, D.B. 1992. Business planning for aquaculture – is it feasible? Northeastern Regional Aquaculture Center Fact Sheet No. 150-1992, University of Massachusetts, Amherst, Massachusetts.

Swann, L. 1993. Transportation of fish in bags. North Central Regional Aquaculture Center Fact Sheet Series No.104, Iowa State University, Ames, Iowa.

General marketing sources

Books

Chisnall, P.M. 2007. *Marketing Research*. 7th edition. McGraw Hill, London.
   Well-established textbook on marketing research techniques and applications.

Curtis, T. 2008. *Marketing for Engineers, Scientists, and Technologists*. Wiley-Interscience, Hoboken, New Jersey.
   This book provides a scientist's perspective on marketing.

Engle, C.R. 2010. *Aquaculture Economics and Financing: Management and Analysis*. Blackwell Scientific, Ames, Iowa.
   Textbook that presents details of financial analysis.

Kottler, P. and K.L. Keller. 2015. *Marketing Management*. Pearson Education Ltd., Harlow, UK.
   This book presents a comprehensive, clear, and informative survey of general marketing.

Wright, L.T. and M. Crimp. 2000. *The Marketing Research Process*. Financial Times Prentice-Hall, Englewood Cliffs, New Jersey.
   Textbook that focuses on the process of market research.

Industrial organization

Books

Bresnahan, T. 1989. Empirical studies of industries with market power. In: Schmalensee, R. and R. Willig, eds. *Handbook of Industrial Organization*. North-Holland, Amsterdam.

Carlton, D.W. and J.M. Perloff. 2005. *Modern Industrial Organization*. 4th edition. Addison-Wesley Longman, Inc., Reading, Massachusetts.
   This book presents the latest theory on the organization of firms and industries and combines it with practical evidence. While it discusses the traditional approach of focusing on structure, conduct, and performance of markets, it also addresses the modern approaches such as transaction-cost analysis, game theory, contestability, and information theory.

Schmalensee, R. 1989. Inter-industry studies of structure and performance. In: Schmalensee, R. and R. Willig, eds. *Handbook of Industrial Organization*. North-Holland, Amsterdam.

Articles

Azzam, A., 1992. Testing the competitiveness of food price spreads. *Journal of Agricultural Economics* 43:248–256.

Azzam, A. and E. Pagoulatos. 1990. Testing oligopolistic and oligopsonistic behavior: an application to the U.S. meat packing industry. *Journal of Agricultural Economics* 41:362–370.

Marion, B.W. and F.E. Geithman. 1995. Concentration-price relations in regional fed cattle markets. *Review of Industrial Organization* 10:1–19.

Menkhaus, D.J., J.S. St. Clair, and A.Z. Ahmaddaud. 1981. The effects of industry structure on price: a case in the beef industry. *Western Journal of Agricultural Economics* 6:147–153.

Muth K.M. and M.K. Wohlgenant. 1999. Measuring the degree of oligopsony power in the beef packing industry in the absence of marketing input quantity data. *Journal of Agricultural and Resource Economics* 24:299–312.

Quail, G., B. Marion, F. Geithman, and J. Marquardt. 1986. The impact of packer buyer concentration on live cattle prices. Working Paper, North Central Project 117, North Central Agricultural Experimental Stations, Madison, Wisconsin.

Schroeter, J.R. 1988. Estimating the degree of market power in the beef packing industry. *Review of Economics and Statistics* 70:158–162.

Schroeter, J.R. and A. Azzam. 1990. Measuring market power in multi-product oligopolies: the U.S. meat industry. *Applied Economics* 22:1365–1376.

Principles of economics

There are a large number of good books that cover the principles of economics. Titles may include "Principles of Economics," "Microeconomics," or "Macroeconomics." A few examples of currently available books include:

Baumol, W.J. and A.S. Blinder. 2011. *Macroeconomics: Principles and Policy*, 12th edition. Cengage Learning, Independence, Kentucky.

Hall, R.E. and M. Lieberman. 2002. *Microeconomics: Principles and Applications*. South-Western College Publishers, Cincinnati, Ohio.

Mankiw, N.G. 2014. *Principles of Economics*. 7th edition. Cengage Learning, Independence, Kentucky.

Pindyck, R.S. and D.L. Rubinfeld. 2013. *Microeconomics*. 8th edition. Prentice Hall, Upper Saddle River, New Jersey.

Rubinfeld, D.L. 2004. *Microeconomics*. Prentice Hall Publishers, Upper Saddle River, New Jersey.

Seafood marketing trade information

Books

Anderson, J.L. 2003. *The International Seafood Trade*. Woodhead Publishing Limited, Cambridge, UK.
This book is an excellent summary of international trade theory, statistics, and issues related to the international market for seafood. It begins with an overview of the worldwide market for seafood with discussions of major importing and exporting nations and net trade flows. It summarizes trade in the major species groups of seafood traded around the world. A chapter on institutions involved in international trade summarizes regulations, and the framework within which trade disputes are resolved among countries.

Annual report on the United States seafood industry. Available at www.urnerbarry.com/.
   This report is put out on annually and provides a good overview of trends in the U.S. seafood
   sector. The report compiles price and quantity data from a variety of sources for both wild-
   caught and aquaculture species and highlights various trends in the retail and wholesale
   sectors that handle seafood.

National Marine Fisheries Service. 2003. *Fisheries of the United States 2003.* National Marine
Fisheries Service, National Oceanic and Atmospheric Administration, Washington, D.C.
   This document summarizes trade statistics for seafood products imported and exported into
   and from the U.S. It includes per capita consumption statistics for countries around the world.

Periodicals

There are several good seafood trade periodical publications.

Seafood Business

   Seafood Business is published by Diversified Business Communications in Portland,
   Maine. It covers the entire seafood market and is giving increasing print space to
   information on aquaculture products. It periodically includes updates by species. A
   retailer survey and restaurant survey are conducted each year to provide updates on
   trends in these two market segments. Feature articles highlight recent newsworthy
   events. A typical issue will include such sections such as: News, Market, Product
   Spotlight, Species Focus, Top Story, Seafood Star, Trend Watch, On The Menu,
   Seafood University, Equipment, and Highlights of the Boston Seafood Show.

Seafood International

   Seafood International is published by Quantum Publishing Ltd. Surrey, UK. Its focus
   is more on European markets for seafood, but it provides a good perspective on
   trends in seafood markets from a different point of view. This company also publishes *Fishing News International* and *Fish Farming International*. A typical issue will
   include: News, Markets, New Products, Publications, Events, Last Bites, and feature articles.

The Catfish Journal

   The Catfish Journal is published monthly by the Catfish Farmers of America. While
   it is the voice of the U.S. catfish industry, it also includes information on processing
   companies, some market developments, and occasional commentary on price
   trends, particularly as these relate to catfish prices.

Surveys

Books

American Statistical Association. 1997.: *What Is a Survey?* ASA series, Alexandria, Virginia.
Blankenship, A.B., G.E. Breen, and A. Dutka. 1998. *State of the Art Marketing Research.* 2nd
   edition. American Marketing Association, NTC Business Books, Chicago, Illinois.
Chisnall, P.M. 1986. *Marketing Research.* 3rd edition. McGraw Hill, London.
Crimp, M. 1990. *The Marketing Research Process.* 3rd edition. Prentice-Hall, Englewood Cliffs,
   New Jersey.

Theories of choice behavior

Books

Deaton, A. and J. Muellbauer, 1980. *Economics and Consumer Behavior*. Cambridge University Press, Cambridge.

Hall, R.E. and M. Lieberman. 2002. *Microeconomics: Principles and Applications*. South-Western College Publishers, Cincinnati, Ohio.

Lancaster, K.J. 1971. *Consumer Demand*. Columbia University Press, New York.

Mankiw, N.G. 1997. *Principles of Economics*. Harcourt Brace and Company Publishers, San Diego, California.

Pindyck, R.S. and D.L. Rubinfeld. 2001. *Microeconomics*. 5th edition. Pearson Prentice Hall, Upper Saddle River, New Jersey.

Rubinfeld, D.L. 2004. *Microeconomics*. Prentice Hall Publishers, Upper Saddle River, New Jersey.

Annotated webliography of sources of data and information for aquaculture marketing

European Union
Common Organization of the Market in fishery and aquaculture products
ec.europa.eu/fisheries/cfp/market/index_en.htm

The EU established a common fisheries policy with foundations laid in 1970. The web site details provisions under the following five categories:

1 common marketing standards;
2 consumer information;
3 producer organizations;
4 price support system based on intervention;
5 arrangements for trade with third countries.

eur-lex.europa.eu/oj/direct-access.html

This site prints all the legislative actions and full texts of regulations coded by number as well as information and notices. This is the site where the Official Journal of the European Union is published. This site provides full text of decisions of the Commission as published in the Official Journal of the European Union. Actions taken on antidumping orders in the EU are published on this site.

www.efsa.europa.eu/

This site is the gateway to the EU and provides overviews of European Community agencies including the European Food Safety Authority. This site contains the standards, logo, and certification program for organic products in the EU.

European Commission's Health and Consumer Protection Directorate General
ec.europa.eu/dgs/health_food-safety/index_en.htm

This is the site of the European Commission's Health and Consumer Protection Directorate General. It contains the full text of the White Paper on Food Safety that contains the major policy provisions for food safety in the EU.

Seafood and Aquaculture Marketing Handbook, Second Edition. Carole R. Engle, Kwamena K. Quagrainie and Madan M. Dey.
© 2017 John Wiley & Sons, Ltd. Published 2017 by John Wiley & Sons, Ltd.

European Food Safety Authority

www.efsa.europa.eu/

This is the site of the European Food Safety Authority (EFSA). The latest opinions and reports of the various scientific panels can be found on this site.

www.eurunion.org/legislat/home.htm

The European Commission's Health and Consumer Protection Directorate General developed a White Paper on Food Safety in 2000 (Commission of the European Communities 2000). The White Paper on Food Safety contained four major initiatives: (1) creation of EFSA; (2) food safety legislation; (3) a framework for monitoring the food supply chain in the EU; and (4) food labeling rules. EFSA was established in 2002 and provides independent scientific advice on food safety. It develops and publishes opinions based on risk assessments of issues pertaining to food safety and works closely with national authorities (European Food Safety Authority 2004). The risk assessments are prepared by scientific panels convened in the following areas: food additives, substances used in animal feeds, plant health and protection, genetically modified organisms (GMOs), dietetic products, biological hazards, contaminants in the food chain, and animal welfare. Food safety legislation in the EU addresses animal feeds, animal welfare, contaminants and residues, food additives, food supplements, organic products, and packaging. The EU's Food and Veterinary Office in Dublin is charged with overseeing and monitoring food safety throughout the supply chain. Food labeling laws in the EU require the following to be included on labels: (1) name, (2) list of ingredients, (3) quantity or categories of ingredients as percentage, (4) the net quantity, and (5) date of minimum durability. There are additional labeling requirements for organic products and genetically modified organisms (GMOs).

Commission of the European Communities. 2000. White Paper on Food Safety. Available at ec.europa.eu/food/food/intro/white_paper_en.htm. European Food Safety Authority. 2004. Annual Report. ar04en.pdf. Available at www.efsa.europa.eu.

France
Ifremer

wwz.ifremer.fr/

Ifremer (French Research Institute for Exploitation of the Sea) publishes market studies on pilot projects, pricing, sector studies, socio-economic studies, and market appraisals in France, Europe, Africa, Asia, and Latin America.

Industry associations
At-sea Processors Association (APA)

www.atsea.org/

This is a web site for the At-sea Processors Association (APA). The APA represents U.S.-flag catcher/processor vessels that participate in the groundfish fisheries of the Bering Sea. Their principal fishery is the mid-water pollock fishery – the largest fishery in the U.S. Members both harvest and process fish at sea.

Efficient Foodservice Response

www.efrcanada.org/

The web site for the Efficient Foodservice Response (EFR) project. EFR is an industry-wide effort to improve efficiencies in the food service supply chain linking manufacturing plants to distribution warehouses to the retail end of the food service industry. It simplifies the flow of products, information, and funds within the food service supply chain. The EFR project is sponsored by five food service industry associations: Canadian Council of Grocery Distributors, International Foodservice Distributors Association, International Foodservice Manufacturers Association, National Restaurant Association (NRA), and Uniform Code Council (UCC).

Electronic Food Service Network

www.efsnetworks.com

The eFS Network, Inc., provides supply chain solutions for the food service industry by combining collaborative workflow technology, hosted application modules, and robust data management services. eFS Network serves suppliers, distributors, and operators as well as other supply chain participants such as sales agencies and carriers. eFS Network's customer base includes Ben E. Keith Company, BiRite Foodservice, Bunn Capitol, Cargill, Inc., Dot Foods, eMac Digital, L.L.C., FoodHandler Inc., Harker's Distribution, Inc., HJ Heinz, Heritage Bag, Kraft Foods, Martin Brothers Distribution Co., Inc., Mattingly Foods, Inc., McCain Foods Limited, Nestlé FoodServices, Performance Food Group, Quality Foods, Inc., Rich Products, Ritz Foodservice, The Schwan Food Company, Sysco Corporation, Thoms Proestler Company, Tyson Foods, Inc., and Ventura Foods LLC.

Euro-Retailer Produce Working Group (EUREP)

www.globalgap.org/uk_en/who-we-are/

The Euro-Retailer Produce Working Group (EUREP) is made up of leading European food retailers. Now developed into GlobalGAP, it is an established mechanism for drawing up production standards for commodities entering the retail trade through their outlets. Extension to the products of aquaculture started in 2001. Products will not enter the retail trade unless they meet the retailers' standard. The EurepGAP program focuses on production process quality, labeling, traceability, and food safety. Third-party verification by an accredited certification body is required.

Foodconnex

www.foodconnex.com

This e-commerce platform is hosted by Integrated Management Solutions, a leading provider of technology solutions to the Food Distribution and Processing Industries. FOOD CONNEX® comes with a software program.

The Global Aquaculture Alliance (GAA)

www.gaalliance.org

The GAA is an international non-profit trade association dedicated to advancing environmentally responsible aquaculture. The GAA program focuses mainly on the management of shrimp farming and processing operations. Third-party verification is required and certified operations can label their products with the GAA logo. The GAA Individual Codes of Practice Food Safety can be found on this site.

International Foodservice Distributors Association (IFDA)

www.ifdaonline.org/

IFDA is a trade organization representing food service distributors throughout the U.S., Canada, and internationally. In 2004, IFDA had 135 members that included broad-line and specialty food service distributors that supply food and related products to restaurants, institutions, and other food-away-from-home food service operations. IFDA advocates the interests of the food service distribution community in government and industry affairs through research, education, and communication.

ISO Programs

www.iso.org/iso/home

This site summarizes the ISO programs and members, and offers copies of a variety of technical summaries and brochures of the more than 14,000 International Standards for business, government, and society.

Ornamental Aquatic Trade Association Worldwide

www.ornamentalfish.org

This site includes marketing and trade statistics for the ornamental fish trade. It also includes a code of conduct for businesses, water quality criteria, and a customer charter.

tanganyika.tripod.com/id101.htm

This site includes contact information for ornamental fish trade companies around Lake Tanganyika, online magazines, books, and photos.

Ornamental Fish International (OFI)
www.ofish.org/
OFI was founded in 1980 and currently has 38 members that represent wholesalers, collectors, breeders, retailers, importers, exporters, plant specialists, airlines, consultants, and manufacturers. It has a code of ethics on the site as well as the *OFI Journal* that is published three times a year.

SECODIP
www.tns-sofres.com
TNS SECODIP specializes in market research related to consumer spending, including consumer panels conducted repeatedly over time to measure changes in consumer spending. SECODIP is considered as the primary source of consumer panel survey data for France. This is the main source of quantitative data on French seafood consumption.

Uniform Code Council Net
www.simplybarcodes.net/
A web site for the Uniform Code Council's (UCC) subsidiary UCCnet™. UCCnet™ makes use of industry standards in the development of powerful tools to synchronize item information and the transfer of information in a business-to-business environment. UCCnet uses standards-based e-commerce to provide non-proprietary collaborative capabilities among trading partners.

The United States Trout Farmers Association
www.ustfa.org
The Trout Producer Quality Assurance Program can be found on this page.

International agencies and associations
The Food and Agriculture Organization of the United Nations (FAO)
www.fao.org
The FAO maintains a web site with the most current global statistics available on aquaculture and fisheries. The site lists datasets from 1991 to 2000 on quantities and values of aquaculture products by groups of species, categories of production areas (inland, marine, etc.), principal species, country, total international trade, international trade by principal importers and exporters.

The FAO web site also includes articles that summarize trends as well as summary statistics. Examples include the following:

Rana, K. and A. Immink. 2003. Trends in global aquaculture production: 1984–1996. Available at www4.fao.org/cgi-bin/faobib.exe?vq_query=A%3DRana, %20K.&database=faobib&search_type=view_query_search&format_name=@ ELMON&sort_name = @SCHR&table = mona&page_header = ephmon&lang=eng.

Yearbook of Fishery Statistics. Available at www.fao.org/fi/search/yearbooks.htm.

The FAO web site further offers two databases:

FishStatJ

A set of fishery statistical databases downloadable to personal computers together with data retrieval, graphical, and analytical software. Available databases for use with FishStatJ are:

- aquaculture production: quantities;
- aquaculture production: values;
- capture production;
- total production;
- fishery commodities production and trade;
- Eastern Central Atlantic capture production;
- Mediterranean and Black Sea capture production.

Fishery Data Collection in FAOSTAT of WAICENT (World Agricultural Information Center)

The FAO web site includes information on fish processing on a variety of different levels. Specifics on fish freezing are included as well as planning and engineering data for fish processing businesses.

- Fish production: This domain presents the volume of fish production (catches and aquaculture) by country, by 50 groups and species of the FAO International Standard Statistical Classification of Aquatic Animals and Plants (ISSCAAP) and 29 FAO major fishing areas.
- Fishery data: This domain contains time series data by country on volume of annual production (catches and aquaculture) from all waters, production of processed and preserved products, and external trade of these groups of products in volume and value. The data are provided for seven aggregates of species and eight main types of product preservation, divided into fishery primary products and fishery processed product.
- On the basis of production utilization and trade data, balance sheets by individual countries are prepared which also provide indications on the role of fish in consumption.

Commodities and Trade Division

The Commodities and Trade Division of the Food and Agriculture Organization of the United Nations includes articles and statistics on trade in aquaculture and fisheries at:

www.fao.org/publications/soco/the-state-of-agricultural-commodity-markets-2015-16/en/. These include commodity notes, tables of apparent consumption, estimated value of fishery production by groups of species, trade flow by region, international exports by species and year (1996–2000), and the relative importance of trade in fishery products in 2000.

FAO-Uruguay Round Agreement on Agriculture
www.fao.org/docrep/003/x7353e/x7353e03.htm
This is a page on the FAO site that includes information on the Uruguay Round Agreement on Agriculture.

Code of Conduct for Responsible Fisheries

www.fao.org

The FAO Code of Conduct for Responsible Fisheries is available on this site. This document lays the foundation for responsible management of aquaculture and fisheries stocks.

FAO Codex Alimentarius

www.fao.org/fao-who-codexalimentarius/en/

This page presents an international regulatory framework for fish safety and quality. It discusses the World Trade Organization (WTO) agreements on the Application of Sanitary and Phytosanitary Measures (SPS) and the FAO Codex Alimentarius.

FAO HACCP

www.fao.org/docrep/005/y1579e/y1579e03.htm

Fact sheet on HACCP from the FAO.

Globefish

www.fao.org/in-action/globefish/en

Globefish is a publications unit within FAO that publishes a wide variety of reports and analyses related to fish and seafood markets around the world, including global overviews, world market reports by species, specific market situation analyses, international trade, fishmeal, and trade barriers.

INFOFISH

infofish.org/v2/

INFOFISH publishes articles on capture fisheries and aquaculture, processing, packaging, storage, transport, and marketing, and includes announcements of upcoming meetings and seafood shows.

International Institute of Fisheries Economics and Trade (IIFET)

oregonstate.edu/dept/IIFET

IIFET is the International Institute of Fisheries Economics and Trade. This organization is an international group of economists, government managers, private industry members, and others interested in the exchange of research and information on marine resource issues. IIFET holds bi-annual meetings, and publishes a newsletter and proceedings of its various meetings. The newsletters and proceedings can be ordered on the web site.

Convention on International Trade in Endangered Species of Wild Fauna and Flora

cites.org

This is the official site of the Convention on International Trade in Endangered Species of Wild Fauna and Flora. It includes species and trade databases, registers, export quotas, reports, contacts, resolutions, and reports of the standing, animals, plants, and nomenclature committees.

World Trade Organization

www.wto.org

Official site of the World Trade Organization, the only global international organization dealing with the roles of trade among nations. The site includes a training package, videos, list of members, publications, calendar of events, news releases, committee reports, and international trade statistics.

World Aquaculture Society (WAS)

www.was.org

The World Aquaculture Society is an international non-profit society founded in 1970 with the object of improving communication and information on aquaculture worldwide. WAS sponsors numerous professional meetings, including the international triennial meetings as well as annual chapter meetings in the U.S., Latin America, Asia, and Europe. The WAS has an extensive publications unit with books, the *Journal of the World Aquaculture Society*, and *World Aquaculture Magazine*.

International trade: theory and background
Suranovic, S.M. 1997–2004. International theory and policy analysis. The International Economics Study Center.

internationalecon.com/v1.0

This site is an introductory course/text on international trade theory and policy. It can be purchased at moderate rates ($49 for the entire file in 2016 for students) for an electronic version. The file is very large and may cause technical difficulties in access, but it is well written. It is easily understood by those with no economics background and is illustrated with a number of clear examples of trade issues, policies, and tools. It lays out clearly the advantages and disadvantages of both free trade and protectionist policy and positions. It is a good starting point for understanding international trade issues.

Deardorff's Glossary of International Economics. Alan Deardorff (UMichigan) collection of citations and definitions regarding international economies.

Organic food certification companies
BioGro

www.biogro.co.nz

BioGro in New Zealand includes organic standards for finfish, shellfish, and crustacean farms, but the aquaculture standards are not a part of BioGro's IFOAM accreditation.

Bio Suisse

www.bio-suisse.ch/en/home.php

Bio Suisse adopted organic aquaculture standards in 2000 for trout and salmon in Europe and *Pangasius* in Vietnam.

KRAV

www.krav.se/krav-standards

KRAV certifies aquaculture, but its aquaculture standards are not accredited by IFOAM. The KRAV program includes salmonids, perch, and blue mussels.

Naturland

www.naturland.de

Naturland is an international association of farmers that promotes organic agriculture. Founded in Germany in 1982, it grew to 40,700 farmers cultivating more than 137,000 ha globally in 2013. Naturland farms raise organic trout (Germany, France, Italy, Spain), organic salmon (Ireland, Scotland), organic shrimp (Ecuador, Peru, Brazil, Vietnam, India, Indonesia), organic tilapia (Israel, Ecuador), organic *Pangasius* (Vietnam), and organic sea bass and gilthead sea bream (Greece, Croatia).

The Soil Association

www.soilassociation.org

In the UK, The Soil Association sets aquaculture organic standards for Atlantic salmon, trout, and arctic char, shrimp, bivalves, and carp.

United States

Aquanic

aquanic.org

Aquanic is the U.S.-based gateway to the world's electronic aquaculture resources. It includes links to:

- discussion groups;
- species;
- systems;
- job services;
- contacts;
- sites;
- publications;
- newsletters;
- media;
- educators;
- news;
- calendars;
- classified ads;
- online courses;
- feedback.

Aquanic includes a page by the Joint Subcommittee on Aquaculture that lists Federal Marketing Services available through the USDA. Not all of these programs directly apply to aquaculture. Aquanic also lists programs of the Agricultural Marketing Service and the Foreign Agricultural Service.

The Economic Research Service (ERS), USDA
www.ers.usda.gov/data-products/.aspx
ERS produces data products in a range of formats, including online databases, spreadsheets, and web files. Data and reports include: farm income, trade, food prices, food markets, diet and health, natural resources, and food consumption trends. The food consumption database includes historical data on the U.S. population and the daily per capita amounts of food energy, nutrients, and food components in the U.S. food supply. The trade data include types of export subsidies, expenditures on export subsidies, and the quantity of subsidized exports during a given year by World Trade Organization (WTO) members. Domestic support data detail the type and amount of support WTO members have provided annually. Market access data contain information on tariff commitments and their implementation by presenting bound tariff levels and tariff-rate quotas agreed to in the Uruguay Round, as well as applied annual tariff rates (i.e., the tariff rates published by national customs authorities for duty administration purposes).

Federal statistics
fedstats.sites.usa.gov//
This site provides statistical profiles of states, counties, cities, congressional districts, and federal judicial districts; comparison of international, national, state, county, and local statistics; descriptions of the statistics on agriculture, demographics, economics, environment, health, natural resources and others, and links to relevant web sites, contact information, and key statistics.

Fishery Market News
www.st.nmfs.noaa.gov/st1/market_news/
The National Marine Fisheries Service (NMFS) in the U.S. has maintained its "Fishery Market News" since 1938 with the objective of providing accurate reports on trade in fish products.
    This web site includes the following under Market News Archives:
- monthly imports of shrimp;
- monthly imports of frozen fish blocks;
- monthly imports of selected fishery products;
- monthly exports of selected fishery products;
- quarterly fish meal and oil production;
- market news abbreviations.

    The NMFS Northeast Region Reports include:
New York: Fulton Fish Market fresh prices (daily)
              weekly New York frozen prices (Friday)

Boston: New England auction prices (daily)
  Boston lobster prices (daily, except Wednesday)
  weekly Boston frozen market prices (Wednesday)
  weekly New England auction summary (Friday)
The NMFS Southeast Region Reports include:
- weekly Gulf shrimp landings by area and species (Monday);
- weekly ex-vessel Gulf fresh shrimp prices and landings (Monday);
- weekly Gulf finfish and shellfish landings (Monday);
- weekly fish meal and oil prices (Thursday);
- monthly Gulf Coast shrimp statistics;
- monthly menhaden purse seine landings.

The NMFS Southwest Region Report includes:
- canned tuna import update;
- San Pedro Market fish receipts;
- Japanese shrimp imports;
- Japanese fishery exports;
- Japanese fishery imports;
- Japanese cold storage holdings;
- Tokyo wholesale prices;
- fish landings and average ex-vessel prices;
- sales volume and average wholesale prices.

The NMFS Northwest Region Report includes:
- Oregon weekly prices with comparison report
- Seattle wholesale producer prices.

This site includes graphs of nominal and real wholesale prices from 1991 to 2001 for clam, cod, crab, croaker, flounder, lobster, oyster, pollock, squid, swordfish, and whiting, annual cold storage reports for 1990–2002, annual foreign trade reports for 1996–2002, and an annual summary of Fulton Fish Market fresh prices 1987–2002, New York frozen wholesale prices, annual summary, 1990–1997.

International Trade Commission

www.usitc.gov

This is the official site of the International Trade Commission. It includes information on antidumping and countervailing duty orders for product group, country, and data. The site lists events from the daily and weekly reports and tariff schedules. It has a database of all existing antidumping duties and countervailing orders in the U.S.

National Agricultural Library, ARS, USDA

www.nal.usda.gov

Seafood Marketing Resources includes postings by the USDA National Agricultural Library related to aquaculture, trade, databases, hearings, legislation, journals, U.S. Government contacts, trade associations and organizations, seafood shows and expositions, and lists of distributors/exporters/importers, both foreign and U.S.

The National Aquaculture Development Act

*www.nmfs.noaa.gov/aquaculture/docs/aquaculture_docs/nat_
aq_act_1980.pdf*

The National Aquaculture Development Act became law in 1980. The Act states that is "in the national interest, and it is the national policy, to encourage the development of aquaculture in the United States." This act indicates that the principal responsibility for the development of U.S. aquaculture lies with the private sector, but assigned USDA, USDOC, and USDI responsibility. In a later inter-agency agreement, USDA was given responsibility for research and support activities for private freshwater aquaculture. The NADA has been re-authorized twice, in 1985, establishing USDA "the lead federal agency with respect to the coordination and dissemination of national aquaculture information" and designating the Secretary of Agriculture as permanent chair of the Joint Subcommittee on Aquaculture (JSA).

Antidumping duties

www.nottingham.ac.uk/shared/shared_levevents/.../collie.pdf

Collie, D.R., H. Vandenbussche. 2004. Anti-dumping duties and the Byrd amendment.

www.heritage.org/research/reports/1992/07/bg906nbsp-a-guide-to-antidumping-laws

A complete guide to U.S. antidumping and countervailing duty law.

National Marine Fisheries Service, U.S. Department of Commerce

www.nmfs.noaa.gov/trade/DOCAQpolicy.htm

This web page outlines the mission statement and the vision of the U.S. Department of Commerce for U.S. aquaculture. The statement outlines the specific objectives by the year 2025.

Rural Development Agency

www.rd.usda.gov/programs-services/all-programs/cooperative-programs

This is a federal government web site that provides information on cooperative programs administered by USDA. It provides information on cooperative spotlights, cooperative data, charts on cooperatives, publication on cooperatives, and funding opportunities for research in cooperatives. You can also obtain an electronic copy of *Rural Cooperative Magazine*, a magazine published every other month that focuses on cooperatives and issues facing cooperatives.

Southern Regional Aquaculture Center

Located on the Aquanic site, this page includes a large number of extension fact sheets on a wide variety of topics related to aquaculture, including marketing and economics fact sheets.

U.S. Census Bureau

www.census.gov/epcd/susb/2001/us/US311712.HTM

The web site provides detailed national statistics for the fresh and frozen seafood processing industry from the 1997 Economic Census. Data provided include number of firms, employees, payroll, and revenue by employment-size of the enterprise.

The site provides statistics of U.S. fresh and frozen seafood processing including: employment-size of enterprise, number of firms, number of plant establishments, number of paid employees, and annual payroll ($1,000). The statistics are from 1998 through 2001.

www.census.gov/manufacturing/capacity/

The web site provides results from the Survey of Plant Capacity Utilization conducted jointly by the U.S. Census Bureau, the Federal Reserve Board (FRB), and the Defense Logistics Agency (DLA). The survey collects data for the fourth quarter and includes number of days and hours worked; estimated value of production at full production capability; and estimated value of production achievable under national emergency conditions. Data is from 1994 through 2015.

U.S. Department of Agriculture – Agricultural Marketing Service (USDA-AMS)

www.ams.usda.gov/rules-regulations/cool

This is the site of the Agricultural Marketing Service, an agency within the United States Department of Agriculture that is handling the Country of Origin Labeling rule. The interim rule posted in October, 2004, can be found at the site. Definitions of terms used in the rule, including specific definitions of "retailer," "food service establishment," "covered commodities," "processed food item," etc., are listed on the site. Copies of related rulemaking efforts, resources related to the COOL rule, talking points, overviews, examples of records that may be useful for the COOL verification process, and copies of news releases can be found on the site.

U.S. Department of Agriculture – National Agricultural Statistics Service (USDA-NASS)

www.nass.usda.gov/

NASS provides statistical information on agriculture that includes publications, charts and maps, historical data, statistical research, and a census of agriculture.

U.S. Department of Commerce
www.commerce.gov
This is the official site of the U.S. Department of Commerce. It provides information on the state of the U.S. economy. This site provides export-related assistance and market information, lists export regulations, and includes summaries of trade statistics.

trade.gov/enforcement/operations/
A federal register notice that includes the regulations on antidumping and countervailing duty proceedings to conform to the Department of Commerce's regulations to the Uruguay Round Agreements Act.

The USDA Economics, Statistics, and Market Information System
usda.mannlib.cornell.edu
The site contains nearly 300 reports and datasets from the economics agencies of the U.S. Department of Agriculture. These materials cover U.S. and international agriculture and related topics. Entering "Aquaculture" in the "Search" box will direct the reader to the various reports available. Data and reports on aquaculture include Aquaculture Outlook (by ERS), Catfish Processing: Dataset (by NASS), Catfish Processing: Report (by NASS), Catfish Production (by NASS), and Trout Production (by NASS).

The Economic Research Service (ERS), USDA
www.ers.usda.gov/data-products/.aspx
The ERS produces data products in a range of formats, including online databases, spreadsheets, and web files. Data and reports include: farm income, trade, food prices, food markets, diet and health, natural resources, and food consumption trends. The food consumption database includes historical data on the U.S. population and the daily per capita amounts of food energy, nutrients, and food components in the U.S. food supply. The trade data include types of export subsidies, expenditures on export subsidies, and the quantity of subsidized exports during a given year by World Trade Organization (WTO) members. Domestic support data detail the type and amount of support WTO members have provided annually. Market access data contain information on tariff commitments and their implementation by presenting bound tariff levels and tariff-rate quotas agreed to in the Uruguay Round, as well as applied annual tariff rates (i.e., the tariff rates published by national customs authorities for duty administration purposes).

U.S. Fish & Wildlife Service
www.fws.gov/injuriouswildlife/
Lists those species listed as injurious under the Lacey Act.

World Outlook Board, USDA
www.usda.gov/oce/commodity/
This board serves as the focal point for economic intelligence on the outlook for U.S. and world agriculture. It forecasts supply and demand for major commodities at the world level, and for livestock products and refined sugar at the U.S. level. The forecasts are in the form of a balance sheet that matches supply (beginning stocks added to the anticipated crop) with demand (how much will be consumed at home, exported, or remain as ending stocks).

National Oceanic and Atmospheric Administration (NOAA), Fisheries
www.noaa.gov
This site provides information on U.S. aquaculture, bycatch, grants, international interests, legislation, permits, and recreational fisheries. It also provides information on the Department of Commerce's Aquaculture Policy, National Aquaculture Act of 1980, NOAA Aquaculture Policy, Policy Paper on the Rationale For a New Initiative in Marine Aquaculture, Department of Agriculture's National Aquatic Animal Health Plan, the Environmental Protection Agency's final aquaculture effluents rule, and a draft Code of Conduct for Responsible Aquaculture Development in the U.S. Exclusive Economic Zone. There are reports on Fishery Market News and Fisheries Statistics including domestic and international trade.

American Statistical Association (ASA)
www.whatisasurvey.info/
The site provides brochures about survey research. The ASA Series includes: "What is a Survey?" "How to Plan a Survey;" "How to Collect Survey Data;" "Judging the Quality of a Survey;" "How to Conduct Pretesting;" "What are Focus Groups?" "More About Mail Surveys;" "What is a Margin Of Error?" "Designing a Questionnaire;" and "More About Telephone Surveys."

U.S. Food and Drug Administration
www.fda.gov
This is the site of the U.S. Food and Drug Administration. It includes the mission statement, summaries of what FDA regulates, and its history.

www.fda.gov/Food/GuidanceRegulation/HACCP/ucm2006764.htm
This is a page on the U.S. FDA web site that deals with seafood HACCP. This site provides an overview of HACCP as it relates specifically to seafood. It includes a summary of the provisions in the rule as well as the final rule, full text, for the seafood HACCP rule.

Marketing plans and strategies

There are quite a few web sites that offer assistance in development of marketing plans and strategies. A simple web search will turn up several. Some offer free services, sample market plans, and templates for developing market plans and strategies, while others offer services for fees, workshops, books, and software. These are dynamic sites, but a few examples are listed here.

www.morebusiness.com

This site includes templates for developing marketing plans, sample market plans, and software for business planning.

www.entrepreneur.com

This site contains a market planning checklist, tools and services to enhance marketing success, marketing tips, business coaches, and business services.

money.howstuffworks.com/

This site discusses how marketing plans work.

www.paloalto.com

This site contains sample market plans and includes tutorials on how to write a marketing plan.

Non-governmental organizations

Marine Aquarium Council (MAC)

www.marineaquariumcouncil.org

The Marine Aquarium Council is an international, not-for-profit organization that brings marine aquarium animal collectors, exporters, importers, and retailers together with aquarium keepers, public aquariums, conservation organizations, and government agencies. The mission is to conserve coral reefs and other marine ecosystems by creating standards and certification for those engaged in the collection and care of ornamental marine life from reef to aquarium.

Global Marine Aquarium Database (GMAD)

eol.org/collections/55230

The United Nations Environment Programme-World Conservation Monitoring Centre along with the Marine Aquarium Council has compiled a database on 2399 species from 45 representative wholesale exporters and importers. The database can be queried by genus, then species, year, and by either imports or exports.

Monterey Bay Aquarium

www.montereybayaquarium.org

The aquarium issues a pocket guide for fish consumers that informs on how sustainable each type of fish is. There is a report on each seafood species available on this site.

SeaFood Business
seafoodsource.com
This site provides a summary of the out-of-court settlement between Great Eastern Mussel Farms of Maine and mussel producers from Prince Edward Island, Canada. The settlement followed the antidumping petition filed by Great Eastern Mussel Farms.

Universities
University of Wisconsin
www.wisc.edu/
This is the University of Wisconsin Center for Cooperatives (UWCC) web site that provides information on all aspects of cooperatives including business principles, organizing cooperatives, cooperative financing, cooperative structure, cooperative management, leadership and governance, and related topics for both agricultural and consumer cooperatives.

Index

Seafood and Aquaculture Marketing Handbook, Second Edition. Carole R. Engle,
Kwamena K. Quagrainie and Madan M. Dey.
© 2017 John Wiley & Sons, Ltd. Published 2017 by John Wiley & Sons, Ltd.

Printed and bound by CPI Group (UK) Ltd, Croydon, CR0 4YY

16/04/2025

14658471-0005